ECOLOGY OF REPTILES

RECENT ADVANCES IN ENVIRONMENTAL ECOLOGY SERIES—15

Ecology of Reptiles

A. P. DIWAN
D. K. ARORA

ANMOL PUBLICATIONS PVT LTD
NEW DELHI-110 002

ANMOL PUBLICATIONS PVT LTD
4374/4B, Ansari Road
Darya Ganj, New Delhi-110 002

First Edition, 1995
ISBN 81-7488-124-7

Printed in India

Published by J. L. Kumar for Anmol Publications Pvt. Ltd., New Delhi and Printed at Amit Enterprises, 518/14, Anup Mkt., Maujpur, Delhi.

Preface

Reptiles are the first vertebrate adapted for life in dry places on land. Some live near water or in water, but for the breeding purpose and egg laying they return to land.

The present title is carefully organised, and clearly written with an aim to bring together in one place and in a simple way the fundamentals of ecology with special emphasis on the reptiles. The book is profusely illustrated with appropriate and understandable drawings. In order to make the reading easy and accessible, the simple language has been followed avoiding grotesque phrases throughout the book.

Any worthwhile criticism and suggestion for further improvement in the book would be thankfully acknowledged.

AUTHORS

Contents

1

Introduction

The study of antipredator mechanisms was a prominent aspect of the post-Darwinian cra in evolutionary biology but, with the notable exception of mimicry, defense against predators remains a topic largely without conceptual foundation. A truly comprehensive theory of antipredator mechanisms would enable us to predict structures and responses of an unstudied animal on the basis of other, known aspects of its bioloy. The extent to which such a general theory is possible and factors that might limit it are addressed in Section VI. The following remarks will orient the reader and provide an interim conceptual framework.

This chapter is restricted to predation in a strict sense, i.e., the killing and consumption of another animal. Parasitism, scavenging, and predation on the eggs of reptiles (see Packard, and Packard this volume; Wilbur and Morin, this volume) are excluded on the grounds that they are not likely to influence the evolution of behavioral and external morphological aspects of organisms. With predation thus defined, defensive mechanisms can be considered in terms of the costs and benefits associated with searching for pursuing, handling, and processing food. These components incorporate the probabilities of locating and capturing prey, the risk of injury or death the predator, and the delay or nonperformance of other essential activities while feeding. An antipredator tactic could work by raising any cost component to unacceptably high levels, or by deceptively indicating higher costs and or lower benefits than a predator would "willingly" accept. Section II reviews predation on reptiles, as background for understanding antipredator mechanisms.

Ethology has traditonally recognized four major issues regarding behaviour, viz., development within on individual, sensory and motivational control, evolutionary history, and ecological consequences (Tinbergen, 1963). Parallel questions also are posed with respect to morphological and physiological features of organisms and they highlight important areas that have been largely overlooked in the study of antipredator mechanisms. Factors affecting the individual expression of defensive attributes and intraspecific variation are discussed in Section III. There follows a systematic analysis of variation among species and higher taxa in Section IV and a discussion of the evolution of antipredator mechanisms in Section V. Records of predation on specific taxa in nature, a survey of defensive behaviour in neonate reptiles, a descriptive catalogue of antipredator mechanisms, and a taxonomic review are provided in the Appendices.

Case studies are emphasized throughout this chapter and, whenever possible, conceptual discussions are biased toward reptiles as examples. Occasional mention is made of unpublished observations, because they influenced my perpective and in hopes that they will inspire more detailed research. Auttomy mimicry and parental care (Shine) are dealt with elsewhere in this volume.

PREDATION ON REPTILES

The kinds and numbers of potential predators involved, the ways in which they interact with prey, and the enegetic and reproductive consequences of those interactions are necessary parameters for understanding the ecological and evolutionary significance of antipredator mechanisms. The interaction itself is a "behavioural couple" that can very irrespective of morphological features of predator and prey. Energetic consequences of the interaction might reflect the roles of morphological and physiological constraints on predators and or prey and might figure heavily in the future fitness of both organisms. Reproductive consequences of an encounter are the ultimate measurse of fitness for both predator and prey.

An evaluation of antipredator mechanisms requires knowledge of actual predation on the prey species of interest, as well as of predator taxa that are not eating it becaus the encounter would be too "expensive." The later information can suggest profitable experiments and appropriate interpretations of the behaviour. For example, coatis (Nasua narica) and peccaries (Tayassu tajacu) were used as experimental predators of coral snake models. Although those sympatric mammals are not known to prey upon venomous coral snakes, they do eat small, cryptic vertebrates, and Gehlbach's data demonstrated that coatis and peccaries actually avoid coral snakes.

Several sources of information on predation are potentially available, although relatively few data can be applied conclusively to questions about the evolution of antipredator mechnisms. These sources are discussed below, and include autecological studies of hunting techniques and diets, dietary analyses of entire communities of predators, comparative surveys of injury frequencies, and natural and manipulative experiments. Specific records of predation on reptiles in nature are given in Appendix A, and brief summaries for higher taxa are in Section IV.

Analysis by Region

Animals and their defensive mechanisms evolve in a community context. Work on marine invertebrates poeciliid fishes and other organisms underscores the importance of studying simultaneously predation and antipredator responses in geographic and temporal terms. An especially pertinent example is the suggestion that the large biomass of herbivores and carnivores depresses species richness and overall abundance of reptiles in Africa relative to the neotropics. That complex proposal was based on indirect evidence and crude measures of reptilian densities at African and neotropical sites. Nevertheless, the hypothesis might explain a major pattern in vertebrate geographical ecology, and thereby illustrates the portential value of comparative studies of predation.

A few surveys examine the diets of several sympatric predators, usually within a narrow taxonomic focus. Amons 29 species of raptors in Surinam, at least six feed frequently on reptiles and at least eight feed occasionally on lizards and snakes. Of those, the laughing falcon (Herpetotheres cachinnans) specializes on snakes and the gray hawk (Buteo nitidus) takes many lizards (Haverschmidt, 1962; Voous, 1969).

In Africa, at least 20 of 89 species of diurnal raptors and 2 of 30 species of owls feed only or in part on reptiles, and their collective diets encompass an impressively large number of species (Brown, 1971; Broadley, 1974). For example. 8 secretary birds (Sagittarius serpentarius) contained 61 reptilian prey, including a tortoise, 57 lizards of 6 genera and 11 species, and 1 each of 3 species of snakes. Forty-four reptiles eaten by brown snake eagles (Circaetus cinerus) included a Chamaeleo dilepis, a Varanus niloticus, an Agama sp., and 41 snakes, among which were numerous large, venomous species (Bitis arietans, Dendroaspis polylepis, Naja mossambicar Broadley, 1974).

Reptiles are uncommon in the diets of most of 15 species of medium and large carnivores in southern Africa, but several of those predators might have important impact on them (Pienaar, 1969; Broadley, in litt.). Land tortoises (Geochelone pardalis, Kinixys belliana) and water turtles (Pelusion sinuatus, P. subrufa) are eaten by lions (Panthera leo) and two species of hyaenas (Crocuta crocuta, Hyaena brunnea), and the aquatic turtles are eaten commonly by Crocodylus nileticus as well. Python sebae is eaten by lions and leopards (Panthera pardus).

In contrast, the smaller African carnivores are regular predators on reptiles. These include a small cat, Felis libyca; all of the canids except wild dogs; all mustelids other than otters; and all viverrids except two insectivorous nongooses. Most predation is evidently opportunistic but sometimes predators gorge on particular species or the inhabitants of particular microhabitats. A badger, Mellivora capensis, had eaten 22 geckos, Ptenopus garrulus, as well as an amphisbaenian, rodents, and scorpions. One mongoose, Paracynictis selousi, contained 18 burrowing skinks, Panasvis wahlbergii; 3 lacertids, Nucras taeniolata; 3 snakes, including 2 Lycopludion capense and an Atractaspis biboni; plus a frog, rodents, and crickets.

Comprehensive dietary surveys of entire assemblages of sympatric predators (rather than taxonomic subsets) are available for only four localities, three of them in Mediterranean-type habitats of Spain (Valverde, 1967), Chile and California (a series of papers by Fitch and colleagues, cited in laksic et al., 1982). Among those sites, the overall importance of reptiles as prey is lowest in Chile, intermediate in California, and highest in Spain. The incidence of predation on a species correlated with its relative abundance in Spain and California, but not in Chile. Behavioural and morphological attributes apparently make some prey species more vulnerable to capture (e.g., a large colubrid snake, Philodryas chamissonis, in Chile) and others less so (*e.g.*, a very fast teiid lizard, Callopistes maculatus, in Chile; a venomous rattlesnake, Crotalus viridis, in California) than their abundances would predict.

In an assemblage of mare than 100 sympatric vertebrate predators at Finca La Selva, Costa Rica, a katydid, 2 frogs, 2 lizards, at least 17 snakes, 10 raptors, and 2 mammalian carnivores feed on reptiles. Preliminary data from that locality indicate that most lizards and snakes, particularly smaller species, face the threat of a wide variety of avian, mammalian, and snake predators (van Berkum et al., 1987; Greene, 1987; Greene, Donnelly, Guyer, and Santana, unpubl).

Analysis by Predator Taxa

1. *Invertebrates*

Although reptiles frequently prey on invertebrates, the reverse is rarely true. Most cases of arthropod predation on terrestrial and aquatic reptiles represent occasional attacks on small individuals by predator species that usually eat other invertebrates, but exceptions exist (see McCormick and Polis, 1982, for a review). Some large centipedes, spiders, and scorpions eat vertebrates, and reptiles are apparently an important part of the diet of a few species. Scorpions are abundant in parts of arid western North Amrica, and blind snakes, Leptotyphlops humilis, sometimes form as much as 10% of their diets (McCormick and Polis, 1982). Spiders prey on small snakes (Groves and Groves, 1978; McCormick and Polis, 1982), a mantid ate a gecko (Wright, 1982), a crab caught a hatchling king cobra (Wall, 1926 a), a water bug ate a garter snake (Drummond and Wolfe, 1981), and centipendes often eat Galapagos lava lizards (Snell, in litt.). The prevalence of large, predatory invertebrates in the leaf litter of lowland tropical forests might account for the paucity of small reptiles in those habitats (Myers and Rand, 1969), but more information on the geographic incidence of arthropod predation on small litter vertebrates is needed to evaluate that hypothesis.

Marine reptiles are also preyed upon by invertebrates. A large octopus was found eating a medium sized sea turtle, Eretmochelys imbricata, and flipper damage to other sea turtles might have been caused by octopi (Buxton and Branch, 1983). Crabs probably are prominent predators on hatchling sea turtles throughout the world (*e.g.*, Stancyk, 1981; Frazier, 1984; Cornelius, 1986).

2. *Fishes*

Relatively few reptiles are aquatic or marine, and predatory fish are probably rarely important in the evolution of reptiles (see Section V.A.). The diet of a chub includes terrestrial iguanid lizards, Sceloporus spp. (Neve, 1976); a large catfish had an African viper, Bitis arietans, in its stomach (Thorne and Hamman, 1981); a 3-kg African fish contained a 2-m long black mamba (Dendroaspis polylepis, Broadley, 1983); and a large Amazonian fish had eaten two blind snakes, Leptotyphlops macrolepis (Goulding, 1980). Four Thamnophis sirtalis and one Storeria occipitomaculata were found in 1724 bass (Micropterus salmoides) stomachs in Michigan (Knapik and Hodgson, 1986). An eel (Anquilla rostrata) regurgitated two hatchling musk turtles, Sternotherus odoratus, that survived the encounter (Pough, in litt.).

Six species of sharks take considerable numbers of sea snakes in the waters off Australia, and also perhaps catch them elsewhere (Heatwole, 1975). On rare occasions marine bony fishes eat Pelamis platurus, a surface-feeding sea snake (Pickwell et al., 1983), as well as other marine elapids (Heatwole, 1975). Sharks might pose a significant threat to sea turtles (Stancyk, 1981; Cornelius, 1986) and Galapagos marine iguanas (Amblrrhynchus cristatus, Carpenter, 1966). A single dolphin contained nine young sea turtles of two species (Witham, 1974).

3. *Amphibians*

Most amphibians are not predators on reptiles. The few caecilians that have been studied feed mainly on invertebrates (Wall, 1922; Bemis et al., 1983); an Ameiva undulata and an Anolis dollfusianus in the stomach of a Dermophis mexicanus are the only records of predation on vertebrates by those limbless, tropical amphibians (Moll and Smith, 1967). The only records of predation on wild reptiles by salamanders are of Siren intermedia on a snake. Regina alleni (Godley, 1982); of Dicamptodon ensatus on on iguanid lizard, Sceloporus occidentalis (Bury, 1972); and of Ambystoma tigrinum on a teiid lizard, Cnemidophorus sexlineatus (Camper, 1986). Dicamptodon ensatus is probably the only salamander in the world that regularly eats terrestrial vertebrates in the field.

Most frogs are insectivorous and too small to prey on vertebrates, but there are some interesting exceptions. Rana catesbeina in North America (Minton, 1949; Clarkson and de Vos, 1986), Leptodactylus pentadactylus (Hoogmoed. 1980) in the neotropics, and Pyxicephalus adspersus in Africa (Brach, 1976) eat snakes, including highly venomous species in the first and last cases. A Bufo terrestris ate a brightly colored colubrid snake, Cemophora coccinea (Brown, 1979); a Rana warschewitchii regurgitated a small iguanid lizard, Anolis humilis (Greene. Donnelly, Guyer, and Santana, unpubl.); and a R. grylio ate an Eumeces laticeps (Lamb, 1984). Asiatic B. melanosatictus commonly eats blind snakes Hahn, 1976), Aquatic turtles occasionally fall prey to frogs (Graham, 1984), as do baby Alligator mississippiensis (Neill, 1971).

4. *Reptiles*

The sole living sphenodontid, Sphenodon punctatus, feeds on a variety of prey, including lizards (Gans, 1983a), but the few sympatric geckos and skinks are not a major part of its diet.

Apparently few turtles consumes other reptiles in the field, but Macroclemys temminckii is an important predator on other turtles (Pritchard and Trebbau, 1984). Captive observations suggest that a few other species also might eat reptiles in nature (e.g., Terrapene ornata, Legler, 1960).

Crocodilians feed on a variety of invertebrates and vertebrates, including turtles, lizards, and snakes. Crocodylus niloticus even eats highly venomous puff adders, Bitis arietans (Cott, 1961). Alligator mississippiensis (Neill, 1971) and C. niloticus (Pienaar, 1969) frequently, eat turtles, and several neotropical species occasionally feed on turtles, lizards. and snakes (Alvarez el Toro, 1974; Medem, 1981). Crocodylus palustris sometimes eats sea snakes and marine turtles (Allen, 1974). No crocodilian specializes on reptiles as prey.

Most lizards feed mainly or entirely on arthropods, but some occasionally eat other vertebrates and a few are reptile specialists (for general discussion, see Greene, 1982). Species that feed frequently on other lizards include two large teiids of the genus

Callopistes (pers. obs,), the iguanid Gambelia wislizeni (Tanner and Krogh, 1975; Parker and Pianka, 1976), the pygopodid Lialis burtonis (Patchell and Shine, 1986); and several species of Australian varanids (*e.g.*, Pianna, 1986; Losos and Greene, unpubl.).

Snakes are probably important predators on lizards in many regions (*e.g.*, Shine, 1977) for Australia; Mc Kinney and Ballinger, 1966, for Taxas; Fitch, 1949, for California; Duellman, 1978, for Ecuador; Henderson, 1984a for a West Indies). Some species specialize on other snakes (*e.g.*, Opiniongus hannah, M. A. Smith, 1943; Cylindrophis rufus. Greene 1988a Murarus fulvius. Greene, 1984) or amphisbaenians (Broadley, 1963 Papenruss, 1982). A few species of snakes occasionally feed on turtles (*e.g.*, Drymarchon carais, Ruthven, 1912; Agkistrodon piscivorus, Burken, 1966) and or crocodilians (*e.g.*, A. piscivorus Allen and Swindell, 1948; Eunemes murinus, Wehekind, 1955; an unidentified Eocene henophidian Greene, 1983a).

5. *Birds*

Birds are often enemies of reptiles throughout the world, and at many sites there exists an array of emphasis ranging from occasional to very frequent predation. Species that eat reptiles regularly, at least at times, include the secretary bird (Sagittarius serpentarius) and several other African raptors. Brown, 1971; Sweeney, 1971; Broadley, 1974)), roadrunner (Geococcyx californicus Bryant, 1916 Meinzer 1983) and numerous tropical hawks (Haverschmidt, 1962; Voous, 1969; Greene, Donnelly, Guyer, and Santana, unpubl.). Herons eat many yonng crocodilians (Alvarez del Toro 1974; Gorzula, 1985), and pelicans can eat moderately large land turtles (Land, 1931). Several species of marine raptors eat sea snakes (Heatwole, 1975) and a variety of birds eat young sea turtles (Stancyk, 1981). Although many diurnal reptiles are probably safe from nocturnal raptors, by virtue of the use of retreats, some owls regularly take nocturnal gekkonid lizards (Demeter, 1982). In addition to specialists on reptiles, many other birds occasionally or frequently eat them, including shrikes, jays, toucans, puffbirds, cuckoos, and others (for old but still useful reviews of bird predation on snakes, see Wall, 1906a; Guthrie, 1932).

6. *Mammals*

Few mammals regularly take reptiles as prey; examples include some viverrids (Broadley, in litt.), neotropical mustelids (Galictis sp., Jackson, 1979; Rodda and Burghardt, 1985), some foxes (*e.g.*, Vulpes macrotus, pers. obs.), badgers (Klauber, 1956; Messick and Hornocker, 1981; Kruuk and Mills, 1983; Broadley, in litt.), and some primates (Gans, 1964, 1965; Lorenz, 1971). Hyaenas in South Africa frequently eat tortoises "easily cracking their shells" (Pienaar, 1969). Several carnivorous bats eat nocturnal geckos and diurnal agamid and iguanid lizards (Goodwin and Greenhall, 1961; Douglas, 1967; Tuttle, 1967; Valdez and LaVal, 1971; Vaughn, 1976; Advani, 1981). Sea turtles, especially small individuals, are prey to a variety of terrestrial and marine mammals (Stancyk, 1981). Numerous other mammals in several orders prey occasionally on lizards, amphisbaenians, and snakes (*e.g.*, wild pigs, Sus scrofa, Valverde 1967), and even adult crocodilians are sometimes taken by lions (Panthera leo, Cott, 1961; Pienaar, 1969) and jaguars (P. onca, Alvarez del Toro, 1974). Nesting females of large arboreal lizards, Iguana iguana, are seasonally important prey for jaguars in northeastern Costa Rica (Greene Santana, and Braker, unpubl.). A bear (presumably Ursus arctos) perhaps tried to eat a Testudo graeca (Reed and Marx, 1959).

Injury Frequencies

Scars and missing bony parts do not necessarily index the intensity of successful predation (Schoener, 1979; Jaksic and Greene, 1984; Arnold, this volume), but they can provide a minimum estimate of successful escape from predators (Greene, 1973a), assuming no injuries due to social encounters, mishaps, or illness (Vitt et al., 1974; Khan and Tasnim, 1986a). Injuries are thus potential indicators of predator-prey interactions, both in terms of the frequency and nature of the wounds.

Predation is widely assumed to be an all-or-none phenomenon for prey, although unsuccessful predation is clearly essential for the evolution of antipredator mechanisms (Vermeij, 1982). Observations of encounters and scars indicate that reptiles often confront formidable enemies, incur injuries, and escape. I have found adult iguanid lizards (Callisaurus draconoides, Iguana iguana) with large, healed body scars. A small colubrid snake (Diadophis punctatus) that survived more than 5 minutes of mouthing by a captive kit fox (Vulpes macrotus) was seemingly healthy 2 weeks later pers. obs.). Recovery from serious wounds has also been reported for Sternotherus odoratus (Ernst, 1986), Terrapene coahuila (Brown, 1974), T. ornata (Legler, 1960), Phrynops zuliae (Pritchard and Trebbau, 1984), Crocodylus niloticus (Cott, 1961), Sauromalus abesus (Camp, 1916; Berry, 1974), Eumeces obsoletus (Fitch, 1955), Coluber constrictor (Brunson, 1986), Pituophis melanoleucus (Brunson, 1986), Xenochrophis piscator (Auffenberg, 1980), and serveral species of sea snakes (Heatwole, 1975).

The abilities of reptiles to survive serious injuries from a predator has relevance for the efficacy and mechanisms of defense. The responses to danger work because the prey push the costs of predation above an acceptable level, or because they deceive a predator about the magnitude of the costs. Because escape from predators can be accomplished despite actual physical contact the dynamics of an encounter might influence the behaviour of a reptile and a predator in subsequent interactions, but the extent to which this occurs in nature is unknown see Section III C).

Natural and Manipulative Experiments

Reptiles have existed on some small, isolated islands in the near or total absence of predation for long periods, and numerous cases demonstrate unusual tameness in those animals (*e.g.*, Anolis agassizi, Rand et. al., 1975). The mechanisms production such behaviour (*e.g.*, selection, habituation) remain unstudied (see Sections III C, V.C.).

Modern introductions provide valuable information on the negative impact of prcdators on population of prey that are not capable of defense against them. The arrival of armadillos (Dasypus novemcinctus) in peninsular Florida during recent decades occured contemporaneously with the disappearance of appropriate predators on those generalist hunters, and the result has been a marked reduction in the abundance of leaf litter organisms, including small reptiles (Carr, 1982 ; see also Redford, 1985; Greene, 1987). Introductions of mongoose and faral cats have had severe effects on the reptilian faunas of several Caribbean islands. In a particularly well documented case, the introduction of dogs and cats on a small islands devasted a population of large iguanid lizards, Cyclura, carinata, in 3 years (Iverson, 1978).

Although widely assumed to be a major source of mortality for many reptiles, the impact of predation on population parameters poorly known (sec Dunham et. al., this volume). Experimental exclusion of predators significantly increases the survivorship of particular phenotypes of Uta stansburiana (Ferguson and Fox, 1984).

Some Generalizations

Reptiles are subject to a wide range of predators, although the risks vary geographically as well as among predator taxa at a locality. Even species that might be expected to be expected to be relatively free of enemies are eaten, such as those that are predominantly nocturnal (many geckos), aquatic (crocodilians, many turtles), fossorial (amphisbaenians, scolecophidian snakes), heavily armored (cordylid lizards, tortoises), or highly venomous (Bitis, Naja). Studies of experimental populations, of reptiles subject of introduced predators, and of island reptiles underscore the widespread potential importance of predation as a factor in the lives of these animals. Brief speculation regarding the kinds of problems with which reptiles must contend—the sensory and prey-handling characteristics of their predators—is possible from the information presented above, but generalizations are necessarily broad and tentative.

Birds and mammals probably often are the most significant potential factors in the evolution of antipredator mechanisms in reptiles, for several reasons. They are by and large visually hunting predators, although some, especially mammals, are also guided by auditory and chemical cues. Many birds posses particularly acute vision, and thus are particularly likely to circumvent crypsis (Section V.A.). Parent birds and mammals often transport immobile, intact prey as food for their offspring, making lack of movement subsequent to capture (*i.e.*, failure to elicit additional prey immobilization tactics) a potentially viable defense option (see Sections V.A., VI.B). Birds and mammals have high metabolic rates and high daily energy needs, such that their consumption rates and impact on a prey population can be high even if the latter constitutes a small part of the overall diet.

Snake frequently prey on lizards and, to a lesser extent, on other snakes. Although some species appear to have good vision, many rely heavily on chemical cues to locate prey Burghardt, 1970) and usually begin consuming prey while it is immobilized (*e.g.*, Klauber, 1956; Greene, 1984). At equivalent densities, the impact of snakes and other ectotherms on prey populations will be far less than that of endothermic predators. Probably most terrestrial vertebrate predators are capable of solving at least moderately complex learning problems (see Burghardt, 1977a, for a review of learning in reptiles).

The available evidence suggests that many birds and mammals take species within a broad prey type (*e.g.*, size, higher taxon) and roughly in proportion to their abundances (Jaksic, 1986), although some might avoid preferentially certain dangerous (*e.g.*, Crotalus viridis, Fitch 1949) of particularly fleet (*e.g.*, Callopistes maculatus, Jaksic et. al., 1982) species. Some snakes also might feed indiscriminantly on abundant reptiles, but most species probably specialize on a few prey taxa (*e.g.*, certain Africa borrowing snakes that eat exclusively amphisbaenians, Broadly, 1963; see also Arnold, 1972; Savitzky 1983).

Some predators are "hunting habitat specialists" that feed frequently on reptiles as well as on any other appropriately sized, palatable organisms that they encounter. The neotripical crane-hawk (Geranospiza caerulescens) systematically searches bromeliads and holes in trees for large insects, frogs, lizards, snakes, and nesting birds (Sutton, 1954; Alvarez del Toro, 1971; Bokermann, 1978). Jamaican cuckoos (Saurothera vetula), crows (Corvus jamaicenis), and becards (Platypsaris niger) hunt for anoles and other prey in bromeliads (Cruz, 1972, 1973, 1975). Nocturnal lyre snakes (Trimorphodon spp.) hunt in crevices for resting diurnal iguanid lizards (pers. obs.), and coral snakes (Micrurus spp.) search for small reptilian prey in leaf litter (Greene, 1973b, 1984). African honey badgers (Mellivora capensis) locate a variety of prey, including many lizards and snakes, by digging at the bases of bushes (Kruuk and Mills, 1983; Broadely, in litt.). Such predators present a constant threat to otherwise hidden, inactive animals (*e.g.*, sleeping iguanids at night, quiescent geckos during the day), implying that selection of retreat site might be an important antipredator mechanism.

In summary, many reptiles, perhaps especially diurnal forms that are active above ground, potentially face a large number of predators with diverse prey detection and handling capabilities. Diurnal species may be vulnerable in the daytime to one set of enemies that perceive them visually and precipitate a chase, and to another set that use chemical cues to locate their retreats at night and simply seize the immobile prey. For example, a diurnal desert lizard, Cnemidophorus tigris, may be eaten by kestrels (Falco sparverius), roadrunners, and coachwhip snakes (Masticophis flagellum) during the day, and dug from its borrows by badgers (Taxidea taxus) and long-nosed snakes (Rhinocheilus lecontei) at night (Greene, 1986a; pers obs.). An analogous, reverse dichotomy of predatory problems probably confronts some nocturnal reptiles. A consequence for potential prey of this spectrum of interactions is that some defense is probly effective in some cases, but almost no defense works all the time.

INTRASPECIFIC VARIATION IN ANTIPREDATOR MECHANISMS

Knowledge of variation within an individual during its lifetime can help elucidate developmental mechanisms that shape the phenotype, such as learning and maturation. Variation among individuals and among populations can potentially help explain the naturc of genetic among populations can potentially help explain the nature of genetic and social transmission in the population, and the substrate upon which selection acts. An understanding of those phenomena requires that factors external to the organisms also be understood, including experience, stimulus control, and temperature effects.

Stimulus Control and Context

Potential predators might be perceived by visual, thermal, auditery, chemical, and/or tactile cues. Those modalities differ in the distance over which they are effective, and in the extent to which they are relevant to particular predators and prey. Ideally, two kinds of information are of interest; the response of reptiles to a controlled series of threatening stimuli, and field observations on the reactions of reptiles to potential predators.

Our nations of what constitutes defensive behaviour are heavily influenced by reactions to human collectors, and perhaps only for what reason does it appear that reptiles respond defensively primarily to visual and tactile cues. This has led to erroneous conclusions whenever routine or cautious handling was not sufficient to elicit reactions, as in the supposedly docile nature of venomous coral snakes (Gehlbach, 1972; Greene and Pyburn, 1973) and long-overlooked, noxious secretions of Lichanura trivirgata (Ruben, 1976). Oligonon violaceus presses a human captor with its tail tip and rarely bites when handled, but strikes vigorously and vibrates the tail when attacked by a mongoose (Wall, 1903; Schmidt, 1927a). In fact, some snakes (Weldon and Burghardt, 1979) and lizards (Prieto and Sorenson, 1975) perceive the proximity of predators by chemical cues, and the significance of that modality is surely widely underestimated (for negative evidence for Sceloporus jarrovi, sed Simon et. al., 1981). Moreover, contextual factors (see below and Section IV.C) might interact with sensory cues to determine whether a response occurs.

Observations and studies on several species demonstrate the role of different stimuli in elicting antipredator behaviour. A Nephrurus asper responds to mice, small snakes, and humans with a dramatic threat posture, vocalization, and lunges, but to large snakes by locomotor escape and crouching (Bustard, 1967, 1979). Similarly, Elgaria multicarinata actively threatens birds predators but flees from snakes (Fitch, 1935). Close approach and tactile contact prompt higher tongue flick rates than resting levels in three species of Nercdia, but tactile cues cause a more dramatic increase (Scudder and Eurghardt, 1983). Thamnophis radix responds more defensively head-hiding etc.) to more severe attacks (tapping the head) and more offensively (*e.g.*, striking) to less severe stimuli (tapping the tail) and at lower temperatures (Arnold and Bennet, 1984; see Fitch, 1965, for similar observations of T. sirtalis). A human finger more readily elicits strikes than a cotton swab presented to neonates of three species of Thamnophis (Herzog and Burghardt, 1986). Experimental presentation of carnivores and humans to rattle snakes (Crotalus viridis, Sistrurus catenatus) demonstrated that defensive responses are elicited by generalized visual cues (large moving objects) rather than particular configurational stimuli (Scudder and Chiszar, 1977).

The situation in Crotalus viridis is further complicated by environmental and social factors. Solitary prairie rattlesnakes threatened when away from cover emphasized locomotor escape, whereas those near cover usually remained immobile. Grouped rattelesnakes near cover most frequently resorted to escaping and threatening movements, suggesting social facilitation of individual responses to a predator (Duvall et al., 1985). Other examples of the role of context are Caiman crocodilus, juveniles of which feign death under water but react aggressively on land (Gorzula, 1978), and Bungarus caeruleus, which emphasizes head hiding in daylight but escape and biting at night (Khan and Tasnim, 1986b).

The anatomy and neurophysiology of hearing have been studied extensively in reptiles (*e.g.*, Wever, 1978), but the roles of air and ground-bone vibrations in elicting defensive responses remain poorly known. A tropical teiid lizard, Cnemidophorus lemniscatus, reactus, reacts to alarm cries of small birds by rapid escape to its burrow (Beebe, 1945), and desert geckos, Ptenopus garrulus, retreat to their holes if one steps on a nearby dry leaf or twig (Huey, in litt.). Snakes are widely reported to respond to ground-borne vibrations (*e.g.*, Klauber, 1956).

Genetics and Ontogeny

The phenotype reflects a complex interaction of genetic heritage, the environment, and factors within the developing organism itself (for general treatments, see Burghardt, 1977a, 1977b, 1978; Alberch *et al.*, 1979). This interaction begins at conception (or cleavage, in the case of uniparental forms), and assumes potentially much larger environmental components after birth or hatching. As a descriptive task, we can inquire whether a particular behaviour is present at hatching or birth, whether the behaviour varies within and among litters, whether the behaviour changes with age, and what factors control those differences and changes.

Failure to exhibit an appropriate response during a first predator encounter could be fatal to a young reptile, and a veriety of species typical defensive behaviour patterns are elicited in a taxonomically dixerse sample of young turtles, lizards, and snakes that lack prior experience with threatening stimuli (Appendix B). Nevertheless, that finding does not imply necessarily the lack of maternal and/or postnatal effects. The widespread survival of reptiles after serious predator encounters (Section II.D), the association of young crocodilians and some lizards with their parents (Shine, this volume) and examples of individual changes in response to humans (see below and Section III. E) all indicate the potential for experiential modification of defensive behaviour.

Most descriptions of defensive mechanisms in reptiles either concern adults of unknown background or congenital responses that closely resemble those seen in adults (Appendix B). A few examples illustrate changes with time or lack thereof, but the information provided usually is insufficient to distinguish experience from other factors (*e g.*, maturation of neuromuscular control mechanisms). Juvenile Chelodina longicollis are more likely to expel musk than are adults (Kool, 1981). Recently hatched Pseudemys scripta are more likely to remain immobile and less likely to flee than are adults (Moll and Ledler, 1971). Large Geoclemys hamiltoni are "belligerent," whereas small ones are "sky and gentle" (Minton, 1966). Large Sceloporus torquatus are more wary then small ones (Schmidt and Shannon, 1947). Small individuals of Cnemidophorus lemniscatus are easily approached and resume foraging within seconds after being frightened, whereas adults flee immediately and subsequently remain immobile for several minutes (Beebe, 1945). Newly hatched Crotaphytus collaris run about erratically when chased, whereas those a week or more in age flee directly to a familiar crevice or burrow (Fitch, 1956). Young Varanus varius are secretive, but adult are formidable adversaries and far less wary (Bustard, 1968c; Section IV.G). Juveniles of a Kalahari Desert lacertid, Eremias lugubris, are brightly colored and use a peculiar, jerky locomotion that resembles sympatric, noxious beetles, whereas adults are cryptic and run rapidly (Huey and Pianka, 1977). A number of lizards (Huey, 1982; Garland, 1985) and snakes (*e.g.*, Pough, 1980) exhibit ontogentic changes in locomotor performance abilities, and these are likely to be relevant in predator encounters.

Young Boa constrictor are quiet, whereas adults hiss loudly (Mertens, 1952a; Donnelly, Santana, and Greene, unpubl.). Adult Antillophis parvifrons flatten thier necks and elevate the head when threatened; behaviour that was not observed in newly hatched young (Sajdak and Henderson, 1982). One-day-old Epicrates angulifer strike aggressively, but only exude a mild musk at an age of 13 days (Sheplan and Schwartz, 1974). Juvenile copperheads, Agkistrodon contortrix, cannnot eject a stream of scent gland liquid until they are 8 days old (Gloyd, 1934) Pseudaspis cana do not begin to hiss and strike until 18 hours after birth (Branch, 1973). Neonates of Ninia sebae exhibit defensive head displays that resemble those of adults from the same population, but are somewhat wobby, a difference perheps attributable to a lack of muscular coordination in the young snakes (Greene, 1975). Young Masticophis lateralis typically go under rocks when chased, but adults of that racer will travel 1 m or more to ascend a bush (Sweet, in litt.). Juveniles and adults of six species of Australopapuan elapida resemble each other generally in defensive behaviour, but differ interspecifically in the presence of certain component postures during ontogeny (Johnson, 1975). Adult Pituophis melanoleucus are less likely than juveniles to hiss, vibrate their tails, and form striking coils (Sweet, 1985). Young Naja naja are more aggressive than adults (Wall, 1913a). Newly hatched Lachesis muta strike and vibrate their tails, but cease those responses at the age of 1 month (Switak, 1969; Solozano, Santana, and Greene, unpubl.); that could be attributed to captive habituation, but is also consistent with the phlegmatic behaviour of adult wild bushmasters from some populations (Greene and Santana, unpubl.). Young Vipera russellii are more alert and more easily provoked into biting than are the adults (Wall, 1907a).

Other potential modiffers of defensive behaviour include habituation, learning, and social transmission (see Burghardt, 1977a, for general discussion). A waning of antipredator behaviour in response to handling by humans and other aspects of capacity has been seen in Chelodina longicollis (Kool, 1981), Gambelia wislizeni (Crowley and Pietruszka, 1983), Heloderma suspectum (Bogert and del Campo, 1956), Heterodon platirhinos (Platt, 1969), Pituophis melanoleucus (Sweet, 1985), and some Crotalus spp. (Klauber, 1956). Conversely, Tropidurus albemarlensis that were initially tame became exceptionally difficult to capture after experience with a noose (Stebbins et al., 1967). Crocodylus porosus that have been caught once by humans at night subsequently avoid boats at distances of more than 100 m (Bustard, 1968a).

Most reptiles are not especially social, but the possibility exists that intra-and interspecific responses to predators are more widespred than currently throught. Wild adult Iguana iguana typically follow suit when one lizard jumps out of a tree in response to a predator. However, members of a group that were habituated to a human observer over a period of days subsequently failed to exhibit escape behaviour in response to that of a nonhabituated, transient iguana at the site (Greene *et al.*, 1978). Parental protection of young crocogilians is well known (Shine, this volume), but recent observations on Caiman crocodilus show that distress vocalizations can attract several conspecifics of various sizes and sexes (Gorzula, 1985). The remarkable observation that tropical whiptali lizards, Cnemidophorus lemniscatus, flee in response to alarm calls of birds (Beebe, 1945) deserves detailed study.

The heritability of antipredator mechanisms in vertebrates is inadequately known (*e.g.*, fishes, Vadasz et al., 1978), and the only estimates for a reptile are based on full-sibling analyses for Thamnophis radix (Arnold and Bennett, 1984). The variation among litters in that garter snake is extensive some of it context dependent, and encompasses much of the defensive behavioural variation seen among other colubrids. Replicate trials show that there are stable behavioural phenotypes for individual T. radix, and heritability estimates ranged from 0.37 for single trials to 0.45 for replicate trials. Selection on defensive responses has not been studied in the field, but the recent work by Arnold, Burghardt, and colleagues suggests that is feasible.

Intra-and Interindividual Variation

Apparent intrapopulational variability could be caused by differences in age, stimulus intensity, thermal effects, and/or other undersected contextual variables. Real individual variability might also be dependent on sex. Unfortunately, almost all reports of individual differences in the defensive behaviour of reptiles are not controlled with respect to one or more of those variables.

Diadophis punctatus, a small colubrid snake, illustrates the problems of assessing behavioural variability. An initial study within a Kansas population found that 74 of 100 snakes tested with visual and tactile stimuli displayed their tails. Of a second 100 snakes, tested when initially encountered in the field, 96 displayed. The sex ratio in that sample was approximately equal, and the response seemed equally well developed in young and adults (Fitch, 1975). The difference between the two samples suggests that handling subsequent to capture might have reduced the tendency to display in some snakes. Another study in the same area found that ca. 66% of small D. punctatus displayed their tails, but only ca. 30% of the large snakes did so (Smith, 1974; see also Burghardt, 1978). Several factors probably are at work here, an this species is a prime candidate for detailed studies.

Two species of hognose snakes in Kansas are similarly likely to feign death when first observed 71% of 70 Heterodon nasicus, 83% of 35 H. palatirhninos; Platt, 1969). However, in Texas significantly fewer H nasicus feign death (52% of 25) than did sympatric H. platirhinos (73% of 51, Kroll, 1977). One species loses this behavior sooner in captivity, judging from the proportions responding when re-examined within 24 hours of capture (6% of 98 H. nasicos, 59% of 41 H. platirhinos, Platt, 1969). Two iddividuals of H. simus deffered in the extent to which various behavioral components were included in their defensive displays (Myers and Arata, 1961).

Four of fifty-six male and three of forty-one female horned lizards, Phrynosoma cornutum, squirted blood from their eyes during a total of 18 encounters in which they were handled and toe-clipped by humans. The frequency of that behavior was unaffected by sex of the lizards or by ten environmental variables (Lambert and Ferguson, 1985). Other indications of individual variability include 5 of 65 (7.7%) P. solare that squirted blood (Parker, 1971), 12 of 80 (15%) Diplodactylus williamsi that ejected caudal gland secretions (Bastard, 1964), 10 of 32 (31.3%) of Thelotornis kirtlandii that bit when captured (Sweeney, 1971) and 1 of 26 (3.9%) neonatal Lampropeltit alterna that feigned death (Tryon and Guese, 1984). The tendency of Terrapene carolina (Nichols, 1939) and of T. ornata (Legler, 1960) to withdraw and close their shells varies individually, but each of 169 T. coahuila did so (Brown, 1974; see also Sections IV. B, V.A).

Defensive responses vary with temperature in several species. At high body temperatures, Agama mutabilis and A. savignii flee rapidly when threatend. At lower temperatures (which reduce sprint speed), the lizards remain stationary, adopt threat postures, and bite readily (Hertz *et al.*, 1982). Several species of iguanids flee from predatores at a greater distance at lower body temperatures (Anolis lineatoputs, Rand, 1964, Sceloporus cyanogenys, Greenburs, 1976; Tropidurus peruvianus, Huey, 1977; Uta Stansburiana, Tinkle, 1967). Leopard lizards (Gambelia wislizeni) are more aggressive and vocalize more readily at lower temperaturs in captivity (Crowley and Pietruszka, 1983) and presumably for that reason are more likely to vocalize earlier in the morning in the field (Jorgensen et al., 1963). The defensive behaviors of Bitis caudalis (Gans and Mendelssohn, 1972) and Thamnophis radix (Arnold and Bennett, 1984) vary with body temperature, and flight distance of Regina septemvittata varies inversely with body temperature (Layne and Ford, 1984). However, there is no relationship between incidence of death feigning and temperature in Masticophis flagellum (Gehlbach, 1970) or Natrix natrix (Heusser and Schlumpf, 1962).

Gravid female Lacerte vivipara allow a humen to approch more closely and flee for shorter distances than do nongravid conspecifics under field conditions, presumably reflecting increased reliance on crypsis in the face of reduced ability to flee during pregnancy (Bauwens and Thoen, 1981). Gravid females of Leiolopisma entrecasteauxi run slower bask more often, and are more likely to be caught by snakes than are mals (Shine, 1980a). Males of three species of Nerodia (Scudder and

Burghardt, 1983) and of Telescopus dhara (Zinner, 1985) are more aggressive toward simulated predators than are females. Old males of Acanthodactylus schreiberi are perhaps less able to flee than younger individuals (Zinner, 1967). Camouflage, age, and sex affect flight distance in certain iguanids (Heatwole 1968; Johnson, 1970). Fatigue alters defensive behavior in some snakes (Gans and Mendelssohn, 1972; Pough, 1978; Arnold and Bennett, 1984) and agamid lizards (Anderson, 1963; Heatwole, 1970). The physiological bases for such differences remain unstudied.

The complexities of individual variation are further illustrated by two recent studies. Evasion distances (from a human) by the adder, Vipera berus, are short during cool spring and fall conditions, but longer during the summer. There are no differences among males, females, and subadults, and all flee at significantly shorter distances when rodents are scarce and adders are undernourished (Andren, 1982). Male Galapagos lava lizards, Tropidurus albemarlensis, are faster and allow closer approach then do females; lizards held in captivity a short time are faster than those held for longer periods; and gravid females are slower than nongravid females. Males from sparsely vegetated areas are faster than those from heavily vegetated areas, and both sexes are more wary in sparsely vegetated areas (Snell *et. al.*, in litt.). [Note added in proof: Hailey and Davies (1986) reported detailed experimental studies on variable defensive behaviour in Natrix maura.]

Antipredator responses that are currently rare within populations deserve further study. Examples include ball forming in Chionactis occipitalis (Mitchell, 1978), and death feigning in Masticophis flagellum (Gehlbach, 1970) and Storeria occipitomaculata (Jordan, 1970). Rare behaviour might represent lack of observations (not likely for the above examples, all of which are of common, widespread North American snakes), chance occurrences that have no consequences for individuals, or the behavioural varation upon which selection can act (see also Section III. C).

Geographic Variation

Individual variation can be translated into geographic variation as a result of selection across environmential gradient (Endler, 1977, 1986b), or because of geographic variation in some other variable (*e.g.*, experience) that is correlated with the expression of an antipredator mechanism. The only examination to date of the latter possibility found no relationship between presence or absence of regenerated tails (presumably an indication of predator encounters, see Section II.D) and wariness in Cnemidophorus tigris (Schall and Pianka, 1980). However, the fact that defensive behaviour can be modifiable and that reptiles show evidence of surviving encounts (Section II D) suggest that this topic deserves further study.

Most reports of geographic variation in defensive behavior are anecdotal and potentially confounded by the unknown roles of temperature, sex, stimulus strength, experience, and other factors discussed in the previous sections. Individuals of Kinosternon subrubrum from peninsular Florida are much more prone to bite in defense than are those from farther noth (Carr, 1940). At three localities in Baja California, Callisaurus draconoides either allows close approach and runs only a short distance; runs far along arroyo bottoms, or runs up steep slopes (Tevis, 1944; Asplund, 1967). Throughout most of its range in western North America this species displays the brightly marked underside of the tail, but does not do so in the southern half of Baja California (Tevis, 1944). Sauromalus obsesus and Sceloporus orcutti are much move wary in Baja California than farther north, and the differences are correlated with conspicuous dark coloration, occurrence on pale granite, and increased abundance of raptors at the southern localities (Welsh, in litt.). Urosaurus ornatus is a very shy lizard at some places in the Guadeloupe Mountains of New Mexico, but quite tame in nearby parts of the same range in Texas (Mosauer, 1932). Eremias barbouri escapes into rocks at one locality in China and into bushes at another (Schmidt, 1927a). Geographic variation is weakly indicated for presence or absence of a tail display in Sphaerodactylus glaucus (Harris and Kluge, 1984) and Phrynocephalus arabicus (Arnold, 1984a).

Each of 19 Ninia sebae from southern Mexico exhibited a head display (Greene, (1975), but only 1 of ca. 15 (6.7%) from Belize did so (Henderson, 1977). Coluber constrictor frequently ascends trees to escape in Georgia and North Carolia, but almost never does so in Florida (Carr, 1940). Masticophis flagellum piceus in southern California typically defends itself by rapid escape and vigorous striking is cornered; animals from farther north in the state, M. f. ruddocki, from a stiff; irregular coil with the head concealed when restrained (Sweet, in litt.). Boomslangs (Dispholidus typus) in a treeless, mountain region of southern Africa are more aggressive than elsewhere(Broadley, 1983). Adults of Lachesis muta are usually mild mannered on the Atlantic versant of Costa Rica (Solorzano, Santana, and Greene, unpubl.), but very aggressive on the Osa Peninsula of that country (Bolanos *et al.*, 1978). Pituophis melanoleucus and Crotalus viridis show concordant geographic variation in the tendency to vibrate their tails, hiss, and from striking coils (Sweet, 1985).

Cnemidophorus tigris, a fast moving teiid lizard found in much of the western United States, exhibits considerable geographic variation in defensive responses (Schall and Pianka, 1980). Escape behavior diversity varies positively with incidence of regenerated tails among 12 sites, suggesting a possible ralation to predation intensity This species relies on high speed, but variation among the frequencies of 12 possible responses to approach by an observer at one site varies from 0.3% (for "run from shrub into open") to 38% (for "run from one shrub to another").

Diadophis punctatus in the western United States, peninsular Florida, and northern Mexico has bright orange or red posterior ventral surfaces and a spectacular defensive tail display. In most of the eastern United States the same species has a yellow venter and lacks the tail display. The two types intergrade in Illinois and northern Florida, where snakes with yellow venters have tail displays (Myers, 1965; Greene, 1973a). Although further studies on stimulus control and ontogeny are needed (see above), this appears to be a case of true geographic variation.

Intraspecific geographic variation in morphological defense mechanisms (other than colour pattern) is evidently uncommon. Among cobras, all Hemachatus haemachatus, Naja mossambica and N. nigricollis eject venom in streams, but "spitting" only occurs in some populations of N. naja in Asia (Bogert, 1943). Short-horned lizards, Phynosoma douglassi, are more spinose in southern paris of the species range, where predators are more common (Pianka and Parker, 1975).

SYSTEMATIC VARIATION IN ANTIPRIDATOR MECHANISMS

This section summarizes the taxonomic occurrence of anti-predator mechanisms in reptiles, and seeks to identify patterns of evolution in those attributes. The latter involves a search for taxon-typical patterns and their assessment in terms of phylogenetic relationships. Given a sufficiently large sample of taxa and a phylogeny for the group in question, inferences can be made about the origins and changes that have occurred in those traits.

The term adaptation is used herein to describe a feature that promotes a performance advantage (relative to the next most ancestral condition) that was associated with the origin of feature. This usage is tantamount to saying that a performance advantage are derived at the same level of descent within a clade, and in the absence of additional data (*e.g.* Arnold, 1983; Arnold and Bennet, 1984; Endler, 1986b), does not imply a particular process such as natural selection in the origin of an adaption (see Greene 1986b). Evidence for performance advantages is given in Section V.A and some general trends are discussed in Sections V.B. and VI.C.

The separate consideration of taxa in this section is according to their assignment to monophyletic lineages, except as noted, and the extent to which they can be conveniently and accurately and characterized in terms of antipredator mechanisms. It summarizes current views on the phylogenetic relationships of major tetrapod taxa. A feature is judged to be derived if it is absent in two successive outgroups, or, in the absence of precise information on relationships, rare in most other closely related taxa (Wiley, 1981; Maddison *et. al.*, 1984). Unless otherwise noted, sources for this section ara listed by specific taxa in Appendices A, B, and D. Detailed descriptions of antipredator mechanisms are given in Appendix C

Testudines

Few predators seem to specialize on turtles, with the possible exception of some birds and crocodilians, alligator snapping turtles (Macroclenemys temminckii), and humans. Turtles of various species, particularly as juveniles, are sometimes eaten by a variety of fishes, frogs, snakes, and mammals (Section II, Appendix A; Auffenberg and Iverson, 1979).

Antipredator behaviour in turtless emphasizes crypsis, flight into water or a burrow, withdrawal into the shell, clocal discharge, postural threats, and biting. Other defense mechanisms are less common, and include chemical deterrents (*e.g.*, Pelomedusa subrufa, Pienaar *et. al.*, 1978; Sternotherus odoratus, Eisner *et. al.*, 1977) aposematic colouration (perhaps juveniles of Chelus fimbriatus, Pritchard and Trebbau, 1984, and Platysternon megacephalum, Campbell and Evans, 1972), death feigning (Kinixys belliana, Schmidt, 1919), and sound production (kinosternids, Gans and Maderson, 1973; P. megacepinalum, Campbell and Evans, 1972). Malacochersus tornieri has an unusually flat in flexible shell for a testudinid, escapes predators by rapid movements into a rock crevices (Proctor, 1922; Ireland and Gans, 1972).

The defensive behaviours of only a few turtles have been described in detail. Chelydra serpentina (Chelydridae) is a large equatic of estern species and southern North Ameria. This commonsnapping turtle is cryptic and moves away if a threat is perceived at a distance, especially if the turtle is in water. If approached closely on land, a snapping turtle elevates the posterior parts of its body, hisses loudly, and gapes. The turtle pivots, sometimes using the tail for additional support, such that the shell and gaping mouth continue to orient towards a moving aggressor. If the threat is nearby, a snapping turtle strikes violently and hangs on firmly if the attacker is seized If handled, a snapping turtle emits an offesive glandular secretion, struggles with its limbs, and continues to attempt to bite (Mertens, 1946; Ernst and Barbour, 1972; Dodd and Brodie, 1975).

Belle's hinge-backed tortoise, Kinixus belliana (Testudinidae), occurs in savana and swampy areas throughout much of Africa. This turtle responds to an approaching person by walking away, and if halted withdraws the limbs into the shell, and some times feigns death. If picked of up, it stratches withdraws the limps into the shell snaps the beak and hisses and sometimes discharge small quantities of a strong smelling fluid from the cloaca. If further harrassed, it expels air from the lungs, draws the forelimbs in over the head, and shuts down the hinged, posterior portion of the carapace (Schmidt, 1919).

Terrapene ornata (Emydidae) is a terrestrial of important relatively open areas in the south-central United States. If an ornate box turtle sees an approaching predator at a distance it freezes with limbs and head extended. If the threat draws near, the turtle either moves away with surprising speed, or withdraws and retracts the plastral elements to form a tightly closed box around its head and extremities. Some individuals struggle violently, bite, expel a pungent secretion from integumentary glands, and urinate copiously if handled. Others simply remain closed up in their shells until danger has passed (Norris and Zweifel, 1950; Legler, 1960).

The shell is a defining characteristic of turtles, and most authors have concluded without explicit arguments that it was a structural adaptation against predators in a semiaquatic ancestor (*e.g.*, Carr, 1952; Pritchard, 1979). Boker's (1935) hypothesis, that the shell functions as a plow in pushing through dense brush, was soon labelled "sheer nonsense" (Schmidt, 1939). A consideration of the phylogeny and ecology of living turtles supports the traditional view and suggests an interaction between feeding and defense in the origin of the shell. Although diversity in the feeding behavior of turtles is substantial. most species in most higher taxa evidently search actively for relatively immobile food such as plant parts (*e.g.*, South American pelomedusids that eat submerged palm fruit) or mollusks and other slow-moving animals (see Ernst and Barbour, 1972; Pritchard, 1979, 1984; Pritchard and Trebbau, 1984; Bury, 1986). The domed shell configuration typical of fully terrestrial turtles is an independently derived feature of testudinids and some emydids (Terrapene), as are the fast active (trionychids, some chelids) and stationary ambush (Macroclemys, Chelus) modes of hunting (for discussions of turtle phylogeny, see Bickham and Carr, 1983; Gaffney, 1984, 1985). These considerations imply that primitive turtles were relatively slow-moving animals; their feeding tactics of sustained searching and or long handling time were correlated with protection by the shell. The turtle shell was evidently a key innovation that facilitated a potentially dangerous feeding mode, and specialized defense mechanisms also typify several other reptilian lineages with those foraging characteristics (Sections V.B. VI C).

Shell-closing mechanisms have evolved at least seven times in turtles, and provide particularly interesting examples of the interactions of feeding, locomotion, and defense. For example, phylogenetic analyses of functional and ecological attributes suggest that anterior plastral kinesis evolved independently in two lineages of kinosternid turtles, in each case apparently associated with opening the shell to permit large head size, durophagous feeding, and defensive gaping (*e g.*, Claudius angustatus; Dodd, 1978) In Kinosternon, the primitive condition has been converted to a shell-closing mechanism, and there is also posterior plartral kinesis. The morphoclines within kinosternids are correlated with shifts away from durophagous feeding, increased emphasis on locomotion and protection of body parts rather than aggressive defense, and life in arid regions (Bramble 1974; Bramble and Hutchinson, 1981; Bramble *et al.*, 1984).

Crocodilia

Adult crocodilians generally are not subject to predation, except by humans (Neill, 1971), although lions (Panthera leo), jaguars (P. onca), and sharks eat large Crocodylus spp. (Section II, Appendix A). These reptiles generally respand to intruders by flight into water or aggressive attacks with their jaws and tails, but there are few detailed descriptions of antipredator behavior by crocodilians. Other responses to predators include death feigning, hissing, vocalization. cloacal discharge, exuding secretions from throat glands, gaping, and writhing.

In Venezuela, individuals of Caiman crocodilus typically swim away from a potential predator. They often dive, and subadults bury in the bottom mud. Hatchlings feign death when handled under water, but struggle, vocalize, and bite when handled on land. Nearby caimans sometimes responds to the distress calls of juveniles by attacking a potential predator (Gorzula, 1978, 1985). Juvenile crocodilians are subject to heavy predation, and perhaps this has played a major role in the evolution of parental care in these animals (Shine, this volume).

Some features of crocodilians used in defense (*e.g.*, stout tails and strong jaws) are undoubtedly coopted for that role, because those attributes function primarily and primitively in locomotion and feeding. However, the distribution and development of bony scutes varies among living crocodilians, and the possibility that these structures are adaptations for resisting injury from predators should be investigated (cf. Seidel, 1979; Ross and Mayer, 1983).

Rhynchocephalia

The sole living member of this group, Sphenodon punctatus, is largely free of native predators, but might be subject to diurnal predation by hawks (Saint Girons *et al.*, 1981). These animals are coloured cryptically and freeze or take refuge in burrows when approached. Responses to handling and/or intraspecific aggression include erection of a dorsal crest, inflating the body, gaping, biting, and hissing (Carpenter and Ferguson, 1977; Gans, 1983a; Gans *et al.*, 1984). Tail autotomy is present, but regenerated tails might reflect the incidence of aggressive social interactions rather than attempted predation.

Iguanian Lizards

This taxon includes the Agamidae, Chamaeleonidae, and Iguanidae. The first two are more closely related to each other than either is to the third (Estes, 1983). Iguanians are generally diurnal, terrestrial or aboreal, small, and insectivorous. They are eaten by a variety of reptilian, avian, and mammalian predators (Section II, Appendix A). As a group they tend to have somewhat stocky bodies, with the notable exception of the laterally flattened chamaeleonids. Evolutionary limb loss is unknown in the Iguania.

Most iguanians are probably cryptic, and initially respond to a threat by immobility and locomotor escape. Flight is often to an adjacent microenvironment where pursuit by a predator would be difficult, such as up a tree or into a burrow. In a few instances it involves a radical shift in habitat, such as into air or water. Morphological specializations for locomotor escape are infrequent, but include features associated with running on sand (*e g.*, fringes in Uma, Pough, 1970; Carothers, 1986), running on water (*e.g.*, fringes in Hydrosaurus, Luke, 1986), gliding (Draco, Colbert, 1967), and blocking burrows or crevices with the tail (*e g.*, Ctenosaura clarki, Duellman and Duellman, 1959).

Some iguanians have exaggerated threat displays, which may be accompanied by morphological adjuncts (*e.g.*, the frill and its erection mechanism in Amphibolurus barbatus. Throckmorten *et al.*, 1985, and the black throats of Gambelia and some species of Agama, Hertz *et al.*, 1982. Most species will bite if handled, and in some this can be a formidable defense (*e.g.*, Gambelia wislizeni, Crowley and Pietruszka, 1983; Cyclura carinata, Iverson, 1979). Less common defensive mechanisms include rigid immobility (Brookesia sp., Bohme, 1974), tail display (Cophosaurus texanus, Dial, 1986; Phrynocephalus helioscopus,

Obst, 1959), lashing with the tail (Iguana iguana, Greene *et al.*, 1978; Uracentron azureum, Greene, 1977; Hydrosaurus pustulosus, pers. obs.), vocalization (Gambelia wislizeni, Crowley and Pietruszka, 1983; Liolaemus chiliensis, Carothers, 1987; Otocryptis wiegmanni, Matuschka, 1978), and group responses (Agama stellio, Zinner, 1967; Iguana iguana, Greene *et al.*, 1978).

Fourteen species of the amuaniagenus Phrynosoma are characterized by flattened, spinose bodies and the presence of sharp, sturdy spines ("horns") on the head. There is ample evidence of the efficacy of those structures in thwarting prenom both direct in the reactions of predators (*e.g.*, Tollestrum 1982, Section V.A.) and indirect in the fact that P. douglassi has hignry returned horns in parts of its range where predation evidently is lower man elsewhere (Pianka and Parker, 1975). Some species of Phrynosomia scrum a stream of blood from the eyes, hitting objects up to 2 m away. Section III.D). This response involves derived aspects of the cernain acculation, themselves modified from a vascular mechanism of head temperature regulation, and contraction of the protrusor oculi muscles (Burneson, 1942; Heath, 1966). Observations on captive predators suggest and blood squirting by horned lizards is specifically effective against minds (Cowles, 1977). The Australian agamid Moloch is curvengence in some respects with Phrynosoma, including a flattened, spurose body and stationary foraging for considerable periods of time ant nest and trails (Planka and Pianka, 1970; Greene, 1982; Section VAMB.

Detailed descriptions or arnuredator mechanisms in iguanians are few, but three case studies illuscare some of the diversity in this group:

Corytophanes cristatus Cguarudae is a slow-moving, cryptic, arboreal neotropical forest iizard, individuals remain stationary unless disturbed, then move away with a hupping gait and frequent changes of direction. When ceasing voluntary incomotion. this helmeted iguana maintains whatever position it happens to be in. If threatened by a snake, the lizard orients perpendicularly to the attacker, elevates its trunk, depresses a throat fan, and lowers the head, thus elevating a dorsal crest. The overall effect is to increase the apparent lateral aspect of the animal approximately four-fold. If the threatening stimulus changes position, the lizard shifts its position to maintain a lateral presentation, and occasionally bobs its head. Sometimes a lunge is made at the threat, the lizard bites fiercely if grasned and law strength is enhanced by elongate adductor muscles (Davis 1958, Schwenk, 1980; pers. obs.).

Iguana iguana (Iguanidae) is a large, cryptic, herbivorous neotropical forest lizard. Juveniles are highly social, and tend to move during periods in which the wind is moving the vegetation around them. They respond to the approach of a predator by mass flight on the ground, into low vegetation, or across the surface of the water. Juveniles thrash and sometimes defecate when handled. Adults leap off their perches in the canopy and either disappear into the vegetation below or take refuge in an adjacent stream or take. Prolonged submergence is facilitated by high tolerance for anaerobic metabolism. If cornered or seized, adults thrash and scratch with their claws, lash with their tails, and sometimes bite (Moberly, 1969; Burghardt *et al.*, 1977; Greene et al., 1978; Meritt, 1981).

Phrynocephalus mystaceus (Agamidae) is a small, cryptic, Asiatic desert lizard. This animal sometimes burrows rapidly in sand with its feet if threatened. If the source of danger is nearby, the lizard jumps, rises high on its legs, and faces the attacker. The tail is elevated and curled repeatedly, both vertically in its entirety and laterally at its curcled tip. The mouth is opened widely, unfolding the fringed flaps at the angles of the jaws. At this time the body is inflated and the buccal tissues are engorged with blood, such that the mouth appears bright red. The tongue is not protruded. If seized, the lizard bites fiercely and hangs on to the predator (Nikolskii, 1915; Mertens, 1952b; Polynova, 1982).

Interfamilial trends among iguanians are few. Chamaeleonids depend heavily on crypsis and aggressive defenses, rather than on flight. Iguanids tend to have lateral displays, whereas agamids more frequently have frontal displays that emphasize enlarged teeth, flaps, fringes, and other cephalic structures. The reasons for this difference are unknown, but it parallels differences in social displays (Davis, 1953; Carpenter and Ferguson, 1977). In any case, agamids perhaps are derived from within the Iguanidae so their frontal displays might represent modifications of a more primitive lateral posture.

Gekkotan Lizards

All recent considerations of squamate relationships agree that gekkonids and pygopodids are sister taxa (together the Gekkota), and in turn the sister taxon of other noniguanian squamates. The Gekkota is of particular interest, in that it clarly involves the evolution of limblessness (Shine, 1986) in a group that distinctive in defense relative to other lizards.

The Gekkonidae is a large cosmopolitan group, currently divided into four subfamilies (Kluge, 1967). Gekkonines and diplodactylines are usually nocturnal, have specialized scansorial pads on their feet, and rely on crypsis and rapid locomotion for initial escape from predators. These geckos are eaten by diurnal predators, such as snakes, that catch them in retreats, and by snakes, owls, and bats at night (Section II, Appendix A). Most species probably adopt elevated threat postures when cornered, and vocalize and bite if seized (*e.g.*, Teratoscincus scincus, Mebs, 1966; Nephrurus asper, Bustard, 1967). Tail autotomy is almost ubiquitous. Less frequent antipredator mechanisms that are derived within the family include epidermal sound production (all species of Teratoscincus. Gans and Maderson, 1973), gliding with the aid of skin flaps (all species of Ptychozoon. Marcellini and Keefer, 1976), skin loss (*e.g.*, Aristelliger hechti, pers. obs.), and glandular secretions from the tail (some Diplodactylus. Rosenberg and Russell, 1980).

Eublepharines are nocturnal, terrestrial geckos that vocalize, flee, and in some cases tail display from an elevated posture when threatened by predators. Sphaerodactylines are diurnal geckos that apparently rely only one on crypsis and flight, although little is known of their responses to potential predators. One species has a defensive tail display (Sphareodactylus glaucus, Harris and Kluge, 1984), and one has large eyespots that possibly frighten predators (S. semasiops, Thomas, 1975).

Exaggerated postures and vocalizations have been reported as antipredator mechanisms in eublepharine, diplodactyline, and gekkonine gekkonids (*e.g.*, Mebs, 1966, 1973; Johnson and Brodie, 1974; Frankenberg, 1975; Bustard, 1977). The Eublepharinae include terrestrial, slow-moving species that are not known to use vocalizations during social behaviour (Marcellini, 1977, 1978). It is the sister taxon of other geckos, and the latter as a group are probably primitively noctutnal, specialized for climbing, and use complex intraspecific vocalizatians, Milton and Jenssen (1979) concluded that vocalizations in certain Anolis function in defense (as they do in a few othe iguanids, *e.g.*, Gambelia wislizeni, Liolaemus chiliensis, see above), and they speculated that such sounds could represent precursors of intrapccific vocal communications. Phylogenetic relationships among gekkonids imply that the evolutionary sequence hypothesized by Milton and Jenssen actually occurred in that family: Vocalizations arose as a defensive adaptation in ancestral geckos and were later coopted and adaptively modified (for increasing complexity; see Marcellini, 1978; Werner *et al.*, 1978) within the noneublepharine geckos for intraspecifice communicative functions.

Pygopodids are limbless or nearly limbless, elongate relatives of geckos. Their predators include bats and snakes (Appendix A). Three species in two genera exhibit threat displays and false strikes that might mimic venomous elapid snakes (Bustard, 1968b; but see Kluge, 1974:161), and also use saltation as an antipredator mechanism (Bauer, 1986). A captive Lialis burtonis writhed and defecated when restrained, but did not bite (pers. obs.). Interestingly, pygopodids are not known to gape in response to predators, a behavioural pattern that is widespread in geckos. [Note added in proof: Kluge (1987) analyzed phylogenetic relationships among higher taxa in the Gekkota, reviewed literature on defensive vocalizations in that group (including pygopodids), and tentatively concludes that response in primitive for tetrapods.].

Autarchoglossan Lizards

1. *General*

The Autarchoglossa are a diverse group of ten families, the precise relatihnships of white to each other and to amphisbaenians and snakes are uncertain (Estes 1983; Estes et al., 1988). This uncertainty restricts the excent to which generalizations are possible, a problem made even more senous by the extreme heterogeneity of body form and life styles among the species in these families.

2. *Xantusiidae*

These secretive lizards are eaten by snakes, other lizards, and perhaps other predators Appendix A). They struggle, defecate, autotomize their tails, and bite flercely when handled.

3. *Sciencidae*

Most skinks probably are usually solitary, territorial during the breeding season, and feed by intensive searching (Stamps, 1977; Regal, 1978). They are eaten occasionally by a variety of vertebrates, bot few predators specialize on skinks (Section II, Appendix A; exceptions include certain snakes see below).

A defensive respose that is nearly ubiquitous for skinks and possibly primitive for the family is strong, effective writhing grasped. Skinks generally have elongate bodies and shiny scales that are underlain by osteoderms, and axial bending makes them even more difficult for predators to manipulate. Some owls (Janzen and Pond, 1976) and snakes (Vinson, 1975) have difficulty holding these lizards, and lizards and snakes that feed predominatly on skinks have specializations for holding them (Greene, 1978; Savitzky, 1981; Patchell and Shine, 1986). The hypothesis that smooth overlaping scales and enhanced axial bending are defensive adaptations is confounded by the fact that they also are functionally associated with lateral undulatory locomotion, which serves a variety of biological roles (Gans, 1975). Most skinks typically flee into cover when chased and squirm effectively, bite, and defecate when handled. Several large Australasian species have threat displays that include gaping, tongue extrusion, hissing, and lunging at a predator (*e.g.*, Tiliqua scincoides, Carpenter and Murphy, 1978).

4. *Lacrtidae and Teiidae*

Species in these families are heterogeneous in terms of size, morphology, and ecology, and the monophyly of the Teiidae is arguable (*e.g.*, Harris, 1985). Among one of the undoubtedly monophyletic lineages, the "macroteiids," at least some species have unusual abilities for sustained locomotion (van Berkum *et al.*, 1986; Garland, 1987). A number of macroteiids are eaten by fast moving predators (Section II, Appendix A), and search widely and intensively for hidden prey (Stamps, 1977; Regal, 1978). Teiids and lacertids typically employ crypsis and rapid locomotion to escape, sometimes into burrows. Holaspis guentheri glides out of trees as a means of site change (Schiotz and Volsoe, 1959), and might use that behaviour to escape predators. Most species probably struggle when handled; large ones use postural threats and bite violently (*e.g.*, Callopistes flavipunctatus, Lacerta lepida, pers. obs.), and C. flavipunctatus feigned death when dug from its burrow (pers. obs.).

5. *Cordylidae*

Cordylids are eaten by snakes, birds, and mammals (Appendix A). They often flee to rock crevices or into burrows, and lash with their tails and bite when handled. Cordylus cataphractus grasps its tail when threatened, forming an immobile circle (Rose, 1962). Gerrhosaurus major runs, takes refuge in a burrow, feigns death, scratches, lashes the tail, inflates its body, and bites (Schmidt, 1919).

6. *Dibamidae*

The two genera in this family are limbless, secretive, and of uncertain relationships (Greer, 1985; Estes *et al.*, 1988). Anelytropsis papillosus of Mexico burrows when uncovered, and bites and defecates when handled (Greer, 1985).

7. *Anguidae*

Alligator lizards and their allies are occasionally eaten by various vertebrates (Appendix A), and glass lizards (Ophisaurus spp.) are often taken by coral snakes in the southeastern United States (Greene, 1984). Anguids typically have elongated bodies and shiny scales uncertain by osteoderms. They are usually secretive, and take over when disturbed. Typical responses to handling include violent axial bending, defecation, open mouth threats, biting, and hissing. Species of Elgaria sometimes seize their own tail and encircle a branch, use their tails as "shields," and vocalize (Fitch, 1935).

8. *Xenosauridae*

The Xenosauridae includes two genera of poorly known lizands and is possibly polyphyletic (Hu *et al.*, 1984). Xenosaurus grandis of Mexico and Central America hides in crevices and will inflate, gape, lunge, and bite when cornered and handled (Alvarez del Toro, 1982). Shinisaurus crodilurus is aquatic, dives into water to escape, and bites readily when handled (Fan, 1931).

9. *Helodermatidae*

The two living species of helodermatids are relatively large, slow moving, venomous lizards. They are evidently rarely eaten by predators. The black and yellow colour pattern of Heloderma horridum is prsmitively cryptic, whereas the black and pink colouation of H. suspectum is derived and probably aposematic. If approached by a potential predator, beaded lizards and Gila monsters attempt to escape, sometimes into a burrow or up a nearby tree. If threatened closely, they hiss, inflate the body, elevate the foreparts and gape, then lunge laterally at the intruder. They struggle, discharge their cloacal contents, and bite hard if handled. Use of the jaws in defense is surely facilitated by toxic secretions, but whether venom injection in helodermatids is primarily an adaptation for feeding or for defense against predators is uncertain (Bogert and del Campo, 1956; Pregill *et al.*, 1986; Lowe *et al.*, 1986).

10. *Varanidae*

Monitor lizards (Varanidae, Varaninae) are eaten by a variety of other vertebrates. They are generally sky, fast animals that run to shelter at the first hint of danger. When approached closely, a varanid typically elevates its foreparts, arches the neck, and inflates the throat while hissing and slowly flicking the tongue. If a potential predator approaches closely, the lizard lashed accurately with its tail and lunges. Some monitors defecate, struggle violently, extrude the hemipenes, and bite when handled; at least three species feign death (*e.g.*, Varanus exanthematicus, Schmidt, 1919). Lanthanotus borneensis, the only living lanthanotine varanid (Pregill et al., 1986), squirms, defecates, struggles violently, and sometimes bites when handled (Proud, 1978).

Amphisbaenia

All amphisbaenians are fossorial (Gans, 1969), but at least some do move above ground in the daytime (*e.g.*, Beebe, 1945; Busack, 1978). They are eaten by burrowing snakes, armadillos, and at least occasionally by diurnal birds and mammals (Section II, Appendix A; Greegor, 1980).

The few species of amphisbanians that have been studied resemble autarchoglossans in that they bite, defecate, and writhe with exaggerated body movements when grasped by a predator. Some species are capable of rapid disappearance by burrowing when uncovered (*e.g.*, Blanus cinereus, pers. obs.). Species of Blanus and Trogonophis sometimes freeze in "pretzel" postures (*e.g.*, Stemmler, 1970; Malkmus, 1982a), and Agamodon anguliceps sometimes lies with its bright pink venter exposed (Gans, (1969). At least two species of Amphisbaena have defensive tail displays, and these might direct attacks to the expendable tail as well as serve as tactile cues for retaliatory biting (Greene, 1973a). The success of the former is demonstrated by the incidence of scars and incomplete tails in those animals (Gans, 1962; Greene, 1973a; Papenfuss, 1982), and the efficacy of the latter becomes obvious in handling them (Gans, pers. comm.; obs.).

The defensive behaviour of Amyhisbaena fuliginosa has been described in some detail (Beebe, 1946; Greene, 1973a). This moderately large, brightly patterned, neotropical species burrows rapidly when uncovered, and waves its elevated, curled tail if restrained. At times the display stops and the tail makes burrowing movements. Intensity of the tail movements increases when the animal is grasped firmly. but if lifted free from the substrate it swings the head back and forth with mouth gaping, and bites readily.

Serpentes

Morphological adjunets are widespread among snakes with defensive displays. and include putative aposematic colouration (Diadophis punctatus, Greene, 1973a; micrurine elapids, Greene and McDiarmid, 1981), tracheal rings that facilitate throat inflation (*e.g.*, Dasypeltis scabra, Gans and Richmond, 1957; Spilotes pullatus, Brongersma, 1957; Rossman and Williams, 1966), a glottal keel that enhances hissing in Pituophis (Martin and Huey, 1971; Saif, 1975), elongate anterior ribs that form a hood when abducted (Naja and other cobra genera), reinforced caudal vertebral structures in some erycine boas that elevate the tail (Greene, 1973a; Section V.C.), modification of the fangs for "spitting" in Hemachatus and some Naja (Bogert, 1943), nuchodorsal glands in some natricine colubrids (Smith, 1938), the rattle (Crotalus, Sistrurus, Klauber, 1956; see below), and perhaps vertebral protrusion in Mchelya and Bungarus (Savitzky, 1985). Conversely, snakes have fewer known locomotor specializations associated primmarily with defense than do lizards. Flying snakes (Chrysopelea spp.) have keeled vantral scales that facilitate gliding (Heyer and Pongsapipatana, 1970; Mertens, 1970), and sidewinding and saltation might utilize neuromuscular specializations that serve a defensive role (*e.g.*, Gans and Mendelssohn, 1972).

1. *Scolecophidia*

Three families of blind snakes are all characterized by slender bodies, short tails, smooth scales, and reduces eyes. All are fossorial, but occasionally emerge and even climb trees (*e.g.*, Vanzolini, 1970). They are eaten by scorpions, fishes, frogs, lizards, snakes, birds, and mammals (Section II, Appendix A).

Leptotyphlopids thrash when handled, sometimes displaying a brightly coloured tail tip *e.g.*, Leptotyphlops goudotii, Greene, 1973a); one species feigns death (L. scutifrons, Visser-1966), and another "takes to water" when pursued (L. longicauda, Pienaar *et al.*, 1978). Typhlopids thrash, discharge the cloaca, and press a captor the tail spine. A few specimens of Rhinotyphlops schlegelii squeaked when handled (Sweeney, 1971).

2. *Henophidia*

This infraorder includes serveral families that retain primitive features relative to caenophidians, but their relationships to each other and to the latter group are unclear (Dessauer *et al.*, 1987). Henophidians are conveniently discussed in terms of two superfamilies, although the monophyly of those groups also is questionable.

Anilioids (Aniliidae, Loxocemidae, Uropeltidae) are burrowing snakes with several lizard-like attributes. Some species respond to predators by thrashing, cloacal discharge, head hiding, and tail display *e.g.*, Cylmdrophis rufus, Greene, 1973a). Snakes and perhaps birds eat uropeltids (Gans, 1976; Appendix A), but nothing else is known of predation. Although most species are probably cryptic, some aniliids and uropeltids have bright, contrasting colours that are perhaps aposematic, mimetic or deflective (*e.g.*, Anilius scytale, Greene, 1973a; Uroreltis blytic, Gans, 1973). Uropeltids also respond to predators by coiling around twigs and other objects, and have reinforced tails that attract dirt and form protective coudal plugs (Gans, 1973, 1976; Gans and Baic 1977).

Booids (Boidae, Tropidophiidac, and perhaps others) are more diverse ecological and morphological than anilioids (Greene, 1983a), and are eaten by carbs, lizards, snakes, birds, and mammals (Section II, Appendix A). Tropidophiids generally respond to threates like anilioids, and some exude blood from the eyes and nostrils (*e.g.*, Tropidophis sp., Hecht *et al.*, 1955). Many boids hiss, strike, and bite fiercely (*e.g* , Boa constrictor, Greene, 1983b).

3. *Caegdphidia*

This infraorder includes more than half of the living snakes, but famitial limits and relationships among them are uncertain despite much attention (*e.g*,, McDowell, 1986; Dessauer *et al.*, 1987).

One species with a particularly complex repertoire illustrates much of the range of antipredator mechanisms in the largest caenophidian family, Colubridae. Heterodon platirhinos is a moderately large, nonvenomous or slightly venomous snake found in the eastern United States. The dorsal pattern typically consists of dark dorsal blotches on a lighter ground colour, but in some areas many individuals melanistic. When first disturbed an eastern hognose snake assumes an exaggerated S-coil, hisses loudly, and makes false strikes if approached closely. The tail is tightly coiled and in some instances elevated. That suite of behaviour is repeated if the snake is grasped. but after repeated harrassment an H platirhinos begins to writhe violently with its month open limply and to defecate. If placed on the ground the snake continurs to writhe for a short time, then assumes an inverted, limp posture, usually with the mouth open and tongue hanging out. If turned onto its ventral surface, the hognose snake immediately inverts the entire body. If the threatening stimulus is removed, the snake slowly rights itself and crawls away Edgren, 1955; Platt, 1969; Kroll, 1977 Sexton, 1979). Controlled observations on two litters demonstrated that the complete repertoire occurs in neonates,' that false strikes habituate more rapidly than noes death feigning, and that stable variation among individuals in a litter is substantial (Burghardt, Greene, and Klinghammer, unpubl; see also Section V.A. Appendix B).

Another widespread colubrid, Coluber constrictor, illustrates the responses of a slender, fast-moving species. A startled racer reacts by violent, undulatory movement that quickly changes to rapid, graceful locomotion for 30 m or more. It readily climbs or enters water in the course of escape. If startled a very close range, a racer sometimes coils with its head hidden and writhes, smearing scent gland secretions over its body. Alternatively, the snake assumes an S-coil, vibrates the tail, and strikes repeatedly. If seized, the racer continues to bite and discharge the cloacal contents, and twists its entire body so strongly that sometimes the tail is broken off (Fitch, 1963; Section III.E).

Atractaspidae includes a single genus of African burrowing snakes with powerful venom and long, movable fangs. The only record of predation was by a mammal (Section II, Appendix A). Species of Atractaspis respond to threats by rapid locomotion, thrashing, tail display, cloacal discharge, and biting. The head is held in a peculiar, ventrally flexed position when these snakes are disturbed; that is probably a prestrike posture, because bitting prey is accomplished by a posteriorly directed, ventral and/or lateral movement (pers. obs. on A. bibroni and A. microlepidotus). Greene (1979) interpreted the similar defensive behaviour of Atractaspis and certain elapids (see below) as convergence, but their response to predators might be homologous if the two families are indeed closely related (Cadle, 1982; McDowell, 1986.

The large family Elapidae includes dangerously poisonous snakes with predominantly neurotoxic venoms and short, relatively immobile fangs. Terrestrial and arboreal elapids are eaten by lizards, other snakes, birds, and mammals (Section II, Appendix A). Most are nervous snakes that respond to predators with rapid, often erratic locomotion. Many elapids bite readily, and one or more major clades within the family consist of species that elevate and flatted the foreparts as a threat. This behaviour and its structural correlates occur to varying degrees in "cobras" (two species of Aspidelaps, two species of Boulengerina, Hemachatus haemachatus, six species of Naja, and Ophiophagus hannah) and some mambas (Dendroaspis sp., Armitage, 1956). Three species of Afaican cobras (H. haemachatus, N. mossambica, N. nigricolls) and some populations of Asiatic N. naja eject streams of venom that are aimed accurately at the face of an enemy (Bogert, 1943). The relationships among cobras are not well understood, so it is not yet possible to evaluate the origins of hoods and spitting, but the phylogenesis of these antipredator mechanisms invites detailed study.

The approximately 50 species of New World elapids are characterized as a group by a suite of antipredator attributes (Appendix D), illustrated by the Sonoran coral snake. Micruroides euryxanthus is a small, crepuscular and rocturnal coral snake found in the southwestern United States and Mexico. Its bright red, black, and white or yellow ringed pattern might be both cryptic and aposematic. A Sonoran coral snake crawls rapidly away if touched lightly. Whed restrained or touched firmly, it writhes rapidly and violently, with the head tucked laterally and the tall elevated into a coiled loop or spiral. As the snake writhes and displays its tail, loops of the body move erratically in vertical and horizontal planes. The cloacal contents are expelled with a series of sharp, popping sounds that are audible to a human ear from a distance of at least 1 m. If released immediately, a Sonoran coral snake crawls rapidly away or freezes for a moment, often with its head tucked under a loop of the body and its tail elavated and coiled. It continues to writhe and bites repeatedly when restrained, and even a small adult is capable of envenemating a large mammal with serious consequences. Eventually a Sonoran coral snake may disgorge a recently swallowed prey snake, sometimes weighing ca. 50% of the weight of the Micruroide (Woodin, 1953, Bogert, 1960; Russell, 1967; Vitt and Hulse, 1973; Greene 1973 and McDiarmid, 1981; pers. obs., Pough, this volume).

Venomous sea snakes probably represent two or more independent marine rodiations from terrestrial, Australasian elapid ancestors (McDowell, 1986). Some specises are evidently inoffensive (*e.g.*, Laticauda spp.), while others bite fiercely (*e.g.*, Aipysurus). Pelamis platurus has noxious fiesh as well as a venomous bite, and the bright tail display is aposematic (Rubinoff and Kropach, 1970, Caldwell and Rubinoff, 1983).

Most species of Viperidae assume an S-coil and strike when when threatened, and many vibrate the tail and hiss (*e.g.*, Agkistrodon contotrix, Fitch, 1960; Azemiops feae, Kardong, 1986); those behaviour patterns are found in all three viperid subfamilies as well as several other families and are probaly retained primitive traits (see Henophidia, above, and. Terrestrial Bothrops asper occasionally gape at threatening humans (pers. obs.). and aquatic Agkistroden piscivorus routinely do that (Allen and Swindell, 1948). Arboreal species in the predominantly terrestrial subfamilies Crotaline (B. schlegeli) and Viperinae (Atheris squamier) also gape defensively (Kratzer, 1969; implying the independent evolution of that display several times. Species of Cerastes and Echis do not hiss (a loos perhaps functionally related to problems of water loss in desert environments, Mendelsohn, 1963, 1965), but instead produce a loud, "hissing" sound by rabbing obliquely keeled scales on adjacent body coils against each other (Gans and Baic, 1974; Similar behaviour and sound production in snakes of the colubid genus Dasypeltis are mimetical convergence (Gans and Richmond, 1957; Pough, this volume). Whether the sound producing mechanisms of Cerastes and Echis result from a single evolutionary event is not resolved (cf. Marx and Rabb, 1965), but the shifting, exaggerated threat displays and loud hissing of other African Viperinae perhaps represent the antecedent defensive response (*e.g.*, Causus resimus.

Rattlesnakes

Approximately 31 species of New World pit vipers in the genera Crotalus and Sistrurus form a monophyletic group characterized by the presence of a rattle. That complex, epidermally derived caudal structure is associated with specialized shaker muscle (Martin and Bagby, 1972), and rattlesnake are prominent creatures in a variety of North American ecosystems.

A particularly widerspread, well studied species illustrate the range of rattlesnake defensive mechanisms. Crotalus viridis is a moderately large, diurnal or nocturnal species that occurs in much of the western half of the United States and southern Canada. These snakes very geographically in colour and are probly cryptic. When approached from a distance by a potential predator, a western rattlesnake is either already immobile or freezes. If approached closely it usually crawls rapidly away from a threat. If cornered or restrained, the rattlesnake assumes a prestrike posture, with its hsad elevated and foreparts drawn into an S-shaped coil. By the stage, the snake is typically rattling vigorously and continuously. If the predator remains in close proximity or touches the rattlesnake, it may strike and bite, hide its head under a body coil, or first hide its head and then suddenly strike from under a coil. If the predator is a kingsnake (Lampropeltis getulus), the rattler forms long, low "body bridges," which are sometimes jerked slightly up and down. C. viridis struggles violently if restrained, attempts to bite, continues to rattle, and discharges the cloacal contents in a fine spray (Bogert, 1941; Klaber, 1956; Carpenter and Gillingham, 1975; Weldon and Burghardt, 1979; Kardong, 1980; Duvall *et al.*, 1985; Sweet, 1985).

The rattle is fundamentally a defensive signaling device today, and it is never regularly used in any other context. Hay (1887) and some subsequent authors favoured an original role in warning large, trampling ungulates of a rattler's presence. Others suggested that the rattling sound originally diverted attention away from the snake's head and toward its tail

(Garman, 1889; Williams, 1966). As Klauber (1956) noted, the first scenario entailed inconsistencies if, as he throught likely, the most primitive living rattlesnakes are small species restricted to high, forested areas on the Maxican Plateau (Klaber's reasons for choosing a particular phylogeny were only vaguely supported in print). Such forms probably would not be susceptible to trampling by ungulates. Klauber (1956) also pointed out that no evidence supports the second scenario either; he marshalled extensive evidence that the rattle is a warming device in all extant specis that possess it, and by my reading, preferred no particular hypothesis beyond that.

Schuett *et al.*, (1984) inferred from variation in size and construction among species that the rattle did not arise in a single step. They concluded therefore that the some stage an incipient rattle was small and inaudible, and could not have served as a warning device. Instead, Schuett *et al.* (1984) proposed that a proto-rattle served as a caudal lute during hunting, because caudal luring is widerspread among viperids and could be more advantageous in two ways with a keratinized caudal appendage. The supposed that a proto-rattle would have increased the optical attractiveness of a lure to potential prey, and that it would have protected a rattler from injuries delivered by the prey.

It is refreshing to have alternative hypotheses regarding the origin of the rattle, but the Schuett *et al.* (1984) hypothesis is not strongly supported by evidence at this time. Whether or not the rattle arose gradually or suddenly is not known, although the occurrence of occasional anomalous individuals demonstrates that it can be lost abruptly (Klauber, 1956; Smith *et al.*, 1985). Nevertheless although rattles vary interspecifically in relative size and shape (Klauber, 1956), all are so similar in structural details that this feature surely has arisen only once. Most Crotalus and Sistrurus make noise with their rattles that is audible to a human ear from a distance of more than a meter (*e.g.*, small C. lepidus, C. mitchelli, and C. pricei in rock slides, pers. obs. Granting there might have been transitional stages not evident in living taxa and prior to the origin of rattlesnakes, but assuming the rattles of Crotalus and Sistrurus are homologous as rattles, their ancestor had a rattle that functioned in a defensive context.

Caudal luring probably is primitive for Viperidae or a more inclusive taxon, because it occurs in numerous crotalines, at least two viperines, one elapid, and a least one boid. However, although the effectiveness of the tail as a lure can be enhanced structurally (*e.g.*, swollen and pink in Bothrops bilineatus, Greene and Campbell, 1972), no convincing case has been made for damage to that appendage by prey. Schuett *et al.* (1984) cited data on tail injuries and mortality in Thamnophis (Willis *et al.*, 1982), but those scars probably resulted from bites of predators rather than from small prey items. Finally, caudal luring occurs in a feeding context and involves relatively slow, sinuous movements (*e.g.*, Greene and Campbell, 1972; Schuett *et al*., 1984, for Sistrurus catenatus). Tail vibration by viperids and many colubrids occurs in a defensive context, is typically rapid, and produces obvious noise on some substrates (*e.g.*, Agkistrodon halys, Maslin, 1950; Bothrops sp., Darwin, 1859; an adult Lachesis muta produces a distinct buzzing noise on dry leaves that can be perceived by the human ear from a distance of several meters, Greene and Santana, unpubl.).

Defensive tail vibration and caudal luring might both be ritualized versions of intention movements to escape, or of conflicting tendencies to attack and escape (cf. Parker, 1965; Greene, 1973a; Radcliffe *et al.*, 1980); whether one evolved from the other is not known. However, defensive tail vibration clearly was the behavioral precursor of rattling and the rattle almost certainly evolved solely in a defensive context. Beyond that, the origin of the rattle remains an important problem, the solution to which has bearing on the adaptive radiation of a major group of snakes. Progress will require a convincing phylogeny of rattlesnakes and their closest relatives, and a survey of variation in the natural history of these animals.

Rattlesnakes vary considerably in their responses to danger (Klauber, 1956) and in venom composition (*e.g.*, Minton and

Weinstein, 1984). Some species are remarkably phlegmatic (*e.g.*, Crotalus ruber), some have particularly dramatic threat displays (*e.g.*, C. durissus), and some have especially toxic venom (*e.g.*, some C. scutulatus, C. tigris). The rattle has been lost independently in C. catalinensis and an island population of C. ruber, although individuals of at least the former still make "rattling" movements of the tail (Radcliffe and Maslin, 1975; pers. obs.). The possible ecological significance of such differences and of other variation remains to be investigated, but rattlesnakes are clearly of great potential interest for studying diversification subsequent to major antipredator innovation.

EVOLUTION OF ANTIPREDATOR MECHANISMS

Function and Biological Roles

1. *Crypsis*

Cryptic animals resemble a random sample of the relevant aspects of their environment (Endler, 1978). A rigorous demonstration of crypsis requires comparisons with a specified hypothesis of resemblance, and consideration of other potential influences. The latter include biological roles in social interactions, mechanical protection, thermoregulation, asposematism, mimicry, and even hunting (*e g.*, Norris, 1967; Ficken *et al.* 1971; Pough *et al.*, 1978; Burtt, 1979; Pough, this volume). Egg-eating snakes (Dasypeltis, Gans, 1961, 1970), gopher snakes (Pituophis melanoleucus, Sweet, 1985), and several vipers (*e.g.*, Crotalus spp., Klauber, 1956; Cerastes and Echis, Gans, 1961) illustrate multiple functions, in that their colour patterns are concurrently cryptic, mimetic, and/or aposematic.

Field reports often emphasize the resemblance, sometimes remarkable, of reptiles to the substrates upon which they are found (*e.g.*, Gans, 1976, for uropeltids). Numerous reports mention parallel geographic variation in substrate and lizard colour, the most impressive of which are those concerned with dramatic differences from nearby habitats (*e.g*, Holbrookia maculata at White Sands, New Mexico, Ruthven, 1907; Uta stansburiana at Pisgah Lava Flow, California, Norris, 1967; the

Amphibolurus decresii species group in Australia, Gibbons and Lillywhite, 1981). Several field studies have documented selection among morphs within a population (*e.g.*, Nerodia sipedon, Camin and Ehrlich, 1958; Beatson, 1976; King, 1986; Thamnophis sirtalis, Sattler and Guttman, 1976), but only a single study to date has quantified in detail the extent to which variation in colour patterns tracks environmental colour variation (Sweet, unpubl). Individuals of Pituophis melanoleucus are substantially less cryptic when displaced from their original capture site to another along a subspecific intergrade transect. Furthermore, a broad range of phenotypes is present among hatchlings at any point on the cline, suggesting that there is a selective gradient that maintains geographic variation in the adults.

A few experimental studies have examined the relationship between color and predation. Models of the black morph of Vipera berus are subject to more distubance by predators than are those of the normal, striped morph in the field, and the continued presence of black vipers in the population is aitributed to a thermal advantage associated with melanism (Andren and Nilson, 1981). Phrynocephalus arabicus that had cryptic coloration disrupted by small paint marks suffered increased mortality (Arnold, 1984a), but survivorship did not differ significantly between paint-marked and normal Sceloporus undulatus (Jones and Ferguson, 1980). A study of S. jarrovi demonstrated no significant differences in mortality between paint-marked and unmarked lizards, even when juveniles and adult males were examined separately (Simon and Bissinger, 1983).

Several kinds of indirect evidence support the hypothesis that cryptic colouration in reptiles is important in avoiding predation. Bright contrasting colors are often used in social signaling by lizards (Carpenter and Ferguson, 1977), but these are almost always on hidden body surfaces that are either protruded (*e.g.*, the dewlaps of Anolis and Sitana, the patagia of Draco) or exposed by rapid body movements (*e.g.*, the bright ventral patches emphasized by the "pushups" of many small iguanids).

Another source of evidence for adaptive colouration is found in the correlations among pattern types and escape responses in snakes. Unicoloured and striped species tend to live in grassy habitats and respond to threat by rapid crawling. Blotched and ringed snakes more often rely on immobility or other responses that emphasize defenses other than flight (Jackson *et al.*, 1976). Other discussions of alternative cryptic color patterns in snakes include Klauber (1939), Brattstrom (1955a), Van Devender (1973), Pough (1976), Bechtel (1978), Dunson (1979), Jackson and Pounds (1980), Kardong (1980), Vitt (1980), and Gregory *et al.* (1983).

Cryptic colouration is often accompanied by behaviour that facilitates the effect. Crocodilians and some turtles possess elevated nostrils and eyes, and typically rest in the water with only those sensory structures protruding (*e.g.*, Neill, 1971; Alvarez del Toro, 1974). Iguanids (*e g.*, Urosaurus graciosus, pers. obs.) and agamids (*e.g.*, Draco maculatus, Schmidt, 1927a) often remain immobile and crouch against the substrate when a predator is perceived nearby. Slender, arboreal snakes sway like a branch in the wind (*e.g.*, Ahaetulla sp., Gans, 1974; Opheodrys aestivus, Goldsmith, 1984; Oxybelis aenus. Fleishman, 1985), and some small lizards conducts many of their movements during periods when wind is causing their surroundings to move (Jackson, 1974; Greene *et al.*, 1978).

2. *Locomotor Escape*

The advantage of movement away from a predator is obvious, but that option is limited for most reptiles by their restricted aerobic capacities (*e.g.*, Bennett, 1980; Christian and Tracy, 1981; Pough, 1983). Exceptions include certain teiids and varanids, which have unusual endurance at high speeds compared to other reptiles. Most species seemingly seek inaccessible retreats (vegetation, burrows, rock crevices, etc.) as soon as possible, and some lizards and snakes apparently know the location of those refuges (*e g.*, Ameiva ameiva, Beebe, 1945; Tropidurus torquatus, Rand and Rand, 1966; Uromastyx hardwicki, Minton, 1966). Many arboreal lizards move to the opposite side of a tree from the pursuer (e.g., Sceloporus magister, Parker and Pianka, 1973). Lizards that take refuge in crevices of rocks and trees sometimes watch potential predators from concealment (*e.g.*, Varanus tristis, Pianka, 1971a), and some venomous snakes turn once they reach shelter and prepare to strike (*e.g.*, Agkistrodon piscivorus, Bothrops asper, pers. obs.).

A common gambit is to escape into a radically different microenvironment that is inaccessible to most predators in the original site. An adult Iguana iguana evaded several coatis (Nasua narica) by diving into a lake in Panama (Greene *et al.*, 1978), a response typical of many aquatic turtles, crocodilians, snakes, and other large lizards when approached by humans. That behaviour is also seen as an occasional tactic in a variety of small and medium-sized lizards (*e.g.*, Hoplodactylus pacificus, Whitaker, 1968; Cnemidophorus sexlineatus, Dillon and Baldauf, 1945; Crotaphytus collaris. Mc Allister, 1983; Eumeces septentrionalis, Black, 1976; Lacerta lepida; Malkmus, 1982b), and a regular escape method in several species of semiaquatic Anolis (*e.g.*, Campbell, 1973) and skinks (*e.g.*, Leiolopisma suteri, Whitaker, 1968). A specialization among several distantly related lineages to run across the surface of water, a behavior that is facilitated by fringe-toed hind feet, and emerge on the opposite bank (*e.g.*, Basiliscus, Hydrosaurus; Rand and Marx, 1967; Laerm, 1973; Luke, 1986). Small lizards that jump into water are susceptible to predation by fish at some sites (Campbell, 1973) and perhaps that risk favoured the evolution of fringetoed water runners in tropical areas.

Two other widespread specialized locomotor modes are running on fine sand, which is often facilitated by toe fringes (Carothers, 1986; Luke, 1986), and climbing relatively smooth, vertical surfaces. The latter task is associated with highly derived and convergent digital scansors in anoline iguanids, geckos, and a few species of skinks (Russell, 1979a; Williams and Peterson, 1982).

An even rarer escape behaviour involves leaping into the air from an elevated perch, then gliding or parachuting to another location (*e.g.*, Draco spp., Colbert, 1967; Anolis carolinensis,

Oliver, 1951; Ptychozoon lionotum, Marcellini and Keefer, 1976; Chrysopelia paradisi, Mertens, 1970; Philothamnus spp., Sweeney, 1971). Gliding and parachuting are significantly more common in southeast Asia than in other rainforest areas, where there are fewer vines than in African and neotropical forests (Emmons and Gentry, 1983). Experiments in which the lateral cutaneous flaps of P. lionotum were tied against the body demonstrated that those structures improve gliding performance (Marcellini and Keefer, 1976). Although relatively uncommon among extant lineages, adaptive modifications for gliding have been a recurrent theme in the evolution of reptiles (Colbert, 1967, 1970).

Field observations suggest that forming a ball of coils might have surprising locomotor consequences in predator encounters for Lichanura trivirgata: twice when these small boas were discovered under rocks, they rolled out of reach into nearby crevices (B. M. McGurty, pers. comm.). Humans might be the only predators that can reach ross boas in rock shelters, but ringtails (Bassariscus astutus) occur in the same habitat and eat snakes (W. P. Taylor, 1954). In any case, that behaviour represents a rare use of a gravity powered "wheel" by a vertebrate (cf. LaBarbera, 1983).

Escape from a predator that has grasped them might be facilitated in a few peciess of gekkonids and scincids by the extreme fragility of their skin (*e.g.*, Aristelliger hechti, pers, obs. Gehyra mutilata, Bustard, 1970; Pachydactylus namaquensis, Huey, in litt.; Phelsuma sp., J. B. Losos, pers. comm;. Teratoscincus scincus. Minton, 1966; a few species of Ctenotus and Lerista, Greer, 1986).

3. *Bluffs and Threats*

Bluffs and threats are signals that differ in the terms of consequences if they are ignored, in that the latter imply potential injury to an attacker and the former are deceptive in that regard. That distinction can be difficult to make in practice or actually variable for an individual because of relative characteristics of the participants (*e.g.*, size). Such signals are present sporadically among several groups of reptiles (*e.g.*, some turtles, some iguanids), and predominate in certain others (*e.g.*, agamids, anguimorphans, geckos, many snakes). Bluffs and threats typically include increases in apparent size, intention movements to bite, explosive noises, and sudden exposure of bright colours or, rarely, fake eyes (*e.g.*, Sphaerodactylus semasiops, Thomas, 1975).

Some observational and experimental evidence demonstrates that bluffs and threats work. A Varanus niloticus kept at bay a full grown, hungry lioness (Panthera leo, Schaller, 1972), and Iguana iguana repulsed several coaties (Nasua narica) with its tail (Chapman, 1938). A Calotes cristatellus repelled a snake (Chrysopelea paradisi, Lim, 1959) and a C. versicolour successfully evaded a crow (Minton, 1966). A "Natrix sp." (perhaps =Rhabdophis tigrinus, a dangerously poisonous colubrid) repeatedly rebuffed an attacking house sparrow with a head display before eventualy gaining the safety of a bush (Sen Gupta, 1969). A 35-kg malamute that effectively defended me against a larger dog was later rebuffed by the vigorous hissing and strikes of a 20-gm Pituophis melanoleucus. An adult P. melanoleucus held two adult raccoons (Procyon lotor) at pay for an hour and then crawled away unharmed (Carr, 1940). The defensive display of Phrynosoma platyrninos (lateral presentation of the tilted, flattened body accompanied by hissing and lunges) is almost certainly a bluff, because leopard lizards (Gambelia wislizeni) will approach and avoid horned lizards of a weight they could easily subdue (based on confrontations with nonspinose lizards, Tollestrup, 1981).

Many defensive displays of reptiles probably are bluffs, in the sense that the consequences for a predator of actually attacking would be trivial. In each of the cases described above, the predator probably could have easily overcome the potential prey had it chosen to "attack anyway." This poses the question, why do those displays work? Perhaps predators simply cannot afford to make mistakes unless they are really hungry, because real danger is sufficiently common, stimulus stimulus generalization is so widespread (Werner, 1983, 1986;

see also Pough, this volume), and size alone is sometimes not an accurate indicator of fighting ability. An Elgaria multicarinata immobilized an attacking adult Masticophis lateralis by biting it on the head, and another held off an adult Coluber constrictor by gaping at it (Sweet, in litt.). Naive mammalian predactors and nonraptorial birds that eat small litter vertebrates avoid coral snake models (Gehlbach, 1972; Smith, 1975, 1977; Greene and McDiarmid, 1981: Note 76), herons avoid yellow-bellied sea snakes (Caldwell and Rubinoff, 1983), and predatory fishes avoid yellow-bellied sea snakes (Rebinoff and Kropach 1970), all situations in which attack might be fatal in nature. Likewise, numerous field observations demonstrate that a variety of vertebrates respond with avoidance to the sound of a rattlesnake (Klauber, 1956).

On the other hand, bluffs sometimes fail because the handing abities of predators are not always strictly size dependent. A shrike (Lanius ludovicianus) readily attacked and flew off with a Phrynosoma platyrhinos that was displaying vigorously (pers. obs.). In this case the bird is capable of reducing a relatively large prey lizard to ingestable chunks, whereas a predatory leopard lizard (Gambelia wislizeni) of equivalent size (ca. 50 gm) must swallow it intact and is limited by gape. Coatis (Nasua narica) are rebuffed by models of venomous coral snakes (Gehlbach, 1972), but other carnivores of smaller or similar size (Grison sp.) readily attack living Micrurus (Jackson, 1979).

4. *Diversion of Attack*

Many defensive responses apparently work by diverting attack, either completely or to an expendable portion of the animal. Absolute diversion is not common. but might include the tail displays of certain iguanids (Cophosaurus texanus, Callisaurus draconoides, Dial, 1986) and agamids (Phrynocephalus sp.; see Arnold, 1984a, 1984b, and this volume). Cophosaurus texanus has a ventrally striped tail, which is sometimes elevated and waved at an approaching predator before the lizard flees. Individuals are significantly more likely to display if they perceive a predator at a long distance, and to flee without displaying if surprised at a short distance. The former behaviour might signal a predator that there is no chance of capture, thereby decreasing the likelihood of an actual chase (Dial, 1986).

Many tail displays in snakes and amphisbaenians probably divert attacks to an espencially sturdy or disposable portion of the body, a role that is facilitated by the tendencies of many avian and mammalian predators to attack the head of a prey (Kaufmann and Kaufmann, 1965; Ewer, 1968; Lorenz, 1971; Curio, 1976; Jackson, 1979). Among certain snakes and amphisbaenians, high incidene of tail scars is associated with vertebral structures that strengthen the tail (Greene, 1973a; Section V.C. but see Nussbaum and Hoyer, 1974, for a feeding role of that modification in Charina bottae). Field observations indicated that a jackal (Canis sp.) attacked an Eryx johnii, but abandoned the sand boa with a bleeding tail but otherwise unijured (Minton, 1966). Lizards that form a ring by seizing a body part in their jaws (*e.g.*, some Elgaria, Fitch, 1935; Cordylus cataphractus, Rose, 1962) are probably abandoned unharmed because a predator cannot locate the head or because the resultant "object" is too large to swallow (Lacerta saxicola, Darevsky, 1967).

Some snakes occasionally miscarry young (uropeltids, Gans, in litt.; Sistrurus catenatus, pers. obs.) or regurgiate recent meals (*e.g.*, Heterodon Platt, 1969; Python sebae, Pienaar *et al.*, 1978) when disturbed, and the latter is especially likely in New World venomous coral snakes (Micruroides, Micrurus) that have eaten relatively heavy meals cf.) Greene, 1984; pers. obs ; Section IV.I). Such behaviour might be an incidentan effect of fright and of the difficulties of struggling or locomoting while containing a developing young or a large prey item, but it might also be advantageous in diverting a predator. A potential predator could be repulsed by the sudden appearance of a large, smelly mass or could consume it. In either case, the predator would be distracted and the snake could escape. This hypothesis could be tested easily with captive predators and appropriate snakes.

5. *Removal of Cues*

Predatory behaviour is typically controlled by particular sign stimuli and some kinds of defensive behaviour might succeed because they remove or obscure those cues from the potential prey's appearance (Platt, 1969; Alcock, 1975). Two classes of stimuli that are particularly widespread in eliciting killing behaviour by vertebrate predators are movement and presence of a "head" (Curio, 1976) Tactics that likely involve removal of one or both of those cues include concealing the head under coils in snakes extended, rigid postures in lizards (*e.g.*, Echinosaura horrida, Leviton and Anderson, 1966) and snakes (*e.g.*, Trachyboa boulengeri, Lehmann, 1970); and death feigning by some lizards (*e g.*, Varanus exanthematicus, Schmidi, 1919; Pienaar *et al.*, 1978) and snakes (see below). The efficacy of those reponses might depend in part upon the ability of a prey the withstand some manipulation by a predator, which seems likely in reptiles (see Section II.D.)

The possible biological role of death feigning is intriguing and controversial. McDonald (1974) documented bradycardia in death feigning Heterodon platirhinos, and noted that most snakes that death feign feed heavily on toads and have enlarged adrenal glands (*e.g.*, Hemachatus, Heterodon). He proposed that the suite of activities involved in death feigning represents a "parasympathetic syndrome," facilitated by typically high levels of circulating catecholamines. McDonald (1974) and other authors recognized no obvious, specific advantage to death feigning, and he suggested that it might be a nonadaptive consequence of toad-eating. That nondaptive hypothesis has yet to be directly refuted, but additional considerations are relevant. Removal of cues that elicit prey-killing behaviour might provide an animal with subsequent opportunities for escape (Alcock, 1975), an event that seems especially likely in the case of predators that carry intact, immobile prey to a nest or den (*e.g.*, raptors, some canids). Furthermore, the response (including slight bradycardia, McDonald, 1974) occurs repeatedly if a threat is rapidly presented and removed from a Heterodon platirhinos

(Burghardt, Klinghammer, and Greene, unpubl.), Hemachatus haemachatus remains alert and ready to bite during death feigning (Broadly, 1983). These observations perhaps imply an evoked behavioural response rather than neurohormonal phenomena.

6. *Novelty, Startle Effects, and Aspect Diversity*

Numerous behavioral responses among reptiles are so bizarre relative to normal behavior that they might delay or prevent attacks by virtue of their novel startling qualities. Possible examples include erratic movements and noises, sudden appearance of bright colors or or eye spots, and behavior or appearance that is novel to the predator relative to other members of the prey population (Gans and Richmond, 1957; Humphries and Driver, 1967, 1970). Colour pattern polymorphism is relatively rare in reptiles (reviewed by Neill, 1963, for snakes), but might represent a response to predation via aspect diversity (see Rand, 1967a; Harvery and Greenwood, 1978). Examples include Hoplodactylus pacificus (Whitaker, 1968), Anolis cuvieri (Rand and Andrews, 1975), Gambelia wislizeni (Montanucci, 1978), Sonora aemula (Nickerson and Heringhi, 1966), S. michoacanensis (Echternacht, 1973), S. seminannulata (Gehlbach, 1972), Lampropeltis getulus (Zweifel, 1981, 1982), and Scaphiodontophis (Henderson, 1984b).

An example of interspecific behavioral diversity as a response to predation is found in whiptail lizards of the western United States. One to five species of Cnemidophorus occur at various sites, and the diversity of total behavioural response is positively correlated with tail break frequencies among assemblages (Schall and Pianka, 1980).

7. *Offensive Defense*

Many reptiles are capable of delivering one or more "noxious consequences" to an attacking predator. Features and displays that emphasize unpalatability or danger (*e.g.*, are likely to be defensive adaptations, because they have no other apparent role, and because the behaviour of the predators in rejecting prey sometimes supports that conclusion. Examples include distasteful or toxic flesh (Pelamie platurus, Rubinoff and Kropach, 1970; perhaps some Eumeces, but sce Cooper and Vitt, 1985) and odoriferous glandular secretions (*e.g.*, Diplodactylus sp., Bustard, 1978; kinosternid and pelomedusid turtles, Eisner *et al.*, 1977, 1978; but see Kool, 1981, for contradictory evidence). Offensive mechanisms involving the digestive system (biting, cloacal discharge) have obviously been coopted for defense against predators, and in some cases adaptively elaborated for that role (*e.g.*, the modified fangs of spitting cobras).

The recognition of "active combat" as an antipredator mechanism can be defficult. Casual observation suggest that aggression by some snakes (*e.g.*, Coluber constrictor, Fitch, 1963; Dendroaspis polylepis, Broadley, 1983) and crocodilians (*e.g.*, Crocodylus niloticus, Jackson, 1962) might be mediated by social and hormonal factors, and thus reprcsent misdirected territoriality rather than a real defensive response. Another problem is that threatening and bluffing movements such as tail lashing, lunging, and even biting can merge into actual active combat (see Part 3, this section). Examples of overl aggression (usually as biting) include snapping turtles (Cheydra serpentina, Section IV.B), most large lizards (*e.g.*, Lacerta lepida Callopistes flavipunctatus, several Varanus, Section IV.G), some amphisbaenians (*e.g.*, Amphisbaena alba, Trogonophis wiegmanni, Section IV.H), and many snakes (*e.g.*, Ophiophagus honnah, Howe, 1931; Section IV.I). A further problem is that aggressive defense tactics easily can be misinterpreted as real attacks, as when venomous snakes suddenly change directions, seek shade, or turn to face a threat (*e.g.*, Dendroapis polyepis, Broadley, 1983; Bothrops asper, pers. obs.; Echis carinatus, Minton, 1966).

8. *Morphological Adjuncts*

Morphological features that facilitate defense include cryptic and aposematic colour patterns, structures that enhance locomotion or sound production, structures that improve the utility of other adaptions as defensive weapons (see above), and structures that are evidently adaptive only in defensive roles. The latter include such things as the turtle shell (which of course has other properties, *e.g.*, hydrodynamic and thermoregulatory), certain glands, body and tail spines, and the rattle.

Shell closure mechanisms have evolved repeatedly in turtles. and are believed to be antipredator adaptations (Bramble et al., 1984; Iverson, 1984; Section IV.B). Field observations on Terrapene coahuila at a marsh in Mexico support that hypothesis, because they indicate that secondarily aquatic box turtle is better able to avoid predation than other sympatric turtles that lack plastral kinesis. A coyote (Canis latrans) caught and began to open a slider turtle, Pseudemys scripta. Along the nearby shoreline were remains of nine other sliders and two soft-shelled turtles (Trionyx sp.), and signs indicated that coyotes had eaeen both species (which, like most turtles, lack plastral hinges). Box turtles were common in the marsh, and one was found with fresh scratches on its closed shell and coyote tracks nearby. It was alive and unharmed, and no carcasses of that specties were found around the marsh (Minckley, 1966).

Experimental work has confirmed the locomotor advantage conferred by fringed toes of lizards on water (Laerm, 1973) and sand (Carothers, 1986), and of cutaneous flape and other specializations for gliding in a lizard (Ptychozoon lionotum, Marcellini and Keever, 1976) and a snake (Chrysopelea ornata, Heyer and Pongsapipitana, 1970). Whether these locomotor adaptations actually reduce predation is not yet documented.

Patterns and Processes

1. *Primitive Reptiles and Other Amniotes*

Crypsis, locomotor escape, struggling, defecating, biting, and perhaps hissing are probably primitive for amniotes, judging from their near ubiquity among the living lineages of raptises (Section IV), their presence in a number of mammals (Ewer, 1968) and the presence of all except hissing in at least some living amphibians (Duellman and Trueb, 1985. The prevalence of crypsis, inaccessability, and brief flight in living reptiles thus reflects retention of effective primitive features, and probably also the potentially high energetic costs of more direct responses (Pough, 1983). Among the latter, threatened or actual biting perhaps is the most frequently observed. There have been no thorough reviews of defense in mammals (see Ewer, 1968; Curio, 1976; Eisenberg, 1983; Wemmer and Wilson, 1983) or birds (see Curio, 1975, 1976; Brunton, 1986), but the antipredator mechanisms postulated as primitive for reptiies are evidently also widespread in those classes. Future attempts to evaluate the phylogenesis and ecological significance of defense in birds and mammals should include considearation of their reptilian relatives.

2. *Major Trends Within Repitilia*

Defence against predators in ancestral turtles included the shell and retained, primitive tetrapod mechanisms. Subsequent evolution involved the occasional gain of new features associated with defense (*e.g.*, plastral kinesis, odoriferous glands, spines, bright colours) and perhaps the repeated gains and losses of biting. Crocodilians and Sphenodon evidently retain primitive amniote responses to predation with little in the way of subsequent antipredator specializations, although large size no doubt affords freedom from most predators in the former.

The greatest diversity of antipredator mechanisms is in squamates, perhaps in part because this is by far the largest clade of living reptiles and because they occupy a greater number of microhabitates than do other lineages. The first major split in extant squamate taxa was between iguanians and scleroglossans (gekkotans plus autarchoglossans, amphisbaenians, and snakes, Estes *et al.*, 1988), and clear-cut defferences in many behavioural and ecological attributes parallel that dichotomy. Scleroglossans are frequently secretive and/or nocturnal solitary and nonterritorial (Stamps, 1977), and cover substantial distances in foraging. Although usually cryptic, they often rely on struggling, defecation, powerful jaws, and/or ritualized displays when discovered by a predator, rather then further rapid flight. Iguanians are diurnal, territorial, sedentary predators, that usually rely on crysis and brief of locomotion to avoid predators. Exceptions exsit for both generalizations, including iguanians with impressive stationary threat displays and some teiids and lacertids that surely rely mainly on crypsis and speed. Less frequent themes among scleroglossans include epidermal sound protection (*e.g.*, Echis, Teratoscincus), limbless lateral undulation (*e.g.*, Elgaria multicarinata, pers. obs.; all snakes), bizarre behaviour (*e.g.*, Cordylus cataphractus), and venomousbites (found only in Heloderma and many snakes). Tail autotomy is widespread in some scleroglossan families, but is absent in xenosaurids, helodermatids, varanids, and most snakes (Arnold, this volume).

Snakes and amphisbaenians retain the antipredator responses of a primitive scleroglossan squamate, but thrashing and cloacal discharge are enhanced by specialized, postcloacal scent glands in the former. Several major subsequent radiations of snakes have evolved innovative antipredator mechanisms, such as hoods in certain colubrids and elapids, aposematic coloration and displays in New World venomous coral snakes, and the rattle in one lineage of viperids. Several amphisbaenians (perhaps most) include erratic movements and novel postures, not seen elsewhere among scleroglossans, in their defensive repertoires.

3. *Adaptation and Convergence*

Two classes of innovations are evident in the evolutionary diversification of reptilian defense against predators. Small changes that solve restricted problems have been widespread among squamates, often in the form of individual, intraspecific, or intrageneric variation. Perhaps because the number of options is limited, convergence is common. Examples include behavioural displays in snakes (Greene, 1979; Myers, 1986), and several pairs of iguanids and agamids, respectively: ftinge-toed water runners (Basiliscus, Hydrosaurus), fringe-toed sand runners (Uma, Phrynocephalus), and spinose bodies (Phrynosoma, Moloch). Such innovations are minor in the sense that they apparently have not led to major radiations (typically a dozen species or less) and usually involve either postural changes or minor morphological modifications (*e.g.*, spines and other elaborated epidermal structures).

A few defense-related changes apparently served as key innovations, by solving a problem in a way difficult for some predators to overcome, and facilitated major radiations. Possible examples include the shell of turtles (ca. 230 living species), shiny scales and body elongation in skinks (ca. 31% of 3400 lizards), scent glands in snakes (ca. 42% of 6000 living reptiles), and the rattle (ca. 16% of 190 living vipers). The total number of species encompassed by these relatively few evolutionary events is potentially very large, suggesting that antipredator problems have played a significant role in the diversification and persistence of reptiles.

Although the relevant data on foraging styles are incomplely major antipredator innovations among reptiles apparently are often associated with a shift to active searching and/or with relatively long handling times under exposed conditions. Recent studies on fishes, birds, and mammals confirm an intuitively appealing proposition, that preoccupation with feeding activities decreases the ability of an animal to watch for predators (Milinski, 1984; Lawrence, 1985; Lima, 1985; Carey and Moore, 1986). Show, careful foraging behaviour, particularly in an open environment, probably makes an animal especially vulnerable to attack by a mobile predator. Interestingly, a desert gekkonid, Ptenopus garrulus, suffered increased mortality when it switched from its usual sit-and-wait to widely foraging tactics (Huey and Pianka, 1981). The problem of increased vulnerability while feeding is evidently surmounted in many birds and some mammals by social vigilance and group defense, but most reptiles are solitary (Wilson, 1975; Iguana iguana provides an instructive exception: Juveniles feed on young leaves and flowers in sustained bouts (pers. obs.), are virtually incapable of active defense, and are among the few highly social squamates (Burghardt *et al.*, 1977; Greene *et al.*, 1978). An implication of these considerations is that the repeated coupling of preoccupied foraging styles, antipredator innovations, and life history traits (Duninam et al., this volume) has been a major theme in reptilian evolution (see Section VI.C).

Fossils, Anachronisms, and Recent Changes

The evolutionary study of antipredator mechanisms in modern reptiles will utlimately profit from information on extinct taxa. Indirect evidence from fossil lineages can suggest new and surprising tactics, as in the bony tail club of some ancient turtles (Gaffney, 1985) and the putative defensive adaptations of extinct archosaurian reptiles (Bakker, 1986). This infoımation might be important to the evaluation of constraints and optimal design, because it delimits the bounds of possible defensive responses (Section VI.B). Fossil members of extant clades can provide further evidence of the homology of defensive features and minimal estimates of their times of origin, as in the heavily reinforced caudal vertebrae of Eocene and Recent erycine (boas (Hoffstetter and Rage, 1972; Greene, 1973a).

The analysis of putative defensive adaptations in living taxa is confounded by the possibility that they arose in concert with conditions no longer present (Gans, 1976; cf. Janzen and Martin, 1982), especially in places where there have been recent changes in the relative and absolute abundances of predators. Conversely burgeoning human populations might affect antipredator mechanisms, via the experience of individual reptiles and/or selection on heritable variation in populations. A few examples will illustrate these problems.

Jaksic *et al*. (1982) compared predation on reptiles in Spain, Chile, and California, and discussed the role of predators in shaping the evolution of reptilian phenotypes (see Section II B). They could not evaluate, however, the former role of grizzle bears (Ursus arctos) as predators, even though that large omnivore was common throughout much of California until the end of the last century (Storer and Tevis, 1955). The jaguar (Panthera onca), a generalist predator on many vertebrates (including tortoises, crocodilians, and iguanas; Schaller and Vosconcelos,

1978; Greene, Baker, and Santana, unpubl.), occurred in the southern United States until recently (Brown, 1983). Even when predators remain present: their abundances probably are far lower than carrying capacities in the recent past. For example, more than 50,000 bobcats (Lynx rufus) were killed in the western United States in one year recently (Schueler, 1980), and for several hundred years the trade in skins of spotted cats has drastically reduced their populations throughout the neotropics (Myers, 1973).

Recent effects of humans on antipredator mechanisms would have potential for elucidating causes of rapid change, but few cases have been suggested and none have been studied as yet. Some lay people believe that rattlesnakes in the vicinity of humans are at a disadvantage if they rattle, and that this has favored a reduced tendency for that behavior in such places (Cope, 1871; Klauber, 1956). No evidence supports that hypothesis, but experimental and comparative field research might be feasible on a wide-ranging, common species the distribution of which encompasses areas of high and low human densities, different attitudes toward snakes, etc. (*e.g.*, Crotalus viridis in the western United States). Such studies should elucidate the roles of selection and individual modification as mechanisms underlying the behavioral shifts (see Section III).

REPTILES AND ANTIPREDATOR THEORY

Reptiles as Study Organisms

Reptiles are becoming widely recognized as convenient organisms for studies of functional morphology (Gans. 1974) and behavioural ecology (Huey *et al.*, 1983), and there are several reasons why the study of their antipredator mechanisms might contribute to those and other broad areas in evolutionary biology. First, searching for food and avoiding being eaten probably are the two most likely tasks facing a reptile on a daily basis, and the systems utilized in those two roles are often mechanically and functionally identical or interrelated. Second, because reptiles have a rich fossil record, they can serve as case studies for as yet unsolved macroevolutionary questions, especially in the context of evolutionary morphology (*e.g.*, Vrba, 1984). For example, why are certain innovations correlated with at least modest radiations in some lineages whereas similar changes are exhibited only by single species in other clades (*e.g.*, there are 14 species of Phrynosoma versus 1 species of Moloch)? Finally, Reptilia encompasses a sufficiently large and diverse group of organisms that its members can contribute to general theory as well as to the understanding of the limits of generality.

Optimality and Limits to Theory

Optimization is an approach to research in which design principles are used to evaluate and predict the functional significance of features of organisms (for general discussions, see Oster and Wilson, 1978; Lauder, 1981; Townsend and Calow, 1981). It proceeds by specifying the task (*e.g.*, rebuff a particular type of predator under certain conditions) and a currency by which to measure performance (*e.g.*, time, energy, mortality rate, clutch size), designing a solution to the task (using engineering or economic criteria, demographic models, etc.) such that benefits are maximized relative to costs, evaluating the role of possible constraints on the system and then comparing the results with reality. Optimization can provide a design hypothesis against which organisms are evaluated, it can lead to general theories for examining the diversity of life, and it can clarily factors that need to be investigated (Greene, 1986b; for reptilian examples, see Gans, 1974). A disadvantage is that optimization inherently invalves a number of assumptions, often unrecognized by the investigator some of which might be inappropriate and misleading (Gans, 1966, 1983b; Levins and Lewontin, 1985).

At this point only a few, very general predictions seem possible. Most birds and mammals possess aerobic exercise abilities far in excess of those of reptiles (Bennett, 1980; Pough, 1983). For that reason, reptiles generally are expected to be cryptic and to rely on brief movements into inaccessible microhabitats for escape, rather than on long distance flight or sustained combat. Defenses against predators that encounter active prey should emphasize crypticity and brief flight if possible, whereas an encounter when the prey is inactive necessarily involves actual contact. Death-feigning will not work for species taken primarily by reptilian predators, because the immobile prey would simply be eaten sooner (cf. Gehlbach, 1970; McDonald, 1974:162; Greene, 1984). Antipredator mechanisms that capitalize on visual, auditory, and chemical perception by a predator might be especially effective in particular cases, but as yet no attempt has been made to predict which of these would obtain.

Theories in biology are rarely both precise and general, and this might be especially true for defensive mechanisms. Thus far only a few generalizations are possible: We can make reasonably good predictions about the antipredator strategy of an unstudied snake, based on a knowledge of its size, coloration, and preferred microhabitat (Jackson *et al.*, 1976; Greene, 1979), but very few about those of a bird, a spider, or even a lizard based on studies of snakes. That situation is in contrast to allometry (*e.g.*, Calder, 1984) and some aspects of life history theory (*e.g.*, Dunham *et. al.*, this volume), for which fairly precise predictions have held true across a wide variety of animals and plants.

Perhaps a general theory of antipredator mechanisms for animals (let alone for animals plus plants!) is not possible, because the component phenomena are so complex. What "works" as a defensive response depends on predictable characteristics of the predator, the prey, the context for the interaction between them, and the particular aspect of feeding (*e.g.*, pursuit, search, etc.) that is being manipulated by each animal. Those components and perhaps stochastic factors offer the potential of tremendous complexity, in turn multiplied by the number of possible interactions, so that the number of possible combinations is very high (see also Endler, 1986a). Because antipredator mechanisms are so often idiosyncratic, their explanation is more likely to be informed by functional biology and natural history than by theories based on considerations of optimality (see Gans, 1974, 1986; Greene, 1986a). As with some other problems in evolutionary biology (Colwell, 1984), we may have to settle for sets of restricted theories of antipredator mechanisms.

Some Generalizations

A few general principles concerning antipredator mechanisms seem likely, given the above caveats:

1. There is extensive evidence for a set of design rules with respect to crypsis, aposematic coloration, and mimicry. They follow from certain principles of psychophysics, and apply across a wide range of taxa, including reptiles (*e.g.*, Cott, 1940; Wickler, 1968; Endler, 1978; Hailman, 1979; Owen, 1982; Pough, this volume). Examples include widespread, particular patterns of counter-shading and disruptive coloration, and the use of only a few hues and patterns in warning displays. It is note-worthy that most considerations of aposematic coloration are biased toward invertebrates, and sometimes start from assumptions that probably do not apply widely to reptiles (*e.g.*, kin selection, see Harvey and Greenwood, 1978; Harvey and Paxton, 1981; Guilford, 1985; Endler, 1986b; Engen *et al.*, 1986; Pough, this volume).

2. Although many descriptions emphasize a single defensive component, without mentioning the stimuli eliciting a reaction, animals typically react to the threat of predation with a hierachy of responses that follows from consideration of relative risks, energetic demands, and intrinsic constraints (*e.g.*, Rand and Rand, 1976; Pough, 1983). Numerous cases suggest that a reptile attempts to avoid detection whenever possible, to fleeunlocated and rapid escape is possible, and to engage a predator directly as a last alternative. Examples of that hierarchy of responses include Corytophanes cristatus (Davis, 1953), Thamnophis radix (Arnold and Bernett, 1984), and Micruroides curyxanthus (see Section IV.I) This generality holds across a wide variety of other taxa (Ewer, 1968, Robinson 1969, 1970; Edmunds, 1974; Alcock, 1984).

There are exceptions to the above pattern. Some large or particularly noxious forms immediately threaten a potential predator rather than flee (*e.g.*, some Lachasis muta, pers. obs.), recourse to threat versus flight some is dependents on type of predator (Section III. B.), times and some aposematic organisms apparently rever cehave cryptically. However, in some contexts even animals as dangerous as Crotalus viridis (Section IV. J), Ophiophaus hannah (Oliver, 1956), and Heloderma horridum (Pregill *et al.*, 1986) exhirtt the general hierarchy described above.

3. Features used in defense by plants and animals are usually adaptations for another biological role, and either coopted outright or modified in minar ways (Janzen, 1981). This appears to be true for reptiles, and the scructures coopted are usually thoes used primarily for feeding and locomotion. Indeed it can be difficult to distinguish primary from secondary roles, as in the cranial modifications of Corytophanes (Section IV. E). Many defensive displays of reptiles are plausibly interprened as ritualized locomotor "intention movements" to attack, or of conficting tendencies to attack and flee (cf. Daanje, 1950; Greene. 1973a; Chiszar, 1978), although those possibilities have not been rigorously ecamined.

4. There is a clear tendency among ectotherms for wide-searching and preoccupied foragers to rely on sudden (or omnipresent), highly effective, and often active defenses (Section V.B). That pattern is reasonably consiscent for ieveral lineages of reptiles, including several higher taxa (*e.g.*, turtles, snakes), and anurans (Taigen *et al.*, 1982). Certain solitary mammals also fit the patten (*e.g.*, tenrecs, porcupines, armadillos, skunls; Ewer, 1968; Eisenberg, 1983), suggesting that the coupling of preoccupied foraging and specialized defense might have been a significant factor in the

evolution of tetrapods and deserves detailed consideration.

5. Predictable functional relationship, more taxonomically restricted than the pattern discribed above (4), exist among lizards and snakes. Lizards that forage widely across diverse microhabitats and use speed to escape have longer tails than species that are sedentary predators, habitat specialists and use crypsis as a predominant antipredator tactic (Huey and Pianka, 1981; Vitt, 1983; Arnold, 1984a). Lizards with high exercise abilities typically rely on speed, whereas those with low aerobic scope and work capacity exhibit more complex, static defense repertoires (Bennett, 1980). Snakes that depend on rapid locomotion for escape are often unicoloured or striped, whereas those with static defenses are usually banded or blotched (Jackson et al., 1976).

6. Earlier considerations suggested that defensive mechanisms should be minor innovations that are closely tuned to local, contemporary problems, and that convergence should be widespread (Alcock, 1975; Greene, 1979; Janzen, 1981). Similar, independently evolved displays are indeed widespread among snakes (Gans and Baic, 1974; Greene, 1979; Myers, 1986; see above), but the broader conclusion that antipredator mechanisms have not played a significant role in reptilian evolution (Pough, 1983) is strongly contradicted by the survey in this chapter. Furthermore, as with the courtship of salamanders (S. J. Arnold, 1977) and the constricting behaviour of some snakes (Greene and Burghardt, 1978), certain defensive adaptations have persisted for very long periods. Why some some attributes of organisms should be tightly correlated with contemporary local conditions and others remain stable throughout ancient adaptive radiations is an intriguing question for evolutionary biology.

2

Mimicry in Reptiles

In 1862 H. W. Bates introduced the term "mimetic analogies" to describe the "resemblances in external appearance, shape, and colours between members of distinct families." In particular, he proposed that edible species of Amazonian butterflies were protected from attack by predators by virtue of their resemblance to inedible species that insectivorous animals had learned by experience to avoid. This short of resemblance is now known as Batesian mimicry.

Although Bates was the first to formulate a coherent explanation for mimicry, the existence of resemblances among unrelated animals had long been recognized. Eighteenth century names of day-flying moths like Apiformis and Ichneumoniformis called attention to the similarity in appearance of the newly described taxon to an unrelated noxicus insect (Remington, 1963). Studies of mimicry have concentrated largely upon insects since the early work by Bates and by A. R. Wallace (1865); but examples of mimicry have been noted among vertebrates, and reptiles were among the first of these to be described. In 1860 E. D. Cope called attention to the similarity of several African colubrid snakes to the venomous Atractaspis and of some New World colubrids to coral snakes. In 1867 Wallace pointed out that species of coral snakes (Micrurus) from different parts of Latin America differed in pattern, and that in a number of localities some colubrid snakes bore the pattern of the local coral snake. Mimetic resemblances of egg-eating snakes (Dasypeltis) to night adders (Causus) and of a water snake, Nerodia fasciata, to the water moccasin (Agkistrodon piscivorus) were proposed by Weale (1871) and Goode (1873), respectively.

The second major step in the understanding of mimicry was F. Muller's (1879) solution to a paradox that had baffled Bates—the mutual resemblance of two unpalatable species of butte-fly. Bates had recorded this phenomenon, but had been unable to accommodate it within his hypothesis of a defenseless species gaining protection via a resemblance to a noxious species. Muller proposed that two noxious species could mutually benefit from a shared pattern because predators would consume fewer of each species in the process of learning to reject all animals with the pattern. Mimetic phenomena of this sort are called Mullerian mimicry.

Bates had observed (1862:511) that "the process by which a mimetic analogy is brought about in nature is a problem that involves that of the origion of all species and all adaptations." That succinct characterization explains the controversy that surrounded the peenomenon of mimicry during the late nineteenth and twentieth centuries. Proponents and opponents of evolutionary theory and adherents of different views of evolutionary mechanisms chose mimicry as a battleground, a phenomenon that continues today—see Turner (1983a, b). Three major lines of argument sought to discredit the view that mimetic resemblances were functional or the result of natural selection (Remington, 1963):

1. The putative selective agents (birds, in most instances) were said not to eat butterflies, or eat them too rarely to have any effect upon butterfly populations and evolution.

2. Alternatively, if the predators do eat butterflies they eat as many noxious ones as palatable species.
3. The similarities attributed to mimicry are in fact the result of some other 'factor; either
   (a) chance,
   (b) common ancestry, or
   (c) a common response to some unknown condition of the physical enviroment—that is, genius loci.

Publication of a review of insect mimicry in 1933 by G. D. H. Copenter and E. B. Ford refuted these objections to mimicry as it was observed in insects. Carpenter and Ford documented the occurrence of predation by birds on butterflies and thereby established the existence of the necessary selective agent and selective pressures. The use of evidence from beak marks on butterfly wings as an indication of actual or attempted predation by birds was a critical link in the chain of evidence. (Predators often release unpalatable insects after tasting them (Boyden, 1976; Wirklund and Jarvi, 1982); thus beak marks on the wings of butterfles provide evidence of rejection by predators as wall as of the inefficiency of predators. For a review of early studies of beak marks and more recent data see Pough and Brower, 1977, and L. Brower 1984.)

In the second half of the twentieth century, studies of mimicry among insects have shifted their focus from the question of whether mimicry occurs to investigations of the mechanisms by which it operates. The importance of different relative frequencies of models and mimics or of different degrees of noxiousness of the model have been investigated experimentally and mathematically with a gratifying degree of agreement among the conclusions drawn from different approaches. (See J. Brower, 1958a, b, c, 1960, 1963; Huheey, 1961, 1964, 1976, 1980a; L. Brower *et al.*, 1963, 1964, 1967a, 1971; Holling, 1963, 1965; O'Donald and Pilecki, 1970; Pough *et al.*, 1973; Estabrook and Jespersen, 1974; Pilecki and O'Donald, 1971; Bobisud and Potratz, 1976; Arnold,

1978; Turner *et al.*, 1984 for examples and additional references.) Fulfilling Bates' prediction, the study of mimicry has made important general contributions to evolutionary biology. Of particular significance have been questions about the origion and evolution of aposematic colors (Boyden, 1976; Turner 1977; Gittleman et al., 1980; Harvey and Paxton, 1981a, b; Jarvi *et al* , 1981a, b; Harvey *et al.*, 1982; Wirklund and Jarvi, 1982). The experimental demonstration of a selective advantage for artificially created mimetic color patterns exposed to natural predators under field conditions has provided data to test the predictions of theoretical studies (L. Brower *et al* , 1964, 1967b; Cook *et al.*, 1969; Waldbauer and Sternburg, 1976; sternburg *et al.*, 1977; Jeffords *et. al.*, 1979, 1980). Further lighthas been shed on evolutionary mechanisms by experiments showing that very small changes in colour and pattern, of the magnitude that could be produced by a mutation at single locus, can confer measurable protection from predators (L. Brower *et al.*, 1963, 1971).

These experimets were pivotal in establishing the validity of mimicry theory because some critics of mimicry maintained that only a very close resemblance of a mimic to its model would confer protection on the mimic (Punnet, 1915; Goldschmidt, 1945). Experimental studies have demonstrated that even poor mimics can receive protection. Predators appear to generalize the salient featues of noxious prey and consequently a very slight resmblance between model and mimic can provide a measurable degree of protection. For example, birds that had learned that two species of orange butterflies were distasteful subsequently rejected a blue and yellow butterfly that was the same size and shape as the unpalatable orange butterflies, but accepted brown butterflies of different size and shape (L. Brower *et al*,, 1963). In a second experiment, birds trained to avoid an unpalatable butterfly also avoided three other species of butterflies with variying degrees of resemblance to the model (L. Brower *et al.*, 1971). Generalization of unpleasant experiences can extend to quite dissimilar prey items. Chickenes trained to reject quinine-treated meal, which was presented on a glass slide with a pattern of yellow and black stripes, subsequently refused to peck a cinnabar month caterpillar with a similar striped pattern (Morgan, 1900). Birds trained to avoid cinnabar caterpillars also avoided wasps that had the same yellow and black pattern but an entirely different shape (Windecker, 1939). This generalization of the characteristics of noxious prey by predators has led to an hypothesis for the evolution of Batesian mimicry by a two-step process, the first step providing an inaccurate resemblance that is subsequently refined by selection (Nicholson, 1927; Sheppard, 1962; Ford, 1964).

An important advance in the study of mimicry is the concept of a "palatability spectrum" (L. Brower *et al*., 1967a). Like many butterflies, the monarch (Danaus plexippus) derives its noxious characteristics from its larval foodplant. Monarch caterpillars feed upon a number of species of milkweeds that contain different mixtures and concentrations of cardenolides. As a consequence of their varied larval diets, adult monarchs vary in their cardenolide concentrations. Some individuals are entirely palatable, others have more than a lethal does of toxin and are extremely emetic to a predator. Butterflies at both extremes of palatability as well as at intermediate points may occur in a population (L. Brower *et al*., 1972, 1978; Brower and Moffitt, 1974). Predators feeding on a population of monarchs can experience widely different stimuli depending upon the cardenolide content of a particular butterfly. In its original form the concept of a palatability spectrum embraced variation in the degree of palatability within a single species, but it can logically be extended to apply to a Mullerian mimicry complex or to a Mullerian complex with Batesian mimics (L. Brower *et al*., 1963; Pough *et al*., 1973; Huheey, 1976; but see Turner *et al*., 1984).

The study of mimicry among reptiles has not shared the progress made recently by studies of insect mimicry. Investigations of reptilian mimicry have been largely descriptive rather than experimental. With a few notable exceptions the topic has remained mired in speculative arguments about potential predators and their hypothetical effects on models and mimics, as did the discussions that characterized the study of insect mimicry at the turn of the century.

Mimicry and Behaviour

Mimicry is only one mode of defense or offense potentially available to an organism (cf. Greene, this volume), and the distinction between mimicry and other mechanisms necessarily blurred. Mimicry grades into other forms of protective or exploitative resemblances and into resemblances that function during intraspecific social behaviour. The tendency of scientists to classify natural phenomena in two-dimensional arrays is severely challenged by the multi-dimensional complexity of mimicy. Even the distinction between matching an inanimate background and imitating a living organism has been a point of repeated controversy. Pasteur (1982) presented an amusing graphic representation of the scope of the definitions of mimicry employed by various authors during the last 150 years.

Classifications of mimicry systems have proven to be short-lived and have often stimulated as much criticism as clarification. The year 1976 saw the inception of two such controversies: J.E. Huheey published the seventh of his "Studies in Warning Colours and Mimicry," drawing responses from Benson (1977) and Sheppard and Turner (1977), to which Huheey subsequently replied (1980b). Also in 1976, R.I. Vane-Wright published "A Unified Classification of Mimetic Resemblances" (slightly amended in Vane-Wright, 1980), which elicited responses by Huheey (1980b), Cloudsley-Thompson (1981), Edmunds (1981), Endler (1981), Robinson (1981), and Rothschild (1981) and a rebuttal of these criticisms by Vane-Wright (1981). These examples illustrate the enthusiasm with which competing views of mimicry are expresed and debated.

Rothschild (1981) suggested that an elastic definition of mimicry would best accommodate new examples and new concepts in a series of categories that grade into each other, and Pasteur (1982) emphasized that too rigid a system of classfic a tion can blind us to examples of mimicry that are not readily accommodated by that particular system. Perhaps this is a time to consider examples of mimicry in all their perplexing charm and not a time to construct pigeonholes

Mimicry systems of the sorts considered in this chapter are tripartite (Wickler, 1968). They include three protagonists—a model, a mimic, and a dupe (Pasteur, 1972, 1982). A model is a "living or material agent" with perceptible characteristics such as colour, odour, or sound. A mimic is a living organism that plagiarizes the characteristics of the model, and a dupe is the organism that is deceived by the mimic.

A major dichotomy in systems of protective resemblance can be based on the response of the dupe to the model (Vane-Wright, 1976, 1980; Pasteur, 1982). Some models are of no interest to the dupe (background objects like leaves or the substrate, and inedible objects like twigs or bird droppings,) whereas other models are of interest (potential prey, predators, or mates, for example). Imitation of an object that is not of interest to the dupe can be called camouflage. The central feature of camouflage is failure of the model to elicit any reaction—attraction or avoidance—from the dupe.

Imitation of models that are of interest to the dupes can be designated homotypy ("assimilation of another type"). The model in these systems is normally a living organism, and it is to the advantage of the mimic to elicit the same reaction from the dupe as does the model. This review will be limited to consideration of example of homotype.

Two major sorts of homotype can be distinguished: In concrete homotypy the model is a definite (actual) species or group of similar species, whereas in abstract homotypy the model is an abstraction of a general category of organism that cannot be identified at the species level (Pasteur, 1982). Abstract homotypy appears to be particularly well represented among vertebrates, especially by mimics of venomous snakes.

Another dichotomy within homotypic resemblances can be based on the relationship between the model and the dupe: The medel may be either attractive or forbidding. This distinction underlies most systems of classification of mimicry (Wickler, 1968; Vane-Wright, 1976; Pasteur, 1982), but can be defficult to detrmine in practice because the response of a dupe to a model may not be stable. The proper allocation of a particular relationship may very in time or in space as population numbers or densities change. Even at one place and time, the role of model and mimic may be different for different species of dupes, or even for different age classes of a single species of dupe. For example, in the absence of alternative prey, hungry predators attack moderately unpalatable prey that had previously learned to reject (Holling, 1965; Sexton *et al.*, 1966; Alcock, 1970b); tolerance of aposematically colored prey may increase during the ontogeny of an individual (Rothschild, 1981); some species of predators are more tolerant of unpalatable prey than others (Alcock, 1971; Brower, 1984); and some species of predators behaviours that allow them to separate toxin-containing portions of prey from edible parts (Calvert *et al.*, 1979; Fink and Brower, 1981; Brower, 1984).

An additional division of homotypic resemblances can be based on the taxonomic relationships of the model, mimic, and dupe (Vane-Wright, 1976). If all three protagonists belong to different species, the system is disjunct. Batesian and Mullerian mimicry are examples of disjunct relationships among the protagonists. When two protagonists belong to one species and the third to a different species, the system is bipolar. Mimicry of female hymenopterans by orchid flowers is bipolar; a dupe (the male hymenopteran) tries to copulate with the mimic (an orchid flower) that resembles the model (a female hymenopteran), and in the process carries pollinia from one flower to another. In conjunct mimicry, all three protagonists are conspecific. For example, male scorpion flies (Hylobittacus apicalis) can obtain mates only by offering insect prey to a female as a nuptial gift. Males obtain these nuptial giftf in one of three ways: by catching them, by stealing them by force from other males, or by mimicking female behaviour so that they are offered a nuptial gift by another male (Thornhill, 1979).

Homotypic resemblances do not operate in an ecological or behavioural vacuum. Protective mimicry is normally part of a primary defense that operates before a predator initiates an attack (Edmunds 1974:1), or of a secondary defense that operates after the prey has responded to the presence of a predator (Edmunds, 1964:136), or both. For example, egg-eating snakes (Dasypeltis) have a cryptic pattern of dark dorsal blotches that is similar to the patterns of several species of sympatric vipers (see Section II.B.2.b). Camouflage is the initial primary defense for both egg-eating snakes and for vipers. If a predator discovers an egg-eating snake, the predator may mistake it for a viper because of its similarity to the viperid pattern. Thus homotypy is a second line of primary defense for the egg-eating snake. If both primary defensive mechanisms fail, the egg-eating snakes engages in an active display that incorporates body movements and sound production in imitation of a similar display by vipers This secondary defense enhances the original visual homotypic resemblance of Dasyreltis to a viper by adding movement to the visual stimulus and by adding sound as an additional stimulus. This progression of synergistic defense mechanisms may be a general rule among reptiles, as it is among other vertebrates (Edmunds, 1974), Homotypic resemblances among reptiles are often associated with protean defense (a sudden, erratic change in appearence or movement (Humphries and Driver, 1970)),deimam behaviour (a frightening display, posture, or noise (Maldonado, 1970)) and thanatosis (feigning death (Edmunds, 1974)).

Two categories of resemblances have posed particularly knotty problems for students of mimicry: Mullerian mimicry and homotypic resemblances that are the result of phylogenetic relationship. In Mullerian mimicry two or more noxious species resemble each other, and both species benefit from the resemblances because fewer individuals of each species are lost in the process of educating predators to avoid animals with the shared pattern. Various authors (Poulton, 1890; Wickler, 1968; Wiens, 1978; Pasteur, 1982) have maintained that this situation involves no deceit because avoidance of equally noxious models and mimics is the correct response for the dupe. Because no deceit occurs, they argue that it should not be called mimicry. However, two or more species of animals are unlikely to be equally noxious for all their potential predators, and Batesian and Mullerian mimicry may form a continuum (Huheey, 1976) or a step function (Turner *et al.*; 1984). This distinction has implications for the direction in which selection may act on the stronger and weaker members of a Mullerian mimicry situation (Benson, 1977; Sheppard and Turner, 1977; Turner. 1977, 1984; Huheey, 1980a).

A different problem is posed by homotypic resemblances that reflect phylogenetic relationships rather than being the product of evolutionary convergence resulting from selection by predators. These situations may involve deceit of predators, but selection for homotypy cannot necessarily be inferred. For example, Gingerich (1975) has suggested that the inoffensive, termite-eating aardwolf (Proteles cristatus) is a Batesian mimic of one or both species of (Hyaena (H. brunnea and H. hyaena), both of which are formidable predators. This proposal has been strongly criticized (Goodhart, 1975) and endorsed (Greene, 1977). One point of controversy has been the significance of the phylogenetic relationship of hyaenas and aardwolves : Both are hyaenids and their similarity in color and pattern may be a shared-ancestral character. (Of course, no information is available about changes through time in the similarity in appearance of the two genera.) If mimicry is viewed as an evolutionary process by which two species become more similar through time as a result of selection by predators, a homotypy that embodies a shared ancestral character does not qualify for inclusion unless it can be shown that selection by predators has increased the resemblance of the mimic to the model. On the other hand, if mimicry is viewed in an ecological context as an antipredator strategy, the important question becomes whether a predator is deceived. A resemblance based on a phylogenetic relationship could qualify as mimicry by that criterion. The question of how to interpret mimicry that derives from ancestral characters is particularly relevant to snakes because many instances of mimicry of vipers probably have their origins in a shared ancestral pattern of blotches (Section II.B.2. b).

The Distribution of Mimicry Among Taxa of Animals

Homotypy and camouflage can occur in any sensory modality to which a dupe is capable of responding, and may involve different characteristics within one modality. For example, visual homotypy may include similarities of reflected or emitted light, pattern and form, body outline, or movement. Similarly, a mimic may duplicate thermal or electrical emissions of a model and its odors or sounds. All of these potential mimicries, except duplication of electrical properties, are known to occur (Pasteur, 1982).

Examples of visual mimicry dominate the literature, probably because humans are primarily visually oriented. Humans as dupes are particularly sensitive to visual homotypies, and visual mimicry is over-represented in our sample (Pasteur, 1982). Furthermore, the diurnality and visual sensitivity of humans make us especially likely to recognize mimetic resemblances that result from selection by other animals that are diurnal and visually oriented. Thus, birds and lizards appear to be the selective agents in most of the well-known cases of mimicry. Like humans, these animals attend to details of colour, pattern and movement, and form searh images. Humans are not sensitive to visual properties of objects at wavelengths in the ultraviolet or infrared portions of the spectrum, but other vertebrates may be. Some mimetic resemblances among insects are as good in the ultraviolet as in the visible part of the spectrum (Silberglied, 1979).

Most animals probably rely on smell and hearing more than on vision, and these species seem likely to exert selective pressures that would favour the evolution of olfactory or auditory mimicries that humans would not perceive (Pasteur, 1982). Various species of predators and even individuals of a species may respond to different characteristics of a model or mimic (Alcock, 1971; Terhune, 1977; Vasconcello-Neto and

Lewinsohn, 1984). Antipredator mechanisms are equally diverse; some deter certain types of predators but may increase the risk of attack by other predators (Pearson, 1985). Mimetic relationships combine the effects of all these different types of selection.

To illustrate his system of classification of mimicry, Vane-Wright (1976) listed 56 examples of mimicry, most of which were drawn from previous summeries (Wickler, 1968; Edmunds, 1974). (I have omitted (from this list the "egg mimics" on the anal fin of the cichlid fish Haplochromis because these are probably not examples of mimicry (Otte, 1974:398-399; Hailman, 1977:309) and all of the mimetic activities described for humans.) Vane-Wright's list was not intended to be exhaustive, and it would probably be impossible to compile a truly complete list of examples of mimicry because so much of the literature on the subject consists of anecdotes without experimental data (Pasteur, 1982). Nevertheless, these 56 examples provide a representative sample of known examples to assess the incidence of mimicry among different groups of organisms. Precisely half the examples refer to mimicry by insects, and two more concern spiders. Relatively few examples of mimicry are known among vertebrates; only 14 were cited by Vane-Wright (see Edmunds, 1974; Pasteur, 1982; and Pough, in press, for a few additional examples). The paucity of mimetic relationships among vertebrates is almost certainly an artifact of a mismatch between the sensory modalities to which humans attend and those in which mimicry of vertebrates is most likely to occur. Carefully designed and controlled experiments are necessary to investigate the possible existence of mimicry in sensory modalities to which we are not attuned.

In addition to the mismatch of sensory modalities of humans and other vertebrates, some kinds of protective mimicry among vertebrates may differ qualitatively from the mimicry of insects that has dominated thinking about homotypy (Pough, in press). These differences may reflect the potentially disastrous consequences for a dupe of mistaking a model for a mimic. The advantage gained by a mimic depends in part on the properties of its model. A mildly noxious model protects good mimics, whereas a very noxious model extends protection even to poor mimics (Brower and Brower, 1965; Duncan and Sheppard, 1965; Goodale and Sneddon, 1977).

Vertebrates as a group may be more noxious models than are insects, partly because vertebrates are larger than insects in absolute terms and relative to the size of their predators. A model that combines an effective defense with large body size may pose a substantial risk for a dupe that mistakenly attacks it. If that is true, homotypy among vertebrates may in general be less precise than the insect homotypy that has shaped much of our thinking, and vertebrate mimicry would be correspondingly harder to recognize. In the terminology of Pasteur (1982), abstract (rather than concrete) homotypies may be the rule among vertebrates. Of all vertebrates, the venomous snakes seem to be the animals most likely to translate risk to a dupe into widespread abstract mimicry.

MIMICRY OF VENOMOUS SNAKES

Theoretical Aspects

1. Properties of Reptilian Mimicry

Venomous snakes provided the earliest examples of mimicry among vertebrates (Cope, 1860; Wallace, 1867, 1870; Weale, 1871; Goode, 1873; Lonnberg, 1902) and have continued to attract attention from biologists. The dramatic defensive displays and conspicuous colors of some snakes emphasize the risk they represent to a predator. Indeed, some snakes (particularly vipers) are so dangerous that certain assumptions, terms, and mathematical models that have been developed for mimetic relationships among insects may be only marginally applicable to the mimicry of venomous snakes.

2. Venomous Snakes as Models

Venom is probably an ancestral character for caenophidian snakes (colubrids, hydrophiids, elapids, and vipers), and venom is widespread among living colubrids (Savitzky, 1980). At our present state of knowledge, even the distinction between venomous snakes and harmless ones is uncertain. The widespread development of toxic oral secretions by a variety of colubrid for and the possibility that these secretions are specialized snakes particular functions make allocations of snakes to categories of risk very speculative (Gans, 1978). For example, the colubrid genus Pliocercus lacks fangs and is considered harmless, but recently an adult human was hospitalized by the effects of a bite from P. elapoides (Seib, 1980). For small foliage-gleaning birds any snake, venomous or not, may be a potential predator and a dangerous model for a mimicry system.

The most specialized venomous snakes are the approximately 500 species in the families Elapidae, Hydrophiidae, and Viperidae. Most of these snakes produce toxins that are potentially lethal to vertebrate predators (see Gans, 1978, for a list of these species). These families of snakes include the species most often cited as models that are mimicked by a variety of colubrid snakes and by a wide variety of other animals, including vertebrates and invertebrates.

The unusual features of mimicry of venomous snakes are the result of three features of these animals: their venom, their size, and the unpredictability of their success in biting a predator.

(a) *Venom*

Considerations of the potential lethality of a bite from a venomous snake have focused primarily on the toxicity of snake venom (Brattstrom, 1955; Hecht and Marien, 1956; Mertens, 1957; Gans, 1961; Wickler, 1968). Many factors interact in determining the severity of a bite from a venomous snake: Venomous snakes can bite without delivering venom or can vary the quantity of venom they inject (Kochva, 1978). Venom displays individual, ontogenetic, seasonal, and geographic variation in toxicity (Gans, 1978). The properties of snake venoms also vary phylogenetically (Mebs, 1978). Neurotoxic symptoms are prominent features of bites from elapids and hydrophiids, and include systemic effects such as depression, paralysis of the hind limbs, and respiratory failure. The venom of viperids produces local discolouration and hemorrhage, followed by symptoms of cardiovascular shock and generalized internal hemorrhage. However, classifying snake venoms as neurotoxic or hemorrhagic is an oversimplification. Elapid venoms may produce cardiovascular failure and hemorrhage, and viperid venoms can lead to paralysis and respiratory arrest. The quantity of venom injected, the species of animal being tested, and the route of envenomation are important variables (Mebs, 1978).

The delayed effects of envenomation have usually been ignored in considerations of mimicry of venomous snakes, but these effects seem likely to pose a substantial risk for mammals and birds that prey on snakes. A snake need not deliver a fatal dose of venom to expose an animal to the debilitating effects of envenomation. Tissue necrosis and bacterial infections of damaged tissues are secondary consequences of sublethal doses of venom, even for animals as large as humans (Russell, 1980). Indeed, these secondary effects can be clinically significant results of untreated snakebites. and the loss of digits or limbs was a common aftermath of snakebite before the introduction of massive antivenin therapy (Russell, 1980).

Tissue destruction, infection, or the potential loss of a limb may be as important as the chance of rapid death in determining the risk of preying on venomous snakes. When the short-term and long-term effects of snakebite are considered, the potential risk for a predator of attacking a venomous snake appears to be substantially greater than the risk associated with eating an unpalatable insect, or even of being stung by most wasps. Only among some marine invertebrates that possess lethally toxic venoms (some coelenterates, gastropods, and cephalopods (Halstead, 1965), do we find potential models that may be as formidable to their predators as are venomous snakes.

(b) *Size*

Snakes are not only larger than insects, but snakes increase in size ontogenetically without changing their appearance, unlike the holometabolous insects that have formed the basis for most considerations of mimicry. A large individual of a particular species of venomous snake may represent a very much greater threat to a predator than a small individual of the same species.

(c) *Uncertainty*

Venomous snakes probably must bite a predator to deliver a noxious stimulus. (Some venoms are not toxic when ingested. The possibility that the flesh of some snakes is unpalatable to predators is discussed in Section II.B.2.a). If a snake does not succeed in biting a predator, the experience of the predator with a model is probably not different from its experience with a mimic. (Stinging insects represent a parallel situation in this respect.)

The probability that a predator will be bitten depends on at least two interrelated factors: How good is the snake at delivering venom, and how good is the predator at killing snakes without being bitten? Nature presents a range of specializations on both sides. Front-fanged snakes, especially vipers, are probably capable of killing most predators if they succeed in biting. Elapids, which have short fangs at the front of the mouth, and colubrid snakes, with fangs at the rear of the mouth, may or may not administer a lethal dose of venom. Some snakes, like the nearly toothless colubrids of the genus Dasypeltis and the blunt-toothed sibynophine colubrids, are probably incapable of deterring a predator by biting it (Gans, 1974; Savitzky, 1983).

The specializations of predators interact with the specializations of snakes in determining the probability that a snake will succeed in biting and the consequences of a bite. Many snakes, both venomous and nonvenomous, are immune to snake venom (Kochva *et al.*, 1983). Some predators, especially snake-eating birds like the neotropical laughing falcon (Herpetotheres cachinanns) and the African secretary bird (Sagittarius serpentarius), probably can prey upon even the most dangerous snakes with little risk to themselves. Less specialized birds and most mammals seem likely to face some risk of being bitten if they attack a venomous snake. Risk to a predator may vary intraspecifically as a result of ontogenetic change in its body size and its experience in capturing prey. Chance will play a role, too. If the snake is partly concealed, is its heat where the predator assumes it is? Which direction will the snake strike when it is first seized? If the snake has an erratic display (Tweedie, 1953; Gehlbach, 1970; Greene, 1973, this volume), will it bite the predator as it thrashes about?

As a consequence of these variations in venom delivery capability of snakes and in predatory capacity of predators, one cannot assign a constant degree of noxiousness to a particular venomous snake all are serving as a model. The outcome of an encounter between a venomous snake and a predator might be death or injury for the snake, death or illness for the predator, or possibly the deaths of both or neither of the participants. This situation is different from that known in insects for which models and mimics (species or individuals) can usually be ranked in order of noxiousness. The closest analogy to this situation among insects is the palatability spectrum (L. Brower *et al.*, 1967a). However, in a palatability spectrum the variation in noxiousness results from differences in the quantity and quality of toxic chemicals present in individual models, and each individual has a definable noxious quality.

3. *Special Features of the Mimicry of Venomous Snakes*

The differences in the properties of snakes and insects as models give rise to special features of the selection exerted on predators by venomous snakes, and the selection exerted by those predators on mimics of the snakes. Conventionally, mimicry has been assumed to act through the ability of predators to learn to avoid noxious prey after one or more experiences. The initial learning process is reinforced by subsequent experiences once the initial conditioned avoidance is forgotten. In that situation a mimic benefits from being treated like its model by the predator. A recurrent theme in discussions of mimicry of venomous snakes has been the paradox of the deadly model—that is, the inability of a dead predator to profit from its experience with the model or to avoid a mimic as a result of that experience. To some extent this problem has been overcome by postulating that empathic learning teaches predators that observe a fatal snakebite to avoid the snake in the future (Gans, 1964; Jouventin et al., 1977). This mechanism appears plausible for social animals but is less appropriate for solitary predators. The demonstration of innate avoidance of venomous snakes by predators (Smith, 1975, 1977; Caldwell and Rubinoff, 1983) supplies a mechanism that creates a selective advantage for mimics of venomous snakes in the face of even solitary predators. This situation differs from the sort of response by predators involved in insect mimicry in that the effect produced by experience with the model is not learning in ecological time, but instead genetically fixed alteration of the behavior of a species of predator in evolutionary time.

4. *Predictions*

Both the magnitude of the potential noxious experience and the mechanism of avoidance of models and mimics have special features in the case of mimicry of venomous snakes that are not present in mimicry of insects. These features should be reflected in various aspects of the mimicry of venomous snakes. Several testable predictions can be formulated on the basis of the assumption.

(a) *Ratio of Models to Mimics*

For insects the degree of protection received by a mimic varies inversely with the frequency of the mimic in relation to the model. A variety of factors interact to determine the precise degree of protection a mimic experiences including the noxiousness of the model and the length of time a predator remembers its experience with the model (J. Brower, 1996; Holling, 1963, 1965; Huheey, 1964; Emlen, 1968; Alcock, 1970a, b; L. Brower *et al.*, 1970; O'Donald and Pilecki, 1970; Pough *et al.*, 1973). The common assertion that models must outnumber mimics is incorrect, but the degree of protection a mimic experiences is sensitive to the frequency with which a predator encounters models and mimics, particularly if the model is not very noxious or if the memory of the predator is short.

Mimicry of venomous snakes may be less sensitive to variation in the ratio of mimics to models than is mimicry of insects. If a single noxious experience with a venomous snake can kill a predator, a frequency of models that exposes a predator to one such experience in its normal reproductive lifespan would be effective in creating a selective advantage for those individuals in the predator species with an innate tendency to avoid the model. Mimics of venomous snakes could be considerably more conspicuous or abundant than their models with little loss of protection from predators.

(b) *Abstract Mimicry*

A second consequence of the extreme noxiousness of venomous snakes as models should be broad generalization of the features of the model by the predator and the avoidance of snakes with even minor similarities to the model. Thus, venomous snakes probably serve as abstract models. Because even a very slight similarity to a venomous snake would be selectively advantageous, mimetic resemblances could evolve readily and functional mimicry may be widespread among snakes. Much of this mimicry is likely to be relatively imprecise provided that, once an inaccurate resemblance to a venomous snake is achieved, the very noxious quality of the model limits the selective advantage of modifications that would perfect the mimicry. If predators generalize the characteristics of venomous snakes broadly, one can predict that areas with coral snakes will have higher proportions of nonvenomous snakes with banded patterns incorporating red, yellow (or white), and black than are found in areas without coral snakes because mimics with only superficial similarities to coral snakes should often be avoided by predators.

(c) *Dual Mimics*

In mimicry of insects, predators learn specific characteristics of one or more models, and mimics are often remarkably precise duplicates of one or another of those models. Dual mimics that combine features of two or more models do not occur as part of normal mimetic assemblages. When such hybrid mimics do occur, they are a result of perturbation of the geographic ranges of different mimetic patterns, and the hybrid mimics appear to be less protected than either of the normal forms. However, a mimic that combined the features of two or more species of venomous snakes would not be expected to lose protection, and might even benefit by having more of the abstract features of venomous models because, even within a species of predator individuals respond to different characteristics of a mimic (Terhune 1977). The occurrence among snakes of mimics combining the features of two or more models would be evidence that mimicry of venomous snakes is different from the single-model mimicry of most insects.

Tests of the Predictions

1. *Mimicry of Feptiles of by Insects*

Examples of mimicry of reptiles by insects are particularly useful in testing the hypothesis that mimicry of venomous snakes can be effective without being precise because most examples of mimicry of insects by other insects are concese. That is, the models are identifiable species or groups of species and the mimicry involves extraordinarily exact duplications of the pattern color, and even body form of the models. (For examples of this precision, see the color plates in Bates (1862), L. Brower *et al.* (1963), Turner (1975), and Papageorgis (1975). If mimicry of reptiles by insects is less precise than this, the reason is probably to be sought in the characteristics of the models, not of the predators.

The larvae, and in some cases the pupae, of butterflies and moths in the families Sphingidae, Notodontidae, Oxytenidae, Noctuidae, Geometridae, and Papiliionidae have markings and behaviors that are reminiscent of snakes. These caterpillars occur in both the Old and New World and represent many independent origins of the phenomenon, probably facilitated by the general similarity of the body forms of snakes and caterpillars. In Santa Rosa National Park in northwestern Costa Rica, caterpillars of some 50 species, belonging to five families, have eyespots and may be mimics of snakes (D. H. Janzen, pers. comm). In general the mimicry consists of a display in which the anterior portion of the caterpillars body is inflated, often revealing eyespots, and waved from side to side. Bates (1862:509) described the display of such a caterpillar in South America:

> The most extraordinary instance of imitation I ever met with was that of a very large Caterpillar, which stretched itself from amidst the foliage of a tree which I was one day examining, and startled me by its resemblance to a small Snake. The first three segments behind the head were dilatable at the will of the insect, and had on each side a large large black pupillated spot, which resembled the eye of the reptile: it was a poisonous or viperine species mimicked, and not an innocuous or colubrine Snake; this was proved by the imitation of keeled scales on the crown which were produced by the recumbent feet, as the Caterpillar threw itself backward.

Other descriptions and photographs can be found in Morrell (1954, 1969), L. Brower (1968, 1971), Edmunds (1974, pl. 7b), Aiello and Silberglied (1978), and Janzen (1980).

The similarity of snake-mimicking caterpillars to snakes is not precise, and it may have a simple genetic basis. Turner (1984) has suggested that the combination of four mutations (Moricaud, Zebra, quail, and Multiluar) can convert the larva of the domestic silk moth (Bombyx) into a "snake mimic."

Snake-mimicking caterpillars are found in areas without arboreal vipers or without any vipers, and such caterpillars appear no more nor less viper-like than species that live in sympatry with arboreal vipers. The Malayan hawk moth (Panacra mydon) has been suggested as a mimic of the juvenile pattern of Trimeresurus wagleri, largely because the caterpillar has a dark somc on its head that resembles the dark line that runs horizontally through the eye of the juvenile viper (Morrell, 1969). However, eyestripes of this sort are usual markings of vipers; among 48 species of vipers, 92% had an eyestripe and simclar marks are common among arboreal colubrids (Pough *et al.*, 1978). Because of the ubiquity of eyestripes in potential models, more evidence is required to support the suggestion that the caterpillar of P. mydon is specifically mimicking T. wagleri.

Third and fourth instar larvae of the neotropical sphingid moth Hemeroplanes triptolemus are said to mimic the vine snake Oxybelis, whereas the last instar is described as a "formidable mimic" of the pit viper Bothrops schlegelii (Haber, 1983). The pupa of the nymphalid butterfly Dynastor darius has a snaks-like apperance (Aiello and Silberglied, 1978). The yellow-and-black banded caterpillar of the neotropical sphingid Pseudosphinx tetrio may be sufficienntly like a coral snake to be considered a mimic, but the caterpillars are probably distasteful and are perhaps better regarded as being aposematic in their own right.

Any caterpillar that looks enough like a snake to startle a foraging bird is likely to benefit—if not from the bird's previous experience with snakes, then perhaps from a tendency for bird to flee from a startling or potentially dangerous stimulus (Coppinger, 1969, 1970). The eyespots displayed by many insects as part of a qimatic display may be analogous to snake mimicry in certain functional respects. Realistic eyespots (with eccentric pupils and an apparent reflective highlight) elicited more escape responses by small birds than a a pattern of simple concentric circles, and the concentric circles were more effective than a single circle (Blest, 1957a). The eyespts on the wings of some moths are very realistic; like the eyespots in Blest's experiments, they contain eccentric pupils and a light mark resembling reflection of light from the moist surface of the cornea (Blest, 1957b). Nonetheless, they are an abstract representation of the eyes of vertebrates, not mimics of the eyes of a particular predator of the birds they frighten (Pasteur, 1982); and four or more false eyes are displayed simultaneously by many animals (Wickler, 1968).

Eyelike markings occur among a wide variety of vertebrates including fishes, birds, snakes, lizards, frogs, and turtles (Cott, 1940; Wickler, 1968; Edmunds, 1974; Owen, 1980), and many such animals have more than one pair of eyespots. Some of these markings may deter predators, but eyelike patterns may be only a particularly conspicuous way of combining contrasting colours.

Other insects may be still less precise mimics of reptiles. The head of the neotropical fulgorid bug Laternaria bears a large hollow structure of vaguely reptilian appearance (Poulton, 1924; Hogue, 1984). Other examples of proposed mimicry of reptles by insects are probably illusory. A mimetric resemblance between markings on the hind wings of the tropical butterflies Caligo and Eryphanis and the heads of hylid frogs and anoline lizards respectively (Stradling, 1976) seems problematic. Such mimicry might protect the butterflies from predation by some birds, but other species of birds would be equally likely to prey on treefrog or anoles. The proposal that Eryphanis is left unmolested.

2. *Warning and Defensive Mimicry Among Snakes*

(a) *The Coral Snake Mimicry Complex*

The problem:

The very conspicucus similarities of the colours and patterns or coral snakes and of number of nonvenomous or mildly vanomous snakes was noted by Cope (1860) before a coherent theory of mimicry had been formulated. In the earliest synthesis of mimicry meory, Wallace (1867), an ardent supporter of Bates (1882) suggestion of mimicry as a defensive mechanism, cited coral snakes and their mimics as the preeminent examples of mimicry among vertebrates Wallace immediately perceived the inadequacy of the bare fact a similarity in appearance to substantiate a claim of mimicry. In particular, in appears to have been the first to appreciate the significance of parallel geographic variation among models and mimics. In 1867 he gave four examples of similarity of color and pattern of coral snakes and their coubrid mimics within limited geographic areas; and, in his 1870 revision of that essay, he added two more examples. Unfortunately Willace's interests in mimicry were centred on butterflies, and he did not continue his investigation of coral snakes. Indeed, after its promising start the study of coral-snake mimicry was largely overlooked until the moddle of the twentieth century. The chronological development of hypothesesin this period has been summarized (Dunn, 1954; Hecht and Marien, 1956; Mertens, 1957; Grobman 1978) and will not be repeated have. Instead I will consider the major questions that have been raised about coral snakes and their presumed mimics.

The hypethesis that underlies what has become known as "the coral-snake mimicry problem" states that nonvenomous and mildly venomous colubrid snakes with bicolor or tricolor banded patterns mimic the patterns of highly venomous coral snakes of the genera Micrurus and Micruroides. In the absence of direct evidence of mimicry, the proposal that the similarity in color and pattern is protective have been received with a skepticism that does credit to the abstract ideal of scientific objectivity, although not all publications on the subject have embodied that objectivity. Discussion of the occurrence and mechanics of mimicry of coral snakes has focused on several aspects of the biology of the snakes and of their potential predators. The possible functions of regularly banded color patterns in snakes have formed another area of speculation.

Are the paiterns of coral snakes cryptic or aposematic?

The patterns of coral snakes vary among species and vary geographically within species. (See Roze (1967), Savage and Vial (1974), Amaral (1978), and Duellman (1978) for illustrations.) A coral-snake pattern is any pattern of "alternating black and red (or yellow or white) dorsally" (Savage and Vial, 1974:334). A complex terminology has been useedt to classify patterns as "bicolour," "tricolour," "tricolour triad," and "tricolour quartet" (Dunn, 1954; Hecht and Marien, 1956).

The earliest papers on coral-snake mimicry were apparently written by people who had no experience with coral snakes in the field. In a museum collection, a coral-snake pattern is so conspicuous that an aposematic function was implicitly assumed and was not a subject for comment (Cope, 1860; Wallace, 1867, 1870). The first suggestion that coral-snake patterns are cryptic rather than (or in addition to) being aposematic came from Gadow (1911:2), who may have carried out the earliest experimental investigation of the function of the coral-snake pattern:

> ...In their natural surroundings...they are.. conspicuous at a close distance, the red catching the eye at once, but a distance, say beyond five yards, they seem to vanish, at least parts of them according to the pattern. It may not appeal to the closet zoologist, but it is nevertheless a most instructive experiment to have some flexible tubes painted with the various patterns and colours of these snakes and to study the effect of those toys when thrown at random into an herbaceous border, upon the grass, into shrubs in bright sunshine or on a dull day. In most cases the effacive effect is surprising, whilst a similar toy painted monochrome, draws attention at once.

Much controversy about the function of coral-snake patterns has stemmed from a tendency to seek a single function for the pattern, but coral-snake patterns are mobably aposematic in some circumstances and cryptic in others. Evidence for an aposematic function is provided by experiments that have emonstrated innate avoidance of coral-snake patterns by birds and posibly by mammals as well (Gehlbach, 1972; Smith, 1975, 1977). Additional evidence suggesting an aposematic function for a coral-snake pattern comes from the Central American colubrid snake Lovinis evnephalus, which is normally solid coloured. However, its defensive display reveals a red pattern on the skin. Because the coral-snake pattern of this species is revealed only as a part of a defensive display it clearly serves a warning function (H W. Greene, pers. conm.).

Cryptic effects of coral-snake patterns include a visual illusion, apparent when banded or blotched snakes move, that depends upon the rate of passage of light-dark transitions across a viewer's visual field. If this rate exceeds the flicker-fusion frequency of the viewer, the eye cannot resolve individual bands and perceives the snake as a solid-colored object that is a blend of the band colors (Jackson *et al.*, 1976; Pough, 1976). This illusion is well illustrated by water snakes: Juvenile water snakes (Nerodia sipedon) are prominently banded, whereas the adults are unicolored. The escape behavior of juveniles, in contrast to that of adults, consists of crawling very rapidly for a few seconds and then stopping abruptly. While the snake crawls, the light and dark bands on its body pass an observer's field of view at a speed above the flicker-fusion frequency of most predators and the snake appears to be a unicoloured light grey. The instantaneous transition from motion to immobility removes the elongate, solid-coloured stimulus a predator had perceived and substitutes one in which the sharply contrasting bands conceal the elongate form. A similar blending of colours of individual bands occurs in coral snakes during their erratic displays (Gehlbach, 1970 and pers. comn.).

In addition to the change in the visual stimulus that a banded snake can present to a predator by suddenly becoming immobile, the banded pattern could conceivably be employed to produce a flickering stimulus that might be unpleasant to a predator, or to create an illusion that the bands are flowing in a direction opposite to that in which the snake is moving (Jackson *et al.*, 1976). These phenomena depend upon the visual properties of particular species of predators as well as the details of the patterns of the snakes (high or low constrast, bi-coloured or tricoloured, sharp borders to the bands or irregular ones) and on the viewing conditions (bright or dim, full solar spectrum or light filtered through vegetation). These variations in the physical and biological systems probably contribute to the high degree of variation in banded patterns of snakes and to multiple functions that probably include both aposematism and crypis.

Can predators of coral snakes perceive the colours ?

Much debate has focused on the twin questions of what predators might kill snakes and whether certain potential predators are capable of perceiving and responding to the bicolor or tricolour patterns of coral snakes. These questions involve complex interactions between the visual sensitivities of the predators and the behaviors of the predators and snakes. Simple answers are not likely to be available. The assertion that coral snakes are nocturnal or fossorial and therefore rarely if ever exposed to diurnal predators originated with Gadow (1911) and has been frequently repeated. Several lines of evidence suggest that the nocturnal and fossorial habits of coral snakes have been exaggerated, at least insofar as they affect the possibility of encounters among coral snakes and diurnal predators. Coral snakes and their mimics have been found in the open during the day. In Costa Rica individuals of Micrurus nigrocinctus have been observed in Santa Rosa and Corcovado National Parks moving slowly across the forest floor during the day, poking their heads into holes and doubling back on their paths (D.H. Janzen and W. Hallwachs, pers comm.). This behaviour suggests that the snakes were foraging for food, not escaping from predators or moving from one retreat site to another. Coral snakes and their mimics are eaten by diurnal birds and discovered by diurnal mammalian predators (see Greene and McDiarmid, 1981). Innate avoidance of the coral-snake pattern by two species of diurnal birds has been demonstrated (Smith, 1975, 19/7), a phenomenon that is hard to explain unless those species of birds encounter coral snakes during their daily foraging. The protean defensive display of coral snakes appears to be directed against predators that can see the snake clearly. In Costa Rica the yellow, black, and red colubrid snake Scolecophis atrocinctus has been reported to thrash voilently from side in its defensive display, even propelling itself into the air (D.H. Janzen, pers. comn.). This display is like that of the sympatric venomous coral snake Micrurus nigrccinctus. The possibility that foraging birds or mammals overtun ground litter and expose hidden snakes (Gehlbaoh, 1972) appears to have been underemphasized. A preponderance of evidence suggests that coral snakes and their mimics are encountered by predators during the day with sufficient regularity to have produced stereotyped reactions to that situation among both the snakes and the predators.

The assertion that most mammals lack color vision because histological examination of their retinae revealed few cone cells (Walls, 1942) directed the search for potential selective agents toward birds (Hecht and Marien, 1956). Recent behavioural and biochemical evidence suggests that most mammals are capable of distinguishing colours and of responding to different colors (Muntz, 1974), although the qualitative aspects of colour vision are unknown (Jacobs, 1982). Colour vision of this sort has been demonstrated in both diurnal mammals (dogs, squirrels, ungulates and nocturnal species (cats, opossums, tree shrews), and colour vision must be assumed to be widespread among mammals. This realization greatly broadens the scope of the potential selective agents for an aposematic function of color patterns of coral snakes and a mimetic function of similar patterns in colubrid snakes. The proclivity of many mammals for searching through the ground litter appears to make them particularly likely to encounter small snakes. Few mammals are specilized for killing venomous snakes, consequently mammals may be the animals most likely to evolve innate avoidance responses to venomous snakes and thus to favour the evolution of mimics of those snakes.

The rective frequency of models and mimics

The idea that models must oumuber their minics if the minimics are to benefit can be traced directly as Bates (1862) and Wallace 1865, (1867). It persists in the mimicry literature despite the existence of both experimental and mathematical demonstrations that mimics can receive protection even when the outnumber models by 9 : 1 (J. Brower, 1960; Holling, 1965;

Emlen, 1968). In practice, the protection received by a mimic depends upon the interaction of many factors: A very noxious model protects a higher razo of mimics to models, and protects less precise mimics, than does a less noxious model (Brower and Brower, 1965; Duncan and Shepare 1965; Goodabe and snedden 1977). The availability of alternate prevenches the protection received by a mimic, and poor mimics receive less protection if a predator can compare them with a model than they do if they are encoutnered separately (Sexton, 1960). These and other conditions that affect the precision and selective advantage of mimicry are discussed by Edmunds (1974:81-135).

Various reports have listed the proportions of croal snakes, rearfanged snakes, and aglyphous snakes in collections from different regions (Bunn, 1954; Mertens, 1966). In Panama 62% of 1175 snakes with the coral-snake pattern were Micrurus whereas in the state of Sao Paulo, Brazil, only 17% of 1227 snakes with the that pattern were Micrurus. When the sample area was expanded to all of southern Brazil, the proportion of Micrurus rose to 29% of 1691 snakes with the coral-snake pattern.

These data may not be directly applicable to a consideration of coral-snakes mimicry because the different foraging modes of various species or predators will expose them to different portions of the snake fauna in a habitat. For example, hawks will see only those snakes that move by daylight on top of the ground. Birds scratching on the forest floor will see those snakes plus some partly concealed beneath small pieces of ground litter. Mammals may dig deeper and move larger pieces of litter than birds and are more likely to be active before dawn or at dusk. Only mammals and perhaps owls are likely to sample snakes at night. Each of these kinds of predator is likely to encounter a different frequency of models and mimics, and their samples are all likely to be different from a sample gathered by human collectors. These considerations make it difficult to know what significance to attach to the frequencies of coral snakes and their mimics that have been reported. Nevertheless, two useful points do emerge from these data: The coral-snake pattern is well represented in the Americas and the proportions of the total number of snakes with that pattern that are comprised by Micrurus, rear-fanged, and aglyphous species are extremely variable.

The number of independent origins of a coral-snake pattern in different lineages of snakes would probably be the best measure of the selective forces favouring that pattern. In the absence of detailed phylogenetic information about snake faunas, this analysis is not feasible. As an alternative, the number of species of snakes in a region with a coral-snake pattern is an index of tne occurrence of the pattern that may be less subject to the sampling errors inhereni in different methods of foraning predators than is the number of individuals. Table 2.1 shows the frequencies of the coral-snake pattern among aniilid, boid, and colubrid snakes of several regions. (For the census method and lists of snakes considered to have coral-snake patterns see the Appendix.) The coral-snake pattern is clearly more common in tha neotropics than in other parts of the world. The average frequency of coral-snake patterns in Central and northern South America is 18%, four times the average for Europe, Asia and Africa. Furthermore, an average of 80% of the neotropical species of snakes with coral-snake patterns include red, whereas the average incidence of red is only 20% for the rest of the world. Conspicuously banded elapids in Asia and Australia (Calliophis, Bungarus, Vermicella, Simoselaps, Pseudonaja) might form the basis for a mimicry complex like that of the neotropics (Hecht and Marien, 1956), but this does not seem to have happened. Coral snake patterns in these regions are largely confined to elapids.

The high proportion of snakes with coral-snake patterns in the neotropics compared to the lower frequency of that pattern in most of the rest of the world suggests that many species of snakes mimic coral snakes, as was predicted by a consideration of the consequences of mimicking a deadly model. The great variation in the proportions of coral snakes, rear-fanged snakes, and aglyphous snakes, among the coral-snake pattern group

TABLE 2.1

The Frequency of Coral-Snake Patterns Among Aniliid Boid, and Coiubrid Shakes[a]

Region	*Coral-snake patterns* Number of species	*Coral-snake patterns* Percentage of total species	*Coral-snake patterns* Percentage with red	*Total number of species classified*	*Reference*
(1)	(2)	(3)	(4)	(5)	(6)
The Americas					
Temperate Canada and United States	12—14	10—12	86	120	Stebbins, 1966; Conant, 1975
Argentina	1—2	4—12	100	24	Abalos *et al.*, 1964
Tropical					
Guatemala	7—9	20—26	78	35	Duellman, 1963
Venezuela	15—21	14—20	86	107	Roze, 1966
Ecuador	7—13	15—28	62	46	Duellman, 1978

(1)	(2)	(3)	(4)	(5)	(6)
Brazil	15–22	10–14	95	155	Amaral, 1978, Freiberg, 1982
Europe and Asia					
Temperate					
Europe and Soviet Union	1—3	2–5	67	63	Banikoy *et al.*, 1971 Steward 1971
Iraq	0—1	0—4	0	27	Khalaf, 1959
Indian Subcontinent	25—31	7–9	26	346	Smith, 1943, Minton, 1966
Japan	1–3	4—12	67	25	Nakamura and Uero, 1963
Taiwan	1—2	3–6	0	31	Kuntz, 1963
Tropical					
Thailand	7—10	6–8	10	121	Taylor, 1965
South Vietnam	1—3	1–4	33	71	Campden-Main, 1970

(1)	*(2)*	*(3)*	*(4)*	*(5)*	*(6)*
Malaya	5—11	5—10	18	105	Tweedie, 1953
Philippines	3—5	3—5	20	89	Taylor, 1922
Africa					
West Africa	1—4	1—3	0	117	Villiers, 1975
Uganda	1	1	0	118	Pitman, 1974
South Africa	1—3	1—2	0	116	Fitz Simsons, 1970; Broadley and Cock, 1975
Australia					
Temperate	4—6	6—10	0	64	Cogger, 1979
Tropical	4—7	6—10	14	67	Cogger, 1979

The incidence of such patterns among Australian elapids is included for comparison.

suggests that, as expected, the advantage for a mimic of a coral snake is relatively insensitive to the proportion of models and mimics in a region.

The paradox of the deadly model and the palatability of coral snakes.

The assumption that a bite from a coral snake is lethal has led several investigators to doubt that coral snakes could serve as models because a predator that dies from the bite of a coral snake has not learned to a void a mimic of a coral snake. As an alternative, it has been suggested that rear-fanged species of snakes with coral-snake patterns are the models, and that the nonvenomous and front-tanged species are, respectively, Batesian and Mullerian mimics (Hecht and Marien, 1956; Mertens, 1957; Emsley, 1966). The name "Mertensian mimicry" has been applied to this hypothesis (Wickler, 1968). Uncertainties about the probability of envenomation of a predator by a snake and the toxicity of different venoms preclude complete evaluation of this aypothesis, but under some circumstances a bite from a coral snake probably can elicit conditioned avoidance by a predator (Greene and McDiarmid, 1981). Colubrid snakes may be weak Mullerian mimics of venomous coral snakes, but they are unlikely to be the species on which the mimicry complex is based.

Nonetheless, additional mechanisms of conditioning predators to avoid corol snakes may opesate. Unpalatability of a prey species is a common basis for discrimination by predators and coral snakes might be unpalatable, W. Hallwachs and D.H. Janzen (pers. comm.) report that in Costa Rica scavengers (mostly birds) that quickly consume road-killed snakes of other species leave dead coral snakes undisturbed. Although the phenomenon has not been widely investigated, other vertebrates may use unpalatability at a predator-avoidance mechanism. Nurse sharks (Ginglymostoma cirratum) rejected the flesh of yellowbellied sea snakes (Pelamis platurus) even when it was concealed inside a piece of squid (Rubinoff and Kropach, 1970). Various species of birds and their eggs may be unpalatable to verteberate and invertebrate predators and scavengers (see Cott, 1969, for a review).

Evidence for an aposematic function of coral snakes patterns and for mimicry of coral snakes: Parallel geographic variation of coral snakes and colubrids.

Because of the improbability of observing rare events like encounters between predators and models or mimics in nature, the evaluation of alleged examples of mimicry is based on indirect evidence. Concordant geographic charge in color or pattern of presumed models and mimics has been particularly useful. Among butterflies of the genus Heliconius, for example, local variation in patterns of Mullerian mimicry complexes may reflect Pleistocene fragmentation of the Amazonian forest (Brown *et al.*, 1974). In the presence of model the butterfly Limenitis is mimetic, but in areas in which the model does not occur, it is cryptically coloured (Platt and Brower, 1968).

Geographic variation and correspondence in the patterns of coral snakes and their mimics were appreciated by Wallace, who cited four examples in his 1867 review and added two more in his 1870 revision. Conversely, the occurrence of snakes with patterns similar to those of coral snakes in areas lacking models has been used to argue against mimicry (Gadow, 1911; Brattstrom, 1955; Grobman, 1978). These lines of reasoning have been complicated by differences in the perception of partieular species as mimics of nonmimics by different authors. (Compare Grobman 1978) and Greene and McDiarmid (1981) for an example of these sorts of differences in perception). Similar differenccs in the perception of mimics probably occur among predators of the snakes. Some of the predators appear to have dichromatic color vision, whereas others (including humans) have tirchormatic color perception (Muntz 1974; Jacobs, 1982). Banded patterns are probably differentially visible in different lighting conditions. Even the determination of sympatry of a mode and mimic is not simple. A mimic and model need not occur syntopically, provided a predator includes the ranges of both the model and mimic in its foraging area. If migratory birds are predators and avoidance of coral snakes is innate, models that occur in the tropical regions in which the birds overwinter could protect mimics in North America where the birds spend the summer.

The colubrid snake Lampropeltis triongulum shows a clinal change from a cryptic pattern of dark blotches in the northeastern United States to a coral snake pattern of red, yellow, and black bands in the southeastern United State where it is sympatric with Micrrus fulvirus (Conant, 1943). Several subspecies of L. triangulum in central and western North America have patterns generally like those of coral snakes in the absence of sympatric models. These patterns, too, become less mimetic as one moves away from the region of sympatry with coral snakes (see colour plates in Williams, 1978). L. zonata, which was probably derived from L. triangulum (Zweifel, 1952), also has a coral-snake pattern. In some parts of the range of the species, the red areas of the pattern are largely obscured by black; this may be an example of an originally mimetic pattern changing to a cryptic one (Hecht and Marien, 1956).

Several species of coral-snake mimics show both close similarity to their models and concordant geographic variation. The patterns of milksnakes (Lampropeltis triangulum) in Latin America vary in parallel with the patterns of coral snakes. In Mexico two subspecies of L. triangulum resemble local species of Micrurus in band width and colar (Zweifel, 1960) : In Guerrero L. t. blanchardi occurs with M. browni and both have red rings that are heavily spotted with black. In contrast, in Nayarit, Sinalo, and southern Sonora, where L. t. nelsoni occurs with M. distans distans, the coral snake and milksnake both have broad immaculate red rings. A second coral snake in his region, Micruroides euryxanthus austrialis, has wider red rings in its area of overlap with Micrurus distans than it does in more northern parts of its range. In Oaxaca L. t. oligozona and Micrurus ephippifer have an encroachment of black into red rings. On the Yucatan Peninsula the crossbanded pattern of L. triangulu" breaks down in some localities where

Micrurus diastema exhibits the same effect, and in Honduras the pattern changes to one of black and orange rings like that of Micrurus nigrocinctus (Greene and McDiarmid, 1981). In Costa Rica and northwestern Panama juveniles of L. t. gaigae and Micrurus alleni have a tricolour pattern that darkens with age. The milksnake become bicoloured red and black and eventually the red also is obsecured leaving a black snake. In the coral snakes the process terminates with a bicolored pattern of black and yellow (Savage and Vial, 1974). (Experiments are needed before the hypothesis that the black color is significant in their moregulation (Williams, 1978) can be accorded credence.)

The parallels in colour patterns of coral snakes and some sympatric colubrids extend even to very local variations (Greene and McDiarmid, 1981). For example, a form of Micrurus limbatus from the eastern shore of Lago Catemaco, Veracruz, Mexico, is orange with a black nape band and a few large dorsal spots. A Pliocercus from that locality and one of its offspring had the same pattern, and a second offspring was banded with black and orange like the normal pattern of M. limbatus from the surrounding region. M. bernadi from Necaxa, Puebla, Maxico, is red with dorsal black cross bars or spots; three of four Pliocercus from that locality had the same pattern. Similar local variation in Pliocercus and Erythrolaprus with local variations in Micrurus has been reported from Guatemala, Honduras, Ecuador, and Peru.

Some coral snake mimics combine features of two or more models. Most striking is the situation in southern Mexica and northern Guatemala where Pliocerus elapoides combines elements of the patterns of two local coral snakes. Another example of a mimetic pattern that combines features of two models is Oxyrhopus trigeminus; some individuals from Peru have bicolour patterns on some parts of the body and tricolour patterns on other parts. Both bicolour and tricolour coral snakes occur in sympatry with the Oxyhopus (Hecht and Marien, 1956). The same authors noted examples of polymorphism in the mimetic patterns of Atractus elaps and A. latifrons that parallel variation in Micrurus and might involve some individuals with patterns that combine features of two models. In Argentina, the colubrid snak Lystrophis dorbignyi has a dorsal pattern of dark blotches and a subcaudal pattern of alternating red and black bands. The defensive beeaviour of this species includes a tail display that exposes the red and black makings and resembles the caudal display of the sympatric coral snake Micrurus frontalis, In addition, L. dorbignyi flattens its body and may spread the posterior tips of the quadrate bones, giving the head a triangular shape and enhancing its resemblance to the sympatric viper Bothrops alternatus (A.A. Yanosky, pers. comm.). Such dual mimies are to be expected if the proposed differences between mimicry of venomous snakes and conventional mimicry are correct.

Concordant variation in the patterns of venomous and nonvenomous snakes permits the inference that coral-snake patterns are aposematic and that the colubrid snakes. are mimics of the coral snakes. A mechanism by which such mimicry can be selectively advantageous is provided by recent experimental studies of predatory behaviour of birds and mammals.

Responses of predators to coral-snake patterns

Two important experimental studies have investigated the responses of probable avian and mammalian predators of coral snakes to afficial models painted in banded patterns that resemble coral snakes and in a variety of control patterns. Zoo-reared coatimundis (Nasua narica) and javelinas (Tayassu pecari) in open enclosures and a group of free-ranging coatis were presented with two rubber replicas of snakes (Gehlbach, 1972). One rubber snake was painted to resemble a species of Micrurus (not Mayruroides, which is the coral snake that occurs in the area of origin of the coatis and javelinas). An olive-green rubber snake was used as control. The snakes could be manipulated by threads to simulate serpentine crawling or the tail display of Micrurus.

Juvenile coatis, both zoo-reared and wild, were startled by the replica of the coral snake giving a tail display. The coati closest to the snake vocalized and ran away and one or more of the other juveniles also ran. Neither the individual that had been startled by the snake nor the others approached the model subsequently. When the olive-green rubber snake was moved in simulation of serpentine locomotion, it was attacked eight times and avoided once. In serpentine locomotion the coral-snake replica was attacked only once and avoided five times. Adult coatis, both wild and zoo-reared, paid little attention to either rubber snake, a response reminiscent of an encounter between coatis and a Micrurus nigrocinctus in the wild (Kaufmann, 1962). Adult zoo-reared javelinas were frightened by the tail display of the coral-snake replica and avoided it after one individual had examined it and run away. Data presented in Table 2.1 of Gehlbach (1972) suggest that a tail display induced fright more often than serpentine locomotion, and that the coral-snake pattern induced fright more often than the control pattern (Greene and McDiarmid, 1981).

In two carefully controlled experiments, Smith (1975, 1977) presented wooden dowels painted in a variety of colours and patterns to hand-reared birds, thus ensuring that the birds, unlike the mammals in Gehlbach's experiments, had no contact with anything resembling the experimental stimuli before they were tested. The species of birds Smith used, motmots (Eumomota superciliosa) and great kiskadees (Pitangus sulphratus), are neotropical ground-foraging predators that capture a variety of small reptiles including snakes.

Motmots attacked dowels painted in solid colors (red, yellow, green, blue), in a neutral banded pattern (green with crosswise blue bands), or in a pattern of yellow with lengthwise red stripes. They fled to the opposite corner of their case and uttered alarm notes when they were presented with a dowel painted yellow with red bands. In a subsequent experiment, kiskadees also attacked solid-coloured models (white, green, red, yellow, and black), neutral patterns (white with green bands), and a pattern of yellow with red stripes. They fled, giving high intensity alarm cries, from a dowel painted to resemble a coral snake (black-yellow-red-yellow-black). A yellow dowel with red bands also caused the birds to flee, but to give lower intensity alarm calls. The response of the birds to a dowel painted with red, yellow, and black stripes was particularly reveal.ing: Only four of the six birds attacked it, and they did so only with great hesitation and adopting a defensive posture as they attacked. The birds gave a total of only seven pecks to the red, yellow, and black striped model compared to 59 pecks on the white and green banded model and 44 pecks on the red and yellow striped model.

Smith also used models that were painted for only one-third of their length—the remaining two-thirds was bare wood. She found that the birds attacked the painted ends of dowels that had patterns they did not fear, and the unpainted ends of dowels with patterns they treated as being like corai snakes. Some Latin American snakes, particularly Scaphiodontophis, have a coral-snake pattern that is sometimes limited to the head and anterior part of the body; the rest of the body is solid colored. If predators respond to these snakes as Smith's birds responded to her models, that pattern should reduce the intensity of predation and deflect attacks to the tail of the snake. Scaphiodontophis has an extremely long tail that is readily broken when it is seized (Henderson, 1984; Arnold, this volume).

Smith's carefully designed experiments show that birds are capable of making very precise discriminations among a variety of solid, striped, and banded patterns but that in practice they generalize their response broadly. The types of alarm calls the kiskadees gave when they were presented with the models models indicate that they made a distinction between the red, yellow, and black banded dowel and the red and yellow banded one, but they did not attack either. Similarly, their responses suggest that they perceived that the red, yellow, and black striped model was less like a coral snake than the banded combination of the same colours. Despite their ability to make that distinction, the birds showed only a slight inclination to attack the striped dowel. The implications of these responses for the establishment and function of coral-snake mimicry are critical. The birds generalized sufficiently to provide a selective advantage for even a poor mimic of the coral-snake pattern—the red, yellow, and black striped dowel, for example. Predation on coral-snake mimics by birds that behaved like Smith's experimental animals would not produce very precise mimicry of the coral-snake pattern because the birds failed to attack even a poor mimic, the red and yellow banded dowel. This is exactly the sort of response predicted from a consideration of the consequences of having a deadly model, and some coral-snake mimics fall into the categories of Smith's dowels. The species of Elapomorphus, Apostolepis, and Scaphiodontophis with the coral-snake pattern limited to the anterior portion of the body are one such group (Henderson, 1984). In some populations, Elapomorphus bilineatus has red, yellow, and black stripes and might derive protection from that pattern in a manner analogous to the red, yellow, and black striped dowel in Smith's experiments.

Predators that respond like motmots and kiskadees would produce a situation in which mimics with even a small resemblance to coral snakes would benefit from reduced predation. This sort of response by predators is probably the reason for the high incidence of snakes with coral-snake patterns in the neotropics. Snakes like Anilius, Dipsas, and Liophis are poor mimics; they do not really look very much like coral snakes, but they look as much like coral snakes as some of the painted dowels that frightened Smith's birds. Apparently broad generalization by predators makes even that degree of resemblance advantageous. The very precise resemblances to the patterns of coral snakes and the parallel geographic variation with coral snakes seen among species of Lampropeltis and Pliocercus are a different situation. Apparently in this case some predators are making fine distinctions that have refined an initial imprecise resemblance of the colubrids to coral snakes. This situation may be the only example among snakes of the second stage of the two-step evolution of Batesian mimicry that may be the norm for insects.

The clear demonstration of innate avoidance of the coral-snake pattern by two potential avian predators of coral snakes offers a mechanism for the evolution of aposematic color and mimicry that avoids the difficulties of requiring a predator to experience the lethal consequences of the noxious qualities of a venomous snake (Greene and Pyburn, 1973). Empathic learning appears to be another plausible method of learning to avoid coral snakes. More than a century after Cope first called attention to the situation, the two thorniest issues of the "coral-snake mimicry problem" appear to be resolved.

(b) *Mimicry of Vipers by Egg-Eating Snakes*

> On comparing the two snakes I was much struck by their general resemblance, although, of course, the Dasypsing has a very dissimilar head; but the curlous way it has of blowing itself out, and distending its neck, and darting at intruders heightened the resemblances. I cannot help thinking that these habits must be serviceable to so harmless a snake (Weale, 1871).

Distinctive or conspicuous colours and patterns are not essential components of the aposematic displays of snakes. Most vipers, for example, are cryptically colared. Terrestrial vipers have cryptic patterns that match the substrate colour and incorporate obliterative shading or disruptive colors. African vipers of the genera Causus, Echis, Bitis, and Cerastes fit this general description (Gans, 1961). They may be unicoloured or have dorsal saddles and lateral bars. The predominant hue of a snake matches the color of the substrate on which it is found. Many of these African vipers have a prominent T-shaped mark on the head or neck. Crypsis appears to be the primary function of these patterns; aposematism is secondary and associated with distinctive behaviour after a snake is discovered by a predator.

Interspecific differences in the components of these displays have been described and intraspecific variation probably occurs as well, but the general aspects can be characterized (Gans. 1961).

Night adders. Causus rhombeatus and C. defilippii inflate and flatten their bodies, coil and strike, sometimes leaping from the ground.

Horned viper. Bitis caudalis rests in a C-shaped coil and strikes repeatedly to the accompaniment of loud hissing.

Saw-scaled viper. Echis carinatus also forms a C-shaped coil, but in this species a hissing noise is produced by rubbing together speciaiized scales that are distributed in a band occupying the lower several rows of lateral scales.

Two species of the nearly toothless egg-eating snakes. Dasypeltis scabra and D. palmarum, appear to mimic both the patterns and behaviours of these vipers. Gans (1961) reviewed this subject and summarized the parallels in geographic variation of the vipers and their mimics. The ranges of the species of Dasypeltis are not mutually coextensive with any of the vipers and consequently the most widespread egg-eating snake, D. scabra, mimics different vipers in different parts of Africa. Through most of sub-Saharan Africa, D. scabra closely resembles local populations of the night adders Causus rhombeatus and C. defilippii. These adders have dark brown saddles and lateral bars outlined with cream on a light brown background and a V-shaped mark on the neck. In western Africa the lateral body markings of C. rhombeatus are fused with the dorsal blotches, whereas in central Africa the lateral blotches alternate with the dorsal marks. This geographic pattern variation is paralleled by D. scabra. In regions where a unicolor phase of C. rhombeatus occurs, D scabra is also unicolored. Causus defilippii has a pattern very like that of C. rhombeatus, Causus is a stouter snake than Dasypelitis and shows an ontogenetic changes in pattern. As the adders grow their patters become less distinct and large adults may be unicoloured. The patterns of egg-eating snakes resemble those of adders with body girth, but not length, equal to those of the egg-eating snakes.

In the horn of Africa, Angola, Egypt, and Arabia, Dasypeltis scabra resembles Echis carinatus. In eastern Africa both egg-eating snakes and saw-scaled vipers have patterns of elongate saddles and dark heads without markings. This is the only part of its range in which D. scabra has a dark, unpatterned head Night adders and egg-eating snakes from western Africa have square or hexagonal blotches and a V shaped marking on the head. This pattern is shared with Causus rhombeatus as well.

In parts of southwestern Africa Bitis caudalis appears to be a model for Dasypeltis scabra. The patterns of the horned vipers and egg-eating snakes change in concert from square saddles in the west to hexagonal blotches in the east. In addition Sternfeld (1908, 1910) has suggested that a north-to-south reduction in the length of the length of the tail of egg-eating snakes is mimicry of the short tail of Bitis caudalis.

In Angola Bitis heraldica has a distinctive dorsal pattern formed by a light middorsal band that encloses diamond-shaped markings that are bordered by dark lines. This middorsal pattern contrasts sharply with the dark colouring adjacent to it on the back and sides of the body. Dasypeltis palmarum appears to mimic the distinctive middorsal pattern of the body of that viper. Mimicry of only a portion of the pattern, one that may be perceived as separate from the rest of the pattern of the snake, may make the difference between the slim body of Dasypeltis and the thick body of the viper less apparent to potential predators.

The mimicry of vipers by egg-eating snakes extends to their aposematic displays. Like Echis, Dasypeltis throws its body into a series of concentric C-shaped coils that are moved in opposite directions so that each coil rubs against adjacent ones. Modified laternal scales produce a hissing sound (Gans and Richmond, 1957; Gans, 1974). Examination of the surface configuration with electron microscopy shows that the sound-producing structures on the scales of the vipers and egg-eating snakes are quite different. In Echis carinatus, E. coloratus, and Cerastes cerastes the scales have central keels that terminate in one or two longitudinal series of rounded knobs, whereas in Dasypeltisthe keels of the modified lateral scales are covered by longitudinal ridges (Gans and Baic, 1974).

As an egg-eating snake rubs its body coils together, it appears as a swirling object, and its body proportions are difficult to judge (Weale, 1871). The snake inflates its trached air sac, lung, and posterior air sac, increasing its girth and causing the dark markings to become more conspicuous and its body form more viper-like (Gans and Richmond, 1957). Inflation is facilitated by the structure of the tracheal air sac, a ventral expansion of the trachea that originates immediately behind the hyoid. The cartilaginous tracheal bars are not joined ventrally; hence they straighten as the tracheal air sac is inflated and press against the flattened esophagus and subvertebral muscles. The inflated tracheal air sac, lung, and posterior air sac may act as resonating chambers and amplify the hissing sound produced by the modified scales.

During this display some egg-eating snakes spread the posterior tips of the quadrates, giving the head a triangular shape and a more viper-like appearance. The head is the only part of the body of the snake that is motionaless and thus readily visible to a predator through most of the defensive display. During the culminating phase of the display, the egg-eating snake gapes its jaws widely and strikes repeatedly but without actually biting (Gans and Richmond, 1957; Gans, 1961).

The combination of a cryptic color and pattern shared by the egg-eating snakes and vipers with an aposematic display mimicked by the egg-eating snakes is a noteworthy example of the multiple uses of colors and patterns by animals. The clearly cryptic function of a blotched pattern in snakes has led some authors to deny the hypothesis of mimicry by egg-eating snakes (Werner, 1909; Mertens, 1955). Convergence on a similar cryptic pattern by vipers and egg-eating snakes would be a plausible hypothesis were it not for the mimicry of the vipers' display. Five of six species of egg-eating snakes have populations with blotched patterns, but Gans (1961) considered that only in D. scabra and D. palmarum was the resemblance close enough to the vipers to be considered mimetic. The remaining species of Dasypeltis do have modified lateral scales and engage in a defensive display similar to that described for D. scabra. If predators generalize the features of vipers as broadly as they appear to do those of coral snakes, all species of Dasypeltis might benefit to some extent from their resemblance to vipers.

The similarity of the patterns of the several species of vipers that are models for Dasypeltis suggests that this group may be an assemblage of Mullerian mimics (Gans, 1961). Evaluation of that hypothesis is complicated by the ubiquity of snake patterns that consist of irregular dorsal markings. Patterns of this sort often characterize relatively slow-moving species of snakes that are exposed to predators in daylight and are capable of defending themselves effectively (Brattstrom, 1955; Jackson *et. al.*, 1976). Blotched patterns are found in many species of boids as well as among colubrids, and blotched patterns may be an ancesstral characteristic of caenophidian snakes. Nonetheless, predators that avoid vipers, whether innately or as a result of experience, might mistake another blotched snake for a viper. Thus, a shared ancestral pattern of blotches could function like mimicry in an antipredator display. Parallel geographic variation of the patterns of presumed models and mimics (as seen among vipers and egg-eating snakes) suggests that predators may have exerted selective pressure on the resemblance. If that is the case, the resemblance would be minetic in an evolutionary as well as in an ecological sense. However, a parallel response by model and mimic to some other feature of the physical or biological environment cannot be dismissed. (See Sweet, 1985, for an example of convergent selection for camouflage among snakes.) Additional similarities of colubrids and vipers, especially stout bodies, large heads, and the ability to spread the ventral tips of the quadrate bones, may represent convergent adaptations for swallowing large prey (Marx *et. al.*, 1982, Pough and Groves, 1983), or they might be mimetic (Werner, 1985).

The viperid models of egg-eating snakes pose at least as much risk to predators as do coral snakes. Empathic learning has been suggested as the primary mechanism by which predators learn to avoid vipers, although Gans (1961) noted that innate responses were also possible. Baboons (Papio cephalus), mandrills (Mandrillus sphinx), and warthogs (Phocochoerus aethiopicus) are among the potential predators of African snakes. These mammals forage in groups and might learns avoidance from seeing another individal bitten by a venomous snake. Foraging baboons overturn ground litter and uproot plants, and they do eat snakes; older individuals may be more cautions than young ones about attacking snakes. In captivity mandrills fed slices of banana dyed different colors learned to avoid the colour made unpolatable by quinine. Empathic learning was more effective than direct experience in this situation. Animals that watched another individual sample the palatable and unpalatable colors made fewer mistakes when tested subsequently than the animals that had done the sampling. Discrimination achieved by observation was retained for at least 2 weeks without intervening reinforcement.

These experiments demonstrate that empathic learning by predators is a feasible basis for selection of mimetic color patterns and behaviours among egg-eating snakes. The possibility that some avian or mammalian predators of snakes have evolved innate avoidance of vipers akin to that shown for coral snakes remains to be investigated. Experiments utilizing naive predators and a variety of models could determine whether innate avoidance of vipers exists, and if so, what features of the appearance or behavior of a viper elicit that avoidance. With this information it would be possible to evaluate the numerous suggestions of mimicy of vipers by other snakes.

(c) *Mimicry of Vipers by Boiga*

The distinctive pattern and defensive display of the saw-scaled viper Echis carinatus seem most reasonably interpreted as a compromise between the cryptic and aposematic functions of a pattern of dorsal blotches. The light lines that cross the dorsum are outlined by black scales. This contrast renders the total pattern of a saw-scaled viper more conspicuous than that of vipers with unicoloured dark blotches on a lighter background. Inflation of the body in the defense display emphasizes the contrast of the crossbands.

In southeastern Iran, the rear-fanged Boiga trigonata melancephala, a larger snake than Echis, has a dorsal pattern and a display that resemble those of the saw-scaled viper. In the region of sympatry with Echis the anterior and midbody of Boiga are crossed by light lines with black margins; the posterior part of the body lacks these brightly contrasting marks. As a result of the distribution of the two sorts of pattern on Boiga, the portion of the body that bears an Echis-like pattern has the same length-diameter ratio as Echis despite the difference in the size and relative girth of the two species. The defensive display of Boiga in Iran consists of using the contrastingly marked portions of the body to form a C-shaped loop that resembles the loop formed by Echis. Unlike the viper, Boiga from Iran are said to hide the head beneath a coil of the body, but in other parts of the range of the genus Boiga may strike during the display.

These similarities of pattern and behavior suggest that Boiga is a mimic of Echis. Both species have extensive geographic range and the situation in Iran may be a local manifestation of a more general relationship between Boiga and Echis. Furthermore, the eight species of Boiga that have overlapping ranges with Echis and have patterns like that of B. t melanocephala may also mimic Echis or form a Mullerian mimicry complex that includes Echis. In Africa, B. pulverulenta, Rhamphiophis multimaculatus, Telescopus variegatus, and Pseudaspis cana may mimic vipers.

(d) *Additional Examples of a Mimicry of Vipers*

Many vipers and colubrids have a pattern of dark blotches on a light background. This similarity has startled observers sufficiently to inspire repeated suggestions of mimicry of vipers by colubrids, but other interpretations are also plausible. A blotched pattern is widespread among caenophidian snakes, and is characteristic of slow-moving terrestrial species (Jackson *et al.*, 1976), so that convergence on a common cryptic pattern cannot be dismissed as an explanation for the similarities of some vipers and colubrids. Furthermore, blotched patterns are common among henophidians, and may be an ancestral character for caenophidians.

Most suggestions of mimicry of vipers by colubrid snakes are anecdotal, and few have considered these alternative hypothesis. For example, the assertion that Coluber ravergieri mimics Vipera palaestinae in Israel (Arbel, 1979) is apparently based on the coincidence of a pattern of dorsal blotches in both species; the photographs that accompany the description do not show any noticeable similarity between them, and geographic variation is not considered. The Great Basin gopher snake (Pituophis melanoleucus deserticola) has been proposed as a mimic of the northern Pacific rattlesnake (Crotalus viridis oreganus) because both species have dark dorsal blotches and coil the body and vibrate the tail when they are disturbed (Kardong, 1980). However, a quantitative comparison of geographic variation in color and pattern of Pituophis melanoleucus and Crotalus viridis in coastal and inland areas of central southern California suggested that camouflage is the primary factor involved, and that mimicry is of no more than secondary importance (Sweet, 1985). Coastal populations of rattlesnakes and gopher snakes occur in different plant associations, do not resemble each other, and are cryptic in different habitats: Coastal Crotalus viridis helleri are matched to their background in chaparral and woodland areas, whereas Pituophis melanoleucus annectans from the coast are cryptic in brushy grasslands. Inland populations of P. m. catenifer and C. v. oreganus from the Carrizo Plain where habitat diversity is less than on the coast, show more overlap of habitat utilization and are more similar to each other in color than are coastal populations. The inland populations of both species of snakes are cryptic in open grasslands. The populations of rattlesnkes and gopher snakes from the Carrizo Plain are also more aggressive than are the coastal snakes. Nearly all the individuals of both species from the inland habitat hissed or rattled the tail when they were disturbed, whereas fewer than 20% of the gopher snakes and only 50% of the rattlesnakes from the coast gave these responses. Crypsis and mimicry are not incompatible mechanisms of predator avoidance for these snakes, but mimicry cannot be inferred from the information currently at hand; convergent evolution of camouflage is a more likely explanation of the similarity of the two species in open grassland habitats (Sweet, 1985).

Colubrin snakes can produce a triangular outline of the head when it is seen in dorsal view by spreacing the ventral tips of the quadrate bones laterally. This behaviour may increase the resemblance of a colubrid to a viper, and a snake may enhance the effect by tilting its head so that the triangular shape is more readily apparent to a predator (Rand and Ortleb, 1969; Greene, 1975; Werner, 1986), and by keeping the head motionless while the body moves. Head triangulation may be combined with coiling, hissing, and striking (Gans and Richmond, 1957; Gans, 1961; Sapwell, 1968; Rand and Ortleb, 1969; Kardong 1980; Werner and Frankenberg, 1982; Werner, 1983, 1985; Sweet, 1985). The neotropical colubrid Xenodon has been reported to resemble a terrestrial species of the pit viper Bothrops. Waglerophis merremii, Tomodon ocellatus, and Lystrophis dorbignyi also may mimic species of Bothrops (Astort, 1983).

Vocalizations by nestling birds sound to some observes like the hiss of a snake and have been described as auditory mimicry that deters predators of the birds (Leroy, 1979), but this hypothesis has not been tested experimentally. In western North America burrowing owls and rattlesnakes both live in tunnels dug by rodents. The defensive hiss of the owl is sonagraphically similar to the rattling of the snake, but similar hisses are produced by two other small North American owls that do not inhabit burrows, and the young of several owl species use hisses to solicit food from adults. Thus, the vocalizations might be an ancestral character of burrowing owls (Rowe, 1984). Wild-caught ground squirrels showed interpopulation differences in their responses to the rattling of snake and the hiss of owl. Squirrels from a population sympatric with rattlesnakes treated both sounds with greater caution than squirrels from a population that has probably not been exposed to rattlesnakes for same 3000 years, raising the possibility that the hiss of the owl may have a mimetic function.

Several lizard species have been reported to mimic vipers. In Pakistan the gecko Teratolepis fasciata has a dorsal pattern of light bands on a darker background. The bands on the head are bent into shallow Vs and the bands continue on the broad tail. The defensive display of the gecko consists of curling and flattening the body and and raising the tail so its dorsal surface is presented to a predator. In this position the gecko is said to resemble the sympatric saw-scaled viper, Echis carinatus (Anderson, 1964). Further disturbance causes the lizard to hiss and strike with its head at an assailant. Legless lizards (Anguis fragilis) are polymorphic; juveniles and some adult females are golden brown with a dark dorsal stripe, whereas some females and all adult males are a uniform grey dorsally (M. Smith, 1969). Retention of the dark stripe by some females has been interpreted as mimicry of the adder, Vipera berus (R.H. Smith, 1974).

Individually most of these accounts leave much to be desired, but in sum they suggest that mimicry of vipers may be a worldwide phenomenon. If other predators share human perceptions of snake patterns, mimics may derive benefit from quite imprecise resemblances to their models. J. R. G. Tunner (pers. comm.) described a color variant of the smooth snake, Coronella austriaca, in England that was "grey with a black V-mark on the head and a series of zig-zag black dots down the back, the whole having a sufficient resemblance to the viper, although the V was the wrong way round and the dots not joined up into a zig-zag, to have us looking at the first one we caught for ten minutes before we dared touch it with bare hands." Hesitation of that sort by a predator may provide an opportunity for a snake to escape. Further investigations, especially controlled experiments with potential predators, are neeessary to define the role of mimicry of vipers in the defensive repertoires of reptiles.

(e) *Mimicry of Other Venomous Snakes*

Coulubrid, elapid, and hydrophiid snakes, although they may be less formidable models than vipers, have been implicated in a number of suggested examples of mimicry. As is the case with mimicry of vipers, the evidence is anecdotal and alternative hypotheses have not been tested. A resemblance has been noted between the boa Candoia aspera and the elapid Acanthophis antracticus (Bustard, 1969). Candoia from Lae, New Guinea, have the dorsal and lateral blotches joined to form bands and are said to be particularly like Acanthophis (McDowell, 1979:59). The terrestrial neotropical turtle Rhinoclemmys pulcherrima has a pattern of alternating red, black, and yellow bands on the ventral surfaces of the marginal scutes of the carapace. (A colour photograph is presented in Janzen, 1980.) These bands bear a strong resemblance to the pattern of a coral snake when the turtle is viewed from beneath, as it would be seen by a predator that overturned it. In addition, the dorsal aspect of the gular scutes of the plastron bears the same colours, and this pattern is revealed in dorsal view when the turtle retracts its head into its shell.

In Malaya the inoffensive pipe snake, Cylindrophis rufus (Aniliidae), is said to mimic the colours and defensive display of the elapid snake Maticora intestinalis (Tweedie, 1953). In this region, the elapid has a dorsal pattern consisting of three naraow yellow stripes bordered with black and separated by the brown ground colour. The ventral surface is boldly patterned with black bands on a white background, and the underside of the tail is bright red with black bars. The defensive behaviour of the snake consists of a display that reveals this pattern:

> When molested this snake markes no attempt to escape, but will often raise its tail, displaying the bright red under suface, and will writhe and tumble about, turning the conspicuously barred belly uppermost. This behaviour is doubtless aposematic, and both it and the colouration are mimicked by the harmless pipe snake, Cylindrophis rufus (Tweedie, 1953:95-96).

A colour photograph of the ventral pattern of a Maticora appears in Fogden and Fogden (1974:114). Tweedie (1953:13) also proposed that the elapid snake Calliophis gracilis, which has a ventral pattern like that of M. intestinalis, is another mimic. Two of the three species of kraits in Malaya (Bungarus fasciatus and B. candidus) are banded with black and yellow or white. Tweedie (1953) suggested that several colubrid snakes with banded patterns are mimics of the banded kraits, particularly the mangrove snake (Boiga dendrophila) and the banded juveniles of some species including the banded wolf snake, Lycodon subcinctus. The third species of krait in Malaya, B. flaviceps, is a solid blue-black with a red head and tail. The same pattern is seen in Maticora bivirgata.

A number of species of the venomous sea snakes (Hydrohiidae) have alternating black and white bands, and sympatric species of share this pattern. These coincidences of pattern have been interpreted as Mullerian mimicry among sea snakes and Batesian mimicry of the sea snakes by the eels (Sternfeld, 1913; Pernetta, 1977), but no feeding trials have been conducted with these banded species. Another sea snake, the pelagic yellow-bellied sea snake Pelamis platurus, has the widest geographic distribution of any species of snake (Minton, 1975). In has beeen recorded from southrn Siberia to New Zealand, Tasmania, and the Cape of Good Hope, and it is the only species of sea snake in the western Indian Ocean and on the west coast of the Americas. Pelamis has a bright yellow venter and a black dorsum. The pattern varies from solid blocks of the two colours to spots of yellow and black. Pelamis are frequently found where converging surface currents create a drift line of floating debris, and aggregations of snakes may number in the thousands. Despite their local abundance, no Pelamis were found in the stomachs of 457 fishes, 214 porpoises, or 1000 sharks examined (Kropach, 1975).

No species of sea snake occurs on the Atlantic side of the Americas, and fishes from the Caribbean have presumably had no experience with sea snakes, at least since the Pliocene when the Isthmus of Panama took its present form. Closely related species of fishes from the Atlantic and Pacific coasts of Panama responded differently to Pelamis in experimental feeding trials (Rubinoff and Kropach, (1970). Atlantic fishes attacked Pelamis in 38 of 383 trials, and in three of those attacks the fish was fatally bitten by the snake. Only one attack on Pelamis was recorded in 316 trials with Pacific fishes, and that fish immediately regurgitated the snake.

Avoidance of Pelamis by predatory fishes from the Pacific coast of Panama appears to provide an opportunity for mimicry of the sea snakes by eels, and ophichtid and muraenid eels from Panama (Murichthys tigris and Priodonophis equatorialis) have some resemblance to Pelamis in pattern and colour. However, if this resemblance is mimitic, fishes do not appear to have been the predators responsible for its evolution because Pacific fishes readily attacked the eels. In contrast, potential avian predators did respond to Pelamis and eels in a manner that would favor mimicry. Naive, hand-reared herons (Butorides striatus) and egrets (Casmerodius albus and Egretta thula) fled from live and dead Pelamis with signs of distress (Caldwell and Rubinoff, 1983). The birds largely ignored an arboreal snake (Imantodes cenchoa), immediately ate a speckled snake eel (Murophis punctatus), and ignored a synbranchid eel (Synbranchus marmoratus) that was too large for them to eat. They alternately approached and retreated from a spotted snake eel (Myrichthys tigris), which has a pattern of creamy yellow with dark brown or black spots reminiscent of the pattern of some Pelamis. Both species of egrets occur on the Pacific coast of Panama where they may frequently encounter Pelamis as they forage on tidal mudflats. The herons tested were from the Atlantic side of the isthmus, but are sufficiently mobile to move between regions with and without sea snakes. However, herons normally feed no land or in shallow water and would not be expected to encounter sea snakes frequently. The high toxicity of sea-snake venom may have favored innate avoidance by even an occasional predator (Caldwell and Rubinoff, 1983).

Many snakes spread the neck as part of their defensive display. The dorsoventral neck flattening of the neotropical colubrid Ninia sebae may enhance its resemblance to sympatric coral snakes, or may be part of a more generalized threat display (Greene, 1975; Henderson, 1977). Some authors have suggested that neck displays by colubrid snakes mimic the neck spreading of cobras (Flower, 1899; Wall, 1913, Gharpurey, 1962; Deorag, 1965; Werner and Frankenberg, 1982). The descriptions on which these assertions are based do not always distinguish between a cobra-like dorsoventral flattening of the neck and a lateral display. Lateral flattening of the neck is a widespread defensive behaviour among colubrids (Mertens, 1946, this volume). A cobra-like display is not limited to species of colubrids that occur in sympatry with cobras as this extract illustrates: "When [the neotropical colubrid Liophis reginae is] angered, the head is raised and the neck flattened and widely expanded, becoming a cobra in miniaturi" (Beebe, 1946). Coincidences of this sort are as likely to represent shared ancestral characters as they are to be the result of selection by predators for mimicry of venomous snakes.

Even colubrids may serve as models. The innocuous neotropical colubrid Rhadinaea fulvivittis aemula takes its subspecific epithet from the suggestion that it mimics the more aggressive colubrid Coniophanes piceivittis (Myers, 1974).

Alternative explanations have been suggested for other allegedly mimietic resemblances of reptiles to venomous snakes. For example, the pygopodid lizard Delma fraseri has a dark head and a light brown body. Hall (1905) proposed that this pattern mimicked the juvenile color and pattern of the unbanded form of the elapid Pseudonaja textilis. The pygopodid lizards Pygopus nigriceps and P. lepidopodus also have dark heads. When threatened they raise the head off the ground and flatten the neck laterally. This behavior has been likened to the defensive display of small elapid snakes like Unechis flagellum and U. gouldii (Bustard, 1968).

Juvenile oak skinks (Tiliqua casuarinae) have black bands on the head neck. Like the pygopodids, oak skinks raise their heads when molested, a behavior described as mimicry of juvenile brown snakes (Pseudonaja textilis) (Rankin, 1973). However, another interpretation is plausible: Nearly 60% of the species of pygopodids have heads that are coloured differently from the body, as do 30% of Australian snakes. These dichromatisms of the head and body are most common among elongate reptiles, and may conceal the location of the head and induce predators to size the body instead (Jackson and Pounds, 1980). Mimicry and crypsis are not mutually exclusive explanations of the high incidence of dichromatism among Australian snakes and pygopodids. The Australian snake fauna is unique in being composed largely of elapids, many of which are probably dangerously toxic to predators. A much lower incidence of venomous snakes in other parts of the world has apparently produced widespread abstract mimicry. What are the consequences for mimicry of having a snake fauna that is composed nearly entirely of venomous species? Experiments with Australian snakes and predators of snakes might reveal hitherto unrecognized aspects of mimicry.

MIMICRY OF INVERTEBRATES BY REPTILES

Three examples of apparent mimicry of invertebrates by reptiles have been described. The geckos Chondrodactylus angulifer and Coleonyx variegatus raise their tails over their backs and wave them from side to side when they detect the presence of a predator; this behaviour has been regarded as mimicry of scorpions Simons and Brain, 1958; Parker and Pinka, 1974.

The contrasting colors of mropetids may function in one or more of several ways. As mimicry of the presumbly aposematic colors of centipedes, which are capable of inficting at least a painful bite upon a predator, the colours may dener attack upon the harmless snakes. If a predator does attack, markings on the tail may act as flash colours, causing the attacker to hestitate, or they may and in deflecting attack from the head. These potential functions are mutually compatible (Gans, 1976).

Juveniles of the lacertid lizard Heicipolus lugubris in the Kalahari Desert of southern Africa appear to mimic abundant, conspicuous oogpister beetles (Anthia spp.) that are protected by a noxious fluid they spray at attackers (Huey and Pianka, 1977. The body colour, posture, and locomotion of the juvenile lizard differ from those of the adults and instead resembles the beetles. Adut H. lugubris from the Kalahari are red-tan and blend with the sand on which they occur, whereas the bodies of the juveniles, like those of the bettles ar black with white markings. The juvenile lizards attain the adult colour and pattern at snout-vent lengths of 40–50 mm these correspond to the maximum body size of the oogpisters. Adut H. ligubris move with the lateral undulations of the body that are characteristic of elongate lizards. In contrast, the juveniles move stiffly and jerkily, holding the body arched and the tail pressed to the sand. They resemble the beetles closely enough to deceive human observers.

The frequency of broken tail among samples of juveniles of three species of acertids in the Kalahari was used as an index of the intensity of predation on the lizards. Juvenile Heliobolus lugubris had the lowest frequency if broken tails (10.5%) and differed significantly from juveniles of Puttoplanes lineocellatus (22.3%) and P. namaquensis (34.8%). Juveniles or the latter two species resemble their adults in color, pattern, and behaviour. The low frequency of broken tails among the beetle-mimicking juvenile H. lugubris was interpreted as evidence that the mimiery ones deter some potential predators of the lizards (Huey and Pianka 1977). However, recent reconsiderations of tail-break frequencies amona lizards cast doubt on their usefulness as indicators of the intensity capredation Schoener, 1979; Schoener and Schoener, 1980; Jaksic and Greene, 1984; Arnold, this volume). Arnold this volume).

INTRASPECIFIC MIMICRY

Deceptive actions or appearances may be directed toward

conspecific individuals and some examples of intraspecific deception may represent conjunct mimicry. For example, the apparent resemblance of certain marcngs on the faces and ears of mammals to horns or teeth has generated an unverified hypothesis that these patterns (called "weapon automimicry") food conspecifics into thinking their possessor a formidable opponent (Guthrie and Petocz, 1970).

A better substantiated example of the phenomenon was reported among lizards by Rand (1967). Male Anolis lineatopus behaved as if they were predicting the outcome of agonistic encounters partly on the basis of the size of their opponent. The larger individual was the victor in 85% of the disputes observed, and encounters between individuals of different sizes were less likely to lead to actual physical contact than were encounters between individuals of nearly the same size. A male Anolis has many behavioural and structural mechanisms for increasing its apparent size during agonistic encounters (see illustrations in Crews, 1978): A lizard orients itself laterally to its opponent, arches its back and flattens its sides, thereby increasing the apparent depth of its body. The gular region of the throat is engorged, making the head appear larger, and a dark spot appears benind the eye. Tissue along the neck and back can be raised into a crest to enhance that effect, and some species of Anolis and oiner lizard have a permanently erect on the back or tail. By the reasoning applied to "weapon automimicry" among mammals, these strucures of lizards would be considered "automimicry." However, the employment of threatening displays and devices to make an individual appear larger and fiercer is nearly universal among vertebrates, and occurs during both interspecific and intraspecific interactions (Edmunds, 1974). Extending the term "mimicry" to include these behaviours robs the word of much of its usefulness without illuminating the phenomena so designated. Colours, patterns, structures, and behaviors of this nature can be included in a category of "intraspecific deception" (Otte, 1975).

A different manifestation of intraspecific deception fits more comfortably into the framework of concrete homotypy: mimicry of females of a species by males of the same species. This form of sexual mimicry is one of many alternative mating strategies of males, a topic that is currently of much interest to students of social behaviour (Austad and Howard, 1984). Mimicry of females may be a facultative response to transient conditions, or (apparently rarely among vertebrates) it may be genetically fixed in some individuals of a population (Dominey, 1984). Reptiles appear to provide examples of both situations.

Some male green iguanas (Iguana igunana) employ female mimicry as a mating strategy when they are small (Dugan, 1982). During the breeding season large adult male iguanas maintain territories from which they display to females. The adult males exclude medium-size males from these territories, but small males that have not developed secondary sex characters are apparently not recognized from a distance as males and are not excluded. These small males adopt female behaviours (or at least they refrain from engaging in conspicusly male behaviors), thereby escaping the notice of territorial males. Fifteen forced copulation attemps by these small males were observed. The small ran to a female, leaped on her back, and tried to secure a grip on the female with his mouth. In all 15 attempts the female tried to escape from the male. On five occasions the territorial male chased the small male away, and in three of the remain ing ten attempts small males apparently copulated successfully Of course, successful copulation does not necessarily ensure fertilization. Nonetheless, the long period before full body size is attained by iguanas (approximately 7 years in the Panama population), the low turnover of territorial males (a minimum tenure of 2½ years), and the apparent lack of prospects for copulation by medium-size males may combine to make even a small chance of successful breeding by small males an important factor in their lifetime reproductive success. Female mimicry may be a normal alternative mating strategy for young males of long-lived species of reptiles in which dominant adult males monopolize access to females. The distinctive characteristic of the social systems of some of the large herbivourous species of iguanine lizards (Stamps, 1983) appear particularly conducive to female mimicry. However, similar behaviors have been reported for small, short-lived species of lizards. Young male Anois informs and A garmani, which are the same sive as adult remains of the species, may occupy sites within the femotory of an adult male Band 1967 Trivers, 1976). These males occasionally succeed in matting with resident females.

POSSIBLE MIMICRY OF UNPROFITABLE PREY BY LIZARDS

Aposematism is only one possible function of conspicuous colors or behaviors of vertebrates. Tail displays during encounters with predators have been interpreted as warning signals directed to conspecifics, or as signal informing the predator that it has been observed by its potential prey. Harvey and Greenwood 1978). Some lizards that live in open habiuts and escape from predators by running have conspicuous dark bars in the ventral side of the rail. These makings are revealed by a display in which the lizard arches its tail over its body and undulates the tail slowly from side to side (Greene, this volume).

Two species of iguanid lizares that occur in xeric habitats in North America, Cophosaurus taxanus and Holbrookia propinqua, allow a distinction between the hypotheses that these signals are directed to conspecifics or to predators (Dial, 1986). C. texanus has a pattern of dark bars on the ventral surface of the tail, whereas H. propinqua lacks markings on the tail. These two species differ in their modes of escape from predators : C. texanus runs rapidly from predators, relying on speed to evade capture, whereas H. propinqua runs only a short distance and then remains motionless, trusting to crypsis to avoid detection. The species differ also in population structure : C. texanus occurs at low population densities, whereas H. propinqua occurs at high densities. Thus, the use of tail displays to war nconspecifics of the presence of a predator would be expected for H. propinqua, whereas a tail display directed toward a predator is morelikely to be seen in C. texanus.

C. texanus employed tail displays in 19 of 35 encounters with a human observer, and in 6 of those encounters the lizards displayed a second time after fleeing whereas none of 31 Holbrookia tested used a tail display. No conspecifics were in sight during tail display by C. texanus, whereas a conspecific was nearby in every trial with H. propinqua. Thus, the tail display of C. texanus is probably directed predators, and no warning of conspecifics can be demonstrated for H. propinqua.

A tail display might inform a predator that its presence had been detected and that the prey was prepared to flee, thereby reducing the chance that the predator would attack. A similar function has been proposed for the alarm calls of birds and mammals (Maynard Smith, 1965; Charnov and Krebs, 1975). An additional hypothesis suggests that predators might learn to associate a tail display with lizards that are difficult to capture and are probably not worth chasing ("unprofitable prey"); this function has been proposed for the bright colors of some birds (Baker and Parker, 1979; Baker and Hounsome, 1983).

If the tail displays of lizards like C. texanus that flee swiftly do identify them to predators as unprofitable prey, a less agile species of lizard might be able to dupe a predator by mimicking the tail display of a fleet-footed species. This hypothesis could be tested in parts of the southwestern United States and Mexico where several species of sand lizards occur sympatrically: Callisaurus draconoides and Cophosaurus texanus are lizards that can run fast and have conspicuous bars on the ventral surface of the tails, the four species of Uma in the United States and Mexico have dark bars on their tails but do not run as rapidly as Callisaurus or Cophosaurus, and Holbrookia maculata and H. propinqua lack dark markings and run only a short distance when a predator approaches. Experiments with natural predators like roadrunners (Geococeyx californianus) that pursue lizards on the ground during the day could test the predictions that predators are more likely to chase and capture slow species than fast ones, and that they are less likely to chase and capture lizards that exhibit a tail display than those that do not display. If those predictions are supported by experimental evidence, the ability of native predators to learn to ignore lizards that exhibit a tail display could be tested, as could the ability of a Uma to deter a predator by giving a tail display like that of Callisaurus or Cophosaurus.

THE USE OF LURES TO ATTRACT PREY

Portions of the bodies of certain reptiles bear a presumed resemblance to a larval insect and may be moved in a manner that has been interpreted as an attempt to attract prey within reach of the reptile. This behaviour has been called "feeding mimicry" (Schuett *et al.*, 1984). In all instances of which I am aware, the homotypy is abstract. That is, the model is a general category of objects (larval insects), but not a particular species of insect.

The tongue of the alligator snapping turtle, Macroclemys, temminckii, bears a bifurcate wormlike appendage on its dorsal surface that is used to attract fishes. The lure can be made to wiggle by contractions of the muscular base on which it rests, and the moving lure attracts fish to the turtle (Allen and Neill, 1950; Carr, 1952:58—59; Drummond and Gordon, 1979). Twelve of nineteen prey captures occurred when a fish swam into the mouth of a turtle and bit the lure, and the remaining seven captures were of fish that swam into the mouth of the turtle without biting the lure (Drummond and Gordon; 1979). In contrast, three attacks on fish that were swimming near the head of the turtle were unsuccessful.

Use of the tail to attract prey has been inferred for a number of species of snakes, but actual success in luring and capturing prey has not often been documented. Putative caudal lures are particularly common among vipers but have been suggested also for boids and elapids (Neill, 1960; Greene and Campbell, 1972; Heatwole and Davison, 1976; Murphy *et al.*, 1978; Carpenter *et al.*, 1978).

In many instances, the tip of the tail is bright yellow and contrasts with the cryptic color of the remainder of the body.

The bright colour and the use of the tail as a lure are generally confined to juvenile snakes, but there are exceptions to that generalization. An adult Bothrops bilineatus captured an Anolis lizard that was attracted within striking distance by the twitching of the snake's tail (Greene and Campbell, 1972), and an adult Cerastes vipera captured a lizard, Acanthodactylus (Heatwole and Davison, 1976). Captive adult antarcticus waved their tails, apparently in response to the presence of mice on the wire screen tops of their cages (Carpenter et al., 1978). The function of the tail as a lure in this instance is inferential; the mice were not in the cages with the snakes and there is no indication that these or other prey were attracted to the tail movements of the snakes. Alternative explanations of tail movements by snakes include various defensive behaviours and an analogy to the tail twitching of stalking felines (Hecht *et al.*, 1955; Greene and Campbell, 1972; Greene, 1973, this volume).

Photographs of juvenile Agkistrodon bilineatus waving their tails appear in Neill (1960). Lizards and toads in the cage with the snakes appear to be orienting toward the tails in some of the photographs. A white mouse and a frog were captured by juvenile Vipera russellii when they moved toward the snakes' bright yellow tail tips (Henderson, 1970). A young Calloselasma rhodostoma successfully lured and captured a frog (Schuett, 1984). Frogs (Rana clamitans and R. sylvatica) approached and snapped at the undulating tails of juvenile Sistrurus catenatus (Schuett *et al.*, 1984). Only after a frog had seized its tail did the snake strike, even though other frogs were within striking range before that. These experiments led the authors to suggest that the raitle of rattle-snakes had its origin as a structure that enhanced the effectiveness of a caudal lure (see Greene, this volume).

The legend that snakes charm birds has been explicitly invoked to describe the defensive display of an African vine snake, Thelotornis, and the coral—snake pattern of the California mountain kingsnake, Lampropeltis zonata (Goodman, 1974; Goodman and Goodman, 1976a, b). These hypotheses suggest that the snakes exploit the responses of birds to potential predators. The ringed pattern of Lamproveltis, it is said, makes the snakes conspicuous and thereby facilitates their discovery by nesting birds. (The term "thelgosematic" was coined to describe this unlikely function of a pattern). The birds, in the process of defending their nest, are said inadvertently to guide the snake to it by a process of "hot and cold". The intensity of the defense by the birds supposedly increases as the snake approaches the nest and decreases when the snake moves away, thereby providing directional cues to the predator. However, an experimental test of this hypothesis showed that nesting birds mobbed a cryptic model snake as often as a brightly banded model (Smith and Mostrom, 1985).

The defensive display of Thelotornis includes inflation of the anterior part of the body. This behavior, it has been suggested, converts the snake to a mimic of a fledgling bird with the result that adult birds are presented with conflicting stimuli. On one hand, they perceive the snake as a predator and simultaneously, it is suggested, the resemblance to a fledgling confuses adult birds with possibly fatal effect (Goodman and Goodman, 1976b). The slow tongue movements of Thelotornis, which are shared by the neotropical vine snake Oxybelis, have been suggested as creating an additional alluring effect. Alternative interpretations suggest a combination of cryptic and chemosensory functions for the slow tongue movements (Beebe, 1946; Keiser, 1975).

3

Ecology of Regeneration

Caudal autonomy, the ability to shed the tail, occurs in many lepidosaurians some salamanders and a few rodents Tails shedding usually takes place if an animal is attacked by a predator and often appears to facilitate escape. Despite its apparent survival value, caudal autonomy is not universal; the ease with which breakage occurs also varies. There are differences among taxa, among ontogenetic stages of the same species, and even within the same individual over short periods of time. Recent selective pressures may influence these differences, but historical factors may also be imporiant absence of autonomy may be due to loss in the distant past rather than through lack of present survival value. As with all selective pressures, both costs and benefits need to by considered. In recent years information about caudal autonomy has increased substatantially and it is now easier to assess the factors controlling its variation.

Autonomy is a deceptive phenomenon; it is often thought of as a simple adaption that enables a potential prey animal to escape from predators. However, autotomy is actually highly complex as are other defensive phenomena, such as mimicry (Pough, this volume). Ramifications extend to many aspects of the animal and even to the proclivities and perceptive capacities of its predators, As such, caudal autonomy gives insight into the complexity of natural selection, the interconnections of parts and activities of animals to each other, and of animals to the situations in which they live. The interactions of which caudal autotomy is a nexus resemble Darwin's fomous evolutionary analogy of the tangled bank (Darwin, 1859), in which many things are interlinked so that proximate effects may have very distant and interwoven causes.

Historically, autonomy has provided support for natural selection as the mechanism of evolution rather than Lamarckian inheritance of characters acquired by successful use. Detached parts of animals, like moving lizard tails, may act to the benefit of their recent owners by actively distracting predators. That tails should act in this way after disconnection is understandable in terms of natural selection, for lizards having such tails will tend to survive and pass on this trait to their offspring. However, such separate organs could not be perfected by Lamarckian systems of inheritance as their success could not be transmitted to the gonads of their owners (Poulton, 1895; Gould, 1983).

Many aspects of caudal autonomy have been recently reviewed elsewhere, including its detailed systematic distribution, histological mechanisms, and evolution history (Arnold, 1984; Bellairs and Bryant, 1985). These three topics are only considered briefly here; but new work, especially on behavioral and ecological aspects of autotomy, has been incorporated.

THE EFFECTIVENESS OF CAUDAL AUTOTOMY

(*a*) The Strategy

Caudal autotomy seems able to benefit lizards and some other reptiles in two quite distinct ways: It lets them break away from a predator that has seized them by the tail and it allows the tail to be used as a distraction, engaging the attention of the predator while the lizard escapes. In distraction the tail is sometimes "offered" to a predator while still attached to the lizard thus diverting attack from the unexpendable head and body. Bright or contrasting makings and movements of the tail (Section III.H) may enhance this process by making the tail more conspicuous. Autonomy usually occurs once a predator has grasped the tail, either after such a display or while a lizard is feeding; the writing tail may then act to distract the attacker from further pursuit of the rest of potential prey.

Shed tails usually move vigorously for some time and seem to attract the attention of the predators (Section III.H). Indeed the rate of caudal movement often increases if the tail is touched and may well increase further if the tail is bitten. Eating the shed tail may in fact be a more beneficial course for the predator than pursuing the lizard itself. Although the tail may sometimes be a poorer potential meal than the rest of the lizard, it has the advantage of being already secured, whereas the rest of the lizard may well evade capture (the tail is like the proverbial birds in the hand worth two in the bush). Furthermore, if the predator temporarily abandoned the moving tail to chase the lizard, the tail mighr attract another animal and be eaten by it. Movement may also increase tae time that may be spent subduing and swallowing the tail, thus allowing the lizard more time to escape (Section III.H). The two strategies, distraction and breaking away, are of course not necessarily complete alternatives; some species use one or the other according to circumstance (see, for instance, Bustard, 1968, on Gehyra variegata).

(*b*) Evidence From Predator Stomachs and Encounters in Capacity

The frequent efficacy of autotomy in allowing escape is well known to anyone who has tried to catch lizards. It is also supported by the recovery of substantial numbers of detached tails, but not their owners, from the stomachs and crops of predators (Cott, 1957). For instance, tails have been reported from the stomachs of other lizards (Milstead, 1957; Dixon and Medica, 1966; Cunningham, 1956), and of snakes (Brooks,

1967; Mount, 1963; Carpenter, 1955; Mckinney and Ballinger, 1966; Fitch 1940; Greene, 1984).

Direct evidence of the important of the tail in escape strategies comes from observation in captivity of controlled encounters among lizards and their enemies. Of 30 geckos Coleonyx variegatus with complete tails that were exposed to the spotted night snake (Hypsigleda ochrorhynicha), 19 were captured while 11 escaped by losing parts of the tail; but all tailless geckos were captured (Congdon *et al.*, 1974). Similarly, tailed geckos (Phyllodactylus marmoratus) were more successful in escaping from the shrew-like marsupial Amechinus swainsoni than were tailless ones. Of 35 tailed individuals contacted, 54% escaped by caudal autonomy, whereas all 15 tailless were captured (Daniels, 1983; other experiments reported by (Daniels *et al*, 1986). Exposure of skinks Sermella lateralis) to the snake Lampropeltis triangulum, resulted in capture of only 8 out of 19 tailed individuals grasped by the predator, whereas all 12 tailless ones were captured (Dial and Fitzpatrick, 1984), Caudal autonomy also significantly enhances escape of hatchling Eumeces spp. from Lampropeltis triangulum elapsoides (Cooper and Vitt, 1985 Vitt and Copper, 1986). The survival value of autonomy to plethodontid salamanders exposed to chickens has also been demonstrated (Labanick, 1984). Although autonomy is clearly advantageous in many circumstances, its benefits vary. Thus the slender tail of Anolis cardinensis, which does not move much after autotomy, is not very effective in distracting cats from pursuing the rest of the lizard (Dial and Fitzpatrick, 1983).

(*c*) The Meaning of Broken Tails in Wild Populations

The frequency of broken and regenerated tails in populations has been interpreted as an indication of the effectiveness of autotomy (Haldane and Huxley, 1927). Each lizard with a damaged tail might be an individual saved from death by predation. Alternatively, the incidence of broken tails has been regarded as an indicator of predation intensity; the more damaged tails in a population the greater the intensity of predation (Rand, 1954; Tinkle, 1972, 1973, 1976; Parker and Pianka, 1973, 1975, 1976; Pianka, 1967; Pianka and Pianka, 1976; Schall and Pianka, 1980). Frequency of damage has also been postulated as an index of age difference among populations (Bustard and Hughes, 1966). In fact a number of independent factors may possibly affect the perceived frequency of damaged tails in a sample. At a trivial level, collecting bias is possible. bias is possible. This is likely to apply especially when museum material is used to compare damage rates among taxa, for there is often a tendency to reject damaged specimens during assembly of specimens for taxonomic and morphological work. Again, samples collected from one locality may not be typical of the taxon as a whole (Schall and Pianka, 1980). Although in many species, regenerated tails are conspicuously different from the originals and consequently easy to recognize, this is not always the case (for instance in Teratoscincus scincus, see Werner, 1967); it is sometimes possible to understimate the number of regenerates as they may closely resemble original tails. However, as the vertebrae in regenerated tails are replaced by a cartilaginous tube, such tails can always be identified by radiography.

Tails, or portions thereof, may be lost in conspecific fights. In Agama agama, such damage occurs when males strike each other with their tails (Harris, 1964). Similarly, territorial fighting has been blamed for high frequency of caudal breakage in males of the iguanid Sceloporus magister (Vitt et al., 1974) and of the xantusiid Xantusia vigilis (Zweifed and Lowe, 1966), although this is disputed in the latter case (Vitt et al., 1977). In some groups, such as lacertid lizards, tail loss rarely seems to result from intraspecific conflicts, even in captivity. The importance of this factor in increasing natural incidence of broken tails can only be assessed properly by field observation of the species concerned.

The probability that an individual lizard will have lost its tail necessarily increases with time, although the increase is often not uniform, and young animals are frequently more prone to caudal autotomy than are adults (see, for instance, data in Cooke, 1979; Blair, 1960). All other things being equal, long-lived species will tend to have higher aggregate levels of caudal breakage than will short-lived ones (Brown and Ruby, 1977; Schoener, 1979; Schoener and Schoener, 1980). This complication can be countered by comparing samples of known age, either recognized by relative size or by previous individual marking.

The frequency of autotomized tails in a population will increase only if tail shedding significantly improves the chances that a lizard will escape. If predation efficiency is high and animals are often caught, broken tails may remain rare in samples even though predation levels are high (Schoener, 1979). On the other hand, incidence of broken tails will be higher in species and populations that show frequent encounters with predators and subsequent escape after autotomy. Moreover, tail-break frequencies can be minimal estimates of escape, because predators striking at tails, sometimes miss their prey entirely (Vitt and Cooper, 1986). Finally taxa appear to differ in the ease with which autotomy takes place.

In principle, it is possible to assess all of these factors; however, the necessary substantial ecological knowledge of the species concerned is rarely available. In particular, predator pressure is hard to estimate without extensive collateral studies of other animals in the communities in which the lizards live. Also, the readiness with which autotomy takes place is difficult to measure realistically. Because of these problems, it is difficult to unravel the contributions of intraspeafic conflict, age, predation efficiency and intensity, and ease of autotomy to the incidence of broken tails. This means that the use of caudal damage as an indirect source of information about any of these factors generally requires additional information. Schoener (1979) and Schoener and Schoener (1980) derive and explore a method of estimating predation parameters from tail break and demographic data in Anolis.

Comparison of geographically close populations of the same species is likely to be most useful, for this controls for potential differences in behavioural trails and inherent longevity; in such comparisons, predation level is the most probable cause of substantial differences in frequence of breakage. The same assumptions can be made, but with less confidence, if widely separated populations of the same species, or closely related species, are compared. For instance. the low level of caudal brealage in found juvenile Heliobolus lugubris, compared with that of the young of related forms in the same area has been attributed to reduced predetor pressure arising from mimicry of noxious beetles (Huey and Pianka, 1977; Pough, this volume). This is not unreasonable, because samples are alike in age and are quite similar in many aspects of behaviour and environment.

In several cases for which predation intensity could be quantified tail break frequency shows some correlation with this. In Cnemidophrus tigris, frequency of regenerated tails increases from north to south and correlates with an increase in the number and variety of predators (Pianka, 1970), a trend also found in juvenile Uta stansburiana (Parker and Pianka, 1975). Yearly variation in frequencies of broken tails among hatching Uta stansburiana from southern Nevada correlates with the density of one of its important predators, Gambelia wislizeni (Turner *et al.*, 1982). In Sceloporus jarrovi in south-east Arizona, a higher freqvency of autotomy at low altitudes correlates with higher numbers of potential predators and increased mortality (Ballinger, 1979). In three taxa of Sceloporus, the one with the highest tail break frequency has the lowest survivourship (Vinegar, 1975). In an analogous experimental case, causal injury frequency in tadpoles of the toad Scaphiopus holbrooki correlates with predation intensity (Morin, 1985). On the other hand, incidence of autotomy in Bahamian Anolis (Schoener, 1979) and in lizards from a site in California, at which diets of predators are known (Jaksic and Greene, 1984), shows no correlation with predation level and appears to reflect predator inefficiency.

It might be thought that tail-break incidence does at least give an indication of the relative success with which autotomy enables lizards of different species to evade predators. This would be correct, if each lizard with a broken tail had survived a predator attack, and if survival had been linked with autotomy in each case. However, in some species autotomy may not involve great costs, such forms may shed the tail at a time at which the risks of capture by a pursuing predator are still relatively low, In such cases, a shed tail is unlikely to represent a lizard saved from certain death by autotomy, whereas this may be more nearly so in species in which lost tails involve a great cost and are given up less willingly. Moreover, as already noted, tail-break incidence gives no evidence of the frequency of predator strikes at tails during which the predator missed the prey entirely (Vitt and Cooper, 1986).

Autotomy levels do often show a broad correlation with particular ecological circumstances. Thus, among geckos occurring in Israel, broken tails are commoner in climbing species than among ground dwelling ones (Werner, 1968), and this is also true among Chilean iguanids (Liolaemus: Jaksic and Fuentes, (1980). Similar correlations exist among European lizards of the genera Lacerta and Podarcis, in communities of lizards in eastern Arabia and in species of Agama in Africa (Arnold, 1984). The causes of this correlation are uncertain, largely because many of the relevant ecological data are lacking. Nevertheless, at present, there are no indications that climbing species are more prone to intraspecific combat, to longer life, or to higher rates of predation. On the other hand, subjective impressions suggest that climbing species relinquish their tails more easily than do ground dwellers. If this impression is valid, it could indicate that tail loss might be loss might be less costly in these forms, perhaps because the tail is less important in the locomotion of climbers. Here, the body is not counterbalanced by the tail, as is possible and necessary in cursorial forms. Also, because the feet are firmly attached to the substrate during climbing, lateral oscillation of the hindquarters is less of a problem.

In summary, tail-break frequencies are often unreliable indices of predation rates, especially in interspecific comparisons that fail to control for age and other factors. However, approaches that incorporate demographic data may yield useful estimates of predation rates (Schoener, 1979).

CAUDAL AUTOTOMY AND THE DETERMINANTS OF ITS VARIATION

(*a*) Mechanisms of Caudal Autotomy and Regeneration

The term, caudal autotomy is used here to describe a fracture of the tail that occurs in a regular and predictable way at a distinct region of weakness. Its mechanisms are described in detail by Bellairs and Bryant (1985) and will consequently only be detail with briefly here. Among modern reptiles, caudal autotomy unknown in the Testudines and Crocodilia but is widespread in the Lepidosauria, including Sphenoam; it is absent in the great manority of snakes. It also occured in some exinct groups (Bellairs and Bryant, 1985).

There are two distinct autotomy meceanisms. The first and more widespread is intravertebral autotomy, in which breakage takes place at preformed areas of weakness—the autotomy or fracture planes—that cross each vertebra transversely. In the soft tissues surrounding the vertebra, the autotomy plane follows the myoseptum that separates adjacent segments of axial musculature and the segments of fat that often lie between these and the vertebra itself. In some cases, there may be clear regions of weakness in the skin corresponding to the autotomy planes (Woodland, 1920; Quattrini, 1954). In the second kind of tail breakage, namely intervertebral autotomy, fractur takes place between vertebrate. Species with intervertebral autotomy do not show any obvious caudal modifications relative to nonautotomizing species. It is difficult to distinguish the two conditions, as the degree of cardal fragility is variable. In most cases of autotomy the tail is regenerated after breakage.

(*b*) How Autotomy Mechanisms May Change

Intravertebral autotomy is probably the ancestral condition in squamate reptiles. Consequently loss or a substantial reduction in the ability to shed all or part of the tail is a derived feature (Section III C). Loss or reduction can take place in a number of ways, the most obvious involving whole or partial fusion of the anterior and posterior sections of each vertebra across the fracture plane, so that its inherent weakness is lessened. This may occur in all the post-pygal vertebrae or only the more distol ones. Whenever such fusion occurs, it proceeds from the tip of the tail forwards and from the top of the neural arch downwards in each vertebra (Etheridge, 1967). Vertebral fusion is easy to detect in skeletal material and often in radiographs as well, so it is widely used as an indicator of restricted autotomy. However, ease of autotomy may be reduced by stronger attachment of the muscle segments to the myosepta and loss of planes of weakness in the skin, if originally present. These modifications usually accompany vertebral fusion but sometimes occur in forms in which intravertebral autotomy planes remain, for intance in the geckos of Pristurus carteri group (Arnold, 1984). Neurological factors also seem to be able to change autotomy thresholds (Section IV.E). Whereas all these mechanisms seem to be capable of reducing the facility with which autotomy takes place, they may not be unidirectional in their action and seversal may be possible, although this would not necessarily apply to fusion of the autotomy planes in the vertebrae once it is established throughout ontogeny.

Of the various mechanisms for changing the facility with which autotomy occurs, all seem able to act on evolutionary and ontogenetic time scales. They then produce differences among taxa and among different ontogenetic stages of the same taxon, but neurological factors are likely to be effective only on shorter time scales.

(*c*) Evolutionary History of Autotomy Mechanisms in the Squamata

The evolutionary history of autotomy mechanisms in the Squamata has been discussed elsewhere (Arnold, 1984); hence it is only summarized here. Intravertebral autotomy with capacity to regenerate appears to be the ancestral condition in the group. This is supported by its presence in Sphenodon, the primitive living sister taxon of the Squamata, by its generally wide occurrence, and by its early appearence in ontogeny. Phylogenies of particular groups, such as the macroteiids (Presch, 1974) and the gecko genus Pristurus (Arnold, 1984), also indicate the intravertebral autotomy is ancestral. Absence of intravertebral autotomy in some members of many holophyletic families indicates that the ability to shed the tail in this way has been lost a number of times. In many instances, this loss may have been relatively recent, for instance, it may only be lacking in a few members of a genus or even of a species (*e.g.*, Amphisbaena angustifrons, Gans and Diefenbach, 1972). On the other hand, its absence from whole superfamilial groupings, such as the Platynota (Helodermatide, Lanthanotidae, and Varanidae) or the Agamidae plus Chamaeleonidae, indicate that its loss may be very long standing.

Intervertebral autotomy seems to represent a re-evolution of tail shedding ability, developed in stocks where intravertebral autotomy was previously lost. Within the Agamidae, comparison of the distribution of intervertebral autotomy with a hypothetical phylogeny of the group (Moody, 1980) suggests that it may have arisen as many as eight times; multiple origins are also likely among the colubrid snakes that posses it. Not all intervertebral autotomizes among the Agamidae are able to regenerate shed portions of the tail (Section IV.B), although regenerative ability appears to have developed or bee reactivated up to five times in the family.

(*d*) Factors Determining Ease of Autotomy : Costs Benefits, and History

If natural selection acted without constraint on totally plastic phenotypes, variations in the ease with which caudal autotomy. takes place, or whether it takes place at all, would represent adaptations to different situations that maximized their fitness. This would apply to differences among taxa and among ontogenetic stages, and to short term changes within the samen individual. Autotomy would be retained whenever it increased the chances that a lizard or other reptile would approach its full reproductive potential; it would be lost whenever it reduced such chances. In autotomizing forms the facility with which the tail is shed should be determined by similar considrations.

The value of caudal autotomy depends essentially on the balance of its costs and benefits; only if the latter exceed the former is autotomy selectively odvantageous. However, the presence or absence of autotomy is frequently explained mainly in terms of variations in cost. Thus, loss or reduction of the ability to autotomize is often attributed to the tail being of especial survival value in other ways (cf. Woodland, 1920; Boring *et al.*, 1948). Variation in benefits must also be taken into account. In some situations, the advantages of autotomy are so great that they justify the sacrifice of even a very valuable tail; conversely, in other situations, the benefits are so low that even the shedding of a tail with little positive function is unjustified.

Futhermore, phenotypes may not be totally plastic, so variations in the autotomy mechanism need not represent precise adaption to different situations. In rapidly altering environments, the phenotype may not change swiftly enough to become fully adapted to current conditions. Also, some kinds of phenotypic change may be less likely than others. The mechanisms of intravertebral atutotomy appears to have been lost independently many times, yet there is no evidence that it has ever been redeveloped in its original form. Functionally analogous intervertebral autotomy appears to have evolved in some of the stocks that previously lost the intravertebral mechanism (Section III.C). Lineages might well differ in the facility with which the ability to autotomize and regenerate the tail is lost and regained. If this is so, absence of autotomy may sometimes result from the phylogenetic history of a taxon rather than from disadvantage of autotomy in the present environment. Therefore, species without autotomy should not always show the predicted excess of likely costs over likely benefits. Consequently, loss of autotomy should be less precisely correlated with excess cost, than should the presence of atotomy with excess benefit. As the environments in which animals and their descendants live probably change substantially through time, the longer since a stock has lost autotomy, the less precise the expected correlation. Because of this, better correlations should be expected at low taxonomic levels than at higher ones; this is the observed condition (Section IV.A).

(*e*) Benefits of Autotomy

The frequently effectiveness of autotomy has already been described (Section II), but its benefits are likely to very in different situations.

(1) Benefits will be slight it predation is very rare.

(2) Simply breaking away from a predator is only beneficial if the lizard can elude further pursuit by speed and agility or otherwise reach a secure refuge. Distraction techniques probably demand less speed; however, the lizard must leave the vicinity of the predator before the shed tail has been consumed, and must deny it the choice between the shed and the rest of the prey. The problem may be aggravated, as tail loss itself may reduce the speed of a lizard (Section III.F). Consequently, autotomy has only slight benefits to very slow-moving lizards.

(3) Autotomy will be of limited importance for lizards that have other means of deterring predators. Instantaneously large species might more effectively defend itself with teeth and claws than with autotomy.

(4) Conversely, autotomy may be more beneficial to small species, and use of the tail to divert attention from the head and body will be more important for delicately constructed forms with limited ability to fight back, than for more robust ones.

(5) The effectiveness of the tail as a distraction device depends on its being an acceptable item of diet, or apparently so. Therefore, tails and tail fragments that are very small, or spiny, or that do not move much are

likely to have effect. As will be seen (Section IV.A), the actual taxonomic distribution of autotomy shows some correlation with these factors.

(*f*) Costs of Autotony

In most lizards, tails have survival value in areas other than the evasion of predators by autotomy. Indeed these uses are so widespread that the simple classification of tails into "actively" and "passively" functional types (Vitt *et al.*, 1977) appears inappropriate. (In this scheme, actively functional tails include these with obvious functions apart from autotomy for predator escape, such as swimming, climbing, defense and blance; passively functional tails include those without an obvious function, other than autotomy.) Nevertheless, the aggregate selective value of the tail differs substantially among taxa, and probably among ontogenetic stages; the cost of losing the tail must show a similar variation. Apart from the transient loss of alternative function, this cost also includes the price of replacing the original tail by regeneration.

In many cursorial lizards, the tail is raised during running and acts as a counterpcise to the head and body. The weight is concentrated over the hind limbs, which are often the main source of forward propulsion. This role is particularly important in bipedal forms; thus the iguanid Basiliscus is unable to run bipedally after losing only the distal third of the tail (Snyder, 1949). The tail is also important as a counterpoise in quadrupedal species, such as the lacertid, Patarcis sicula, and loss of the tail results in more weight being transferred to the front legs (Arnold, 1984). The tail may function as an inertial damper in reducing lateral swing of the hind quarters. In most cursorial lizards the legs move in more or less horizontal arcs and the rumb tends consequently to oscillate from side to side; the presence of a large tail with substantial inertia limits this by inducing bending within the limbs (Arnold, 1984).

In a number of cases, experiment has shown that maximum speed is reduced after caudal amputation. Thus, amputation reduces the velocity of the teiid Cnemidophorus sexlineatus by 36% over distances of 3 m (Ballinger *et al.*, 1979); similar results are obtained for the iguanids Dipsosaurus dorsalis (Pond, 1981), Cephosauris taxanus and Uma notata (Punzo, 1982) and for the lacertid Pedarcis snula (Arnold, 1984)˙ Such changes make lizards more vulnerable to predators. Tailed Scincella lateralis are more adept at evading capture by snakes than are tailless ones; 11 out of 30 intact S. lateralis escaped without being contacted by the predator, compared with none of 12 that lacked tails (Dial and Fitzpatrick, 1984). However, tail loss does not always reduce speed; Phyllodactylus marmoratus, in which the robust tail perhaps plays no part in locomotion, runs faster after the tail is removed, at least for distances of 10—30 cm (Daniels, 1983). In the iguanid Sadoporus merriami, maximal sprint speed on a 2-m track is unaltered by the loss of even two-thirds of the length of the tail (R. B. Huey and A. E. Dunham, personal communication).

The tail contributes directly to locomotion in many long-bodied squamates that have very small legs or no legs at all. In many of these species, the tail makes up a large proportion of their total length, and its loss may significantly reduce the ability to move.

In climbing forms, the tail may be important in balancing. Polychrus marmoratus can stand erect on a branch whenever reaching upwards to grasp a twig because the long tail hanging downwards promotes stability (Cott, 1926). The tail is also important in balancing Anolis carolinensis, and its loss reduces ability to perch in unstable situations (Ballinger, 1973). Urosaurus graciosus also use their tail in balancing (Vitt and Ohmart, 1975).

Lizards that climb in flimsy vegetation or travel across its surface often distribute their weight over as large an area as possible via a long, often slender tail. Among lacertids this occurs in such forms as Lacerta viridis, Psammodromus algirus, and especially among the East Asian grass runners, Takydromus.

Tails may be coiled spirally around twigs or stems to maintain position in vegetation. This is best known and particularly developed in the chameleons but occurs in other forms such as the iguanids Polychrus (Cott, 1926; Vitt and Lacher, 1981), Phenacosaurus (Schmidt and Inger, 1957) and Chamaeleolis, the geckos Naultinus. Heteropholis, and Aeluroscalabotes, and even such unspecialized climbers as Lacerta viridis and Podarcis muralis. The gecko Lygodactylus has an adhesive pad on the caudal tip, similar in structure to those on the digits, and uses it as a fifth point of attachment, for instance during jumps (cf. Vitt and Ballinger, 1982).

Many kinds of lizards use the tail in swimming, and in the more regularly aquatic forms, it is laterally compressed. Such forms, include the iguanids Anolis barkeri (Robinson, 1962), Amblyrhynchus cristatus and Basiliscus, the agamids Lophosaurus and Physignathus, the teiids Crocodilurus and Dracaena, the xenosaurid Shinosaurus crocodilurus, some Varanus species and the skink Sphenomorphus quoyii. In the latter species, loss of the tail severely restricts swimming ability (Daniels, 1985b), as it does in Amblyrhynchus (Bartholomew *et al.*, 1976).

In at least some arboreal lizards that can glide, such as the gecko Ptychozoon (Cantor, 1847; Tweedie, 1950; Marcellini and Keeper, 1976) and the lacertid Holaspis guentheri (Schiotz and Volsoe, 1959). the tail promotes aerodynamic stability and contributes to lift. Loss of the tail in these autotomous forms may well reduce gliding ability.

The tail is sometimes used in intraspecific signalling (Carpenter and Ferguson, 1977), for example in reckos of the genus Pristurus (Arnold and Gallagher, 1977; Arnold, 1980), and tail loss sometimes decreases social status in encounters with conspecifics. This has been demonstrated experimentally in juvenile. Uta stansburiana, in which such decreases may impose a social handicap on successful home range acquisition, thereby increasing the risk of death (Fox and Rostker, 1982). A similar loss of social status has been reported for tailless Anolis carolinensis (Hennig, 1974).

These reserves are apparently mobilized whenever reading does not occur, such as during hibernation in Lacerta vivapara (Avery, 1970) and Eumeces fasciatus. Viability can be reduced it lizards are deprived of caudal fat by autotomy. Juvenile Lacerii vivipara survived less well in hibernation after the tail had been lost and only partly regenerated (Bauwens, 1981). Scincella lateralis may survive about 35 days if starved but only 24 days after the tail has been lost. In Phyllodactylus marmoratus, animals with original tails survived 90.4 days, those with regenerates 142.4 days and tailless animals only 49.2 days. Tail loss negatively affects growth in Uta stassuriana but not in juvenile Eumecus.

Tail loss may also affect reproductive capacity (Vitt and Cooper, 1986). In Coleonyx brevis, energy reserves account for 53% of reproductive energy in females with tails but only 29% in ones that have autotomized. The latter females produce eggs are significantly lower in mass (17%) and energy content (19%). Dail and Fitzpatrick, 1981). In the skink, Morethia boulengeri tail loss results in reduced egg mass, whereas in Hemiergis peronii, tail loss reduces clutch size (Smyth, 1974).

If a substantial part of the tail is shed, the tail cannot be used as an effeetive antipredator device until regeneration takes place. This applies whether such use involves further autotomy or some other strategy (Section III.G). A significant proportion of the short-lived iguanid Uta stansburiana regenerate the tail more than once, indicating that predatory attacks are probably quite frequency (Tinkle, 1967). This suggests that such lizards may stand an appreciattic risk of meeting a predator before the autotomized tail has been fully replaced.

Fragile tails may sometimes be lost inadvertently. Also, they can be damaged whenever used as weapons in intraspecific combats, for instance in Agama agama (Harris, 1964), or they may possibly be broken off by the jaws of a rival, although it is uncertain how common this is (Section II.C). Similar loss may also occur by accident, and some predators may even take advantage of caudal fragility to crop tails of lizards that are otherwise too large to eat. For instance, Varanus caudolineatus appears to harvest the tails of Gehyra that are too large to be subdued in their entirety (Pianka, 1969). In captivity at least, the lacertid Podarcis lilfordi often pulls the tails off similarly sized lizards including conspecifics. Finally, autotomy exposes bone cnd soft tissue to infection and attack of invertebrate predators.

Regenerating a new tail must divert energy and material from other bodily functions or, alternatively, involve increased food intake. For instance, up to 50% of total growth energy may be allocated to caudal regeneration in the skink, Eumeces gilberti (Vitt *et al.*, 1977). Energy contained in a regenerated tail of the gecko Coleonyx variegatus has been calculated to be equivalent to 47% of the energy in one egg of this species; underestimates the actual cost of regeneration because it ignores losses of metabolic energy (Vitt et al., 1977). In plethodontid salamanders, autotomized tails are regenerated at some expense to future reproduction (Maiorana, 1977).

Alternative Defensive Uses of Tails

Loss of the tail may also prevent it being used for defensive purposes other than autotomy. These are discussed elsewhere (Greene, this volume) and will be only summarized here. In some lizards, the tail is used as a weapon in intraspecific combat. Thus, males of Agama agama strike each other about the head with their tails (Harris, 1964), and this sort of behaviour may also be directed at potential predators, as occurs in the iguanids Iguana (Schmidt and Inger; 1957) and Ctenosaura, in a number Varanus species (cf. FitzSimons, 1943; Auffenberg, 1981), and in some forms with thick spiny, club-shaped tails such as several species of the agamid genus Uromastyx. Some geckos of the genus Diplodactylus can expel a very viscous exudate from subcutaneous glands in the tail that probably deters predators (Zietz, 1914; Bustard, 1964, 1970; Rosenberg and Russell, 1980).

The agamid Uromastyx thomasi has a disk-shaped tail with a spiny upper surface. Whenever it enters burrows, the tail is turned downwards so that any predator for following the lizard will be presented with an effective blockage that prevents further progress (Arnold, 1980). This also occurs in U, princeps (Laurent and Gans, 1965), and Agama (Xenagama) battilifera and A. (X.) taylori probably use their rather similar tails in this way. The gecko Diplodactylus conspicillatus, which inhabits the abandoned holes of trap-door spiders, also plugs entrances with its expanded tail (Bustari 1970).

The tail may also give defense through mimicry (Pough, this volume). The boid Crurina bottae uses its tail to distract and deter female mice while eating the young (Nussbaum et. al. 1983). In some lizards the tail is very similar to the head in shape and colouring and may mimic it. Such similarity occurs, for example, in the skink Tiliqua (Trachydosaurus) rugosus and in at least some amphisbaenians. In some Amphisbaena, the tail is also moved conspicuously whenever the animal is disturbed; this may direct the attention of a predator away from the more vulnerable heat to the expendable tail (Greene, 1973). Another area of defense in which the tail may be involved is crypsis. The flattened, fringed, lem-shaped tails of such geckos as Phyllurus and Uroplatus improve the match of these lizards with the surfaces on which they usually occur.

It has been suggested that various lizards that habitually raise the tail in the presence of a predator are scopion mimics, for example Chondrodactylus (FitzSimons and Brain, 1958), Coleonyx variegatus (Parker and Pianka, 1974), and Phrynocephalus. However, in at least the latter two taxa, caudal taxa movement is not scorpion-like (Section III.H; also Pough, this volume).

Member of the geko genus Teratoscincus have the dorsal caudal scales modified, so that they produce a hissing noise as lateral waves of movement pass down the tail (Mertens, 1946; Mebs, 1966). This behaviour occurs in the presence of predators and might deter them (although this is not documented perhaps being similar to the noise produced by the vipers of the genus Echis, which are sympatric with Teratoscincus. The vipers generate the noise by rubbing their lateral scales together during threat displays.

The brilliant tails of fire juveniles of some North American species of Eumeces have been thought to indicate that their owners were unpalatable (Clark, 1971), but experiments with predators suggested they are not actually distasteful (Cooper and Vitt, 1985). Bright caudal markings may be "flash coloring": In lizards in which the underside of the tail is conspicuously marked exposure and the sudden concealment of this pattern by a running animal may fix the attention of a predator on the spot at which the bright colouring was last seen, even though the otherwise cryptically coloured prey has moved on (Arnold, 1984) or become hidden in litter as a refuge. Finally, movement of conspicuous tails might warn a predator that it has been seen and that pursuit will be expensive (Arnold, 1980). This strategy has been demonstrated in Cophosaurus texanus (Dial, 1986).

Modifications That Increase the Effectiveness of Autotomy

As observed earlier (Section I), caudal autotomy in reptiles is often a much more elaborate phenomenon than mere possession of a breakable tail refinements increase the effectiveness of autotomy, either by enhancing its efficiency or by reducing its costs.

There is evidence that ease of shedding may be adjusted so tails are are shed or retained in accordance with the balance between the degree of danger and the value of the tail (Section IV.E). Also, the amount of tail shed seems to be related to the kind of escape strategy employed (Section IV.C). In the breaking-away strategy, the minimal amount of the tail necessary to escape is shed, whereas in the distraction strategy, most of the organ is lost.

Tail size appears to alter the effectiveness of autonomy. Thus, the relatively and absolutely smaller tails of juvenile Phyllodoctylus marmoratus are less efficient than those of adults at attracting the attention of the marsupial predator, Antechinus swaisonii (Daniels *et al.*, 1986). High risk of predation may favor the evolution of relatively long tails (Vitt and Congdon, 1978) as this would increase the chances that a pursuing predator would grasp the appendage rather than the lizard. In the Kalahari desert, species of widely foraging lacertid lizards appear appear to be at greater risk from predators and have longer tails than do sit-and-wait species (Huey and Pianka, 1981). A similar correlation occurs in lizards from northeast Brazil (Vitt, 1983), and members of the amphisbaenian genus Bipes have longer tails whenever they coexist with coral snakes, Micrurus, which are one of their main predators (Papenfuss, 1982).

Tails of lizards with intravertebral autonamy have a high energy content, containing large amounts of fat, which could be interpreted as increasing their attractiveness to predators familiar with lizards. However, the caudal fat of the lizard may simply be part of its energy rssources that cannot be stored anywhere else (Dial and Fitzpatrick, 1983); perhaps it is also a necessary component of the intervertebral autonomy mechanism. This last hypothesis receives some support because squamates with intervertebral autotomy or no autotomy lack well developed fat deposits in the tail.

Shed tails are always eaten by the predatars that caused them to become detached. In such circumstances, the individuals that autotomized may return and eat their tail enabling them to resoup some of its contained material and energy. This behaviour is known in Scincella lateralis (Clark, 1971) and has been demonstrated in captive Eumeces and Xantusia Vitt *et al.*, 1977). Experiments suggest that postautotomy movements propels tails of Scincella through leaf litter, so that they move out of sight, facilitating their retrieval and eating by the owner (Clark, 1971).

By modifying its behviour until the tail is least partly regenerated, a lizard could avoid some of the likely functional costs of autotomy. For instance, by stopping close to cover, it could reduce the risk of being attacked by a predator at a time at which caudal autotomy could not be used successfully and locomotory ability is impaired. Anolis carolinensis may compensate for tail loss in thls manner (Hennig, 1979). The skink, Sphenomorphus quoyii usually escapes by diving into water and swimming away using its tail; after autotomy it changes its escape strategy, either by fleeing onto land or by hiding on the bottom of water courses (Daniels, 1985b). Phyllodactylus marmoratus appears to have some automatic bethavioral compensation for tail loss (Daniels, 1983). After the tail is lost, the fleezing lizard almost doubles its speed. On the other hand, not all behavioural changes are likely to decrease predator contact. In captivity, autotomy of the tail by female Coleonyx brevis is followed by a substantial increase in the rate of food ingestion; apparently they compensate for the lost energy reserves of the tail. In natural conditions, this would increase foraging time and exposure of the lizard to predators during a period when primary defense mechanisms are impaired (Dial and Fitzpatrick, 1981).

The cost of tail replacement can be avoided by not regenerating it, or spread by doing so only slowly. However, this is likely only if tail shedding does not involve substantial net long-term costs.

VARIATIONS IN THE AUTOTOMY PROCESS AND THEIR CAUSES

Total Loss of Intravertebral Autotomy

If the loss of intravertebral autotomy frequently resulted from environmental selective pressures similar to those still acting on the taxa concerned, its distribution should correlate with particular modes of life in which the Ekely costs of autotomy exceed its likely benefits. Appraisal if likely costs is difficult, even in an informal qualititative way, as none of the numerons possible components has been fully assessed for any ipecies. Variations in likely benefits are perhaps easier to discern, as power factors are liable to be significant, and these are easier to estimate. In the whole the magnitude of benefits may be more important than that of costs in determining whether autonomy is lost, as the former vary more. Costs tend to be substantial. In most cases, regeneration of the tail requires considerable metabolic input, and during this time some significant caudal functions are lost. On the other hand, costs are rarely extremely high. Presumably, most lizards survive the transient effects of oil loss, which may be of quite short duration, as regrowth is often swift and functional losses may be substantially reduced by the time that the tail is partially regrown. Further more, behavioral considerations during puts period, such as stopping closer to cover or changing escape strategy, may sometimes reduce the risks of subsequent predation (Section III. H). In contrast, benefits of autotomy range from great (*i.e.*, pre lizard escapes) to nonexistent. The latter involves species toh show to escape from predators. Most lizards that do not autotomize fall into three groups, in each of which the potential benefits of autotomy appear to be low.

The first group is made up of large forms and includes the iguanids Auchlyranchus, Condophus, Ctenosaura, Cyclura and Iguana, and the teiid Timnamics. In some Conolophus. and Ctenosaura, Iguana iguana and Twinamivis, the ability to autotomize is lost during ontogeny. Adults of these forms are often sapable of active defense against predators by bring, clawing and tail lashing. Several of the species occur on small isands, on which predation level are likely to be low, at least for adults. Both these factors would be expected to reduce the benefits of tail shedding.

The second group comprises slow-moving forms with unpalatable tais including the iguanids Hoplocercus, Leiosaurus, Phrymosoma, and Unicentron, the gecko Nephrurus laevis, and the skins Egernia depressa, Tiiaua (Trachdosaurus) rugosa, and Tiliqua scincoides. In the latter two species, the ability to autotomize is lost only during ontogeny. Most of these species move slowly and the tail is either very small compared win the body (for instance in Phrynosoma and Nephrurs) or spiny (Homlocercus, Uncentron; Egerina depressa). Inability to flee very effectively and tails that are unlikely to capture the attention of a predator for long again restrict the possible benefits of caudal autotomy.

The third group is made up of forms that climb in vegetation, such as the iguanids Anisoleris, Anolis latiforns series, Apty cholaemus, Basiliscus, Chamaeleolis, Chamaelinorops, Corythophanes, Enyalius, Enyalioides Laemanctus, Phenacosaurus, Polychroides, Polychrus, and Urostrophus and the skink Corucia zebrata. This assemblage includes a range of lizards that appear to spend all, or a substantial part, of their time among small branches, twigs and leaves or other complex vegetation. Rapid escape from predators, such as birds and snakes, is often difficult in such an environment; surfaces are discontinuous and refuges rare. Many of these lizards are sluggish at least at times, and frequently adopt slow gaits. Some are capable of hopping from branch to branch (for instance Corythophanes, Davis, 1953; Laemanctus, McCoy, 1968; Polychrus, Cott, 1926, Boker, 1935), but even this mode of progression is not especially fast. Antipredator mechanisms may involve aggressive displays or postures that increase apparent body size (Davis, 1953), rather than immediate resort to flight.

Caudal autotomy is likely to have reduced valu in enabling such lizards to break away from predators or distract them. tails of such lizards do not show much postautotomy movement (Vitt, 1983) and are often very slender rather than stout. Hence, they are unlikely to deflect the attention of an attacker. Experiments with Anolis carolinensis, which approaches this group in many caudal features, show that its tail is ineffective in deflecting predator attacks (Dial and Fitzpatrick, 1983). This line of explanation is less convincing for the basiliscines (Basiliscus, Corythophanes, and Laemanctus); many of their species are known to descend from vegetation and are capable of bursts of fast locomotion on horizontal surfaces (Duellman, 1963; McCoy, 1968; Davis, 1953). Although they could theoretically benfit from autotomy, these lizards run bipedally and, at least for Basiliscus, the tail is essentially for maintaing running posture (Section III. F).

Loss of autotomy also occurs in some other forms, namely Gambelia, some Ophisaurus, many Agamidae, Chamaeleonidae, Helodermatidae, Lanthanotidae, and Varanidae. The bipedal ground dweller Gambelia may have lost autotomy for the same reasons as the basiliscines, but it is not obvious why loss should have occurred in some species of Ophisaurus. Many Agamidae, Chamaeleonidae, and Varanidae are ecologically similar to species from other families that lack autotomy mechanisms. Thus many varanids are large and capable of actively deterring some predators, whereas chameleons and some agamids generally live in vegetation, are slow moving, and rely on their tails for support whenever moving among twigs; however, numerous forms are ecologically analogous to members of other families that retain intravertebral autotomy. For instance, some of the small Australian species of Varanus are ecologically similar to some autotomizing skinks, teiids, and lacertids. In the same way, the agamid, Leiolepis is quite like the iguanid. Dipssaurus, and the small ground dwelling agamids, Phrynocephalus and Tympanceryptis, superficially resemble such iguanids as Holbrookia. This lack of correlation tends to support the view (Section III. C) that absence of intravertebral autotomy in the Platynota, Agamidae, and Chama-eleonids due to loss early in the history of these groups rather than to its being universally disadvantageous to present species.

It might be asked why intravertebral autotomy has not been lost more widely, given the apparent facility with which this trait disappears. This may simply be because the great majority of species retaining it do not have the sort of life mode favouring loss. For example, among fully autotomous geckos, pygopodids, lacertids, teiids, xantusiids, cordylids and agamids, few species are large, very slow, or possess unappetizing tails; also the great majority of these forms do not inhabit the sort of vegetation in which many nonautotomous iguanids occur .

Re-evolution of Autotomy

Tail brakage by intervertebral fracture occurs in some agamids and snakes and is the only kind of autotomy occurring in these groups. Breakage in agamids was thought to take place across the centra of the more distal vertebrae, which are narrow in many species, and not between them (Siebenrock, 1895). This impression is incorrrect and probably arose because the last vertebra remaining after autotomy undergoes some ablation before healing or regeneration, so its posterior section is absent (Arnold, 1984). Intervertebral autotomy appears to represent a re-evolution of tail-shedding ability, involves relatively slight modification of caudal structure and has apparently arisen a number of times (Section III. C). Even so, in the Agamidae it is by no means universal and is almost entirely confined to two ecomorphological types of lizard. The first are climbing forms, nearly all of which live on rock surfaces and includ Agama (Agama), Agama (Pseudotrapelus), Agama (Stellio), and Psammophilus. The second includes terrestrial species and some that climb in vegetation. They have relatively long hind legs and very long tails that become extremely slender distally. A number of these forms are known or suspected to be bipedal (Russell and Rewcastle, 1979) and autotomy is usually restricted to the distal parts of the tail. They include some Amphibolurus species, (A. caudicinctus A. cristatus, A. isolepis, A. maculatus, and A. pictus), Diporiphora bilineata, Gonocepalus subcristatus, Lophognathus temporalis, Otocryptis wiegmanni, Physignathus lesueurii, and Sitana ponticeriana. Neither of these groups appears to be holophyletic.

Thus, although apparently quite easily evolved, intervertebral autotomy has not developed in a variety of fast, ground-dwelling agamids that, by analogy with intraverebrally autotomizing iguanids (Section IV.A) might be expected to have it. One possible reason for this is that the evolution of autotomy and regenerative ability in agamids may be independent events. Forms like Diporiphora, Otocryptis, Psammophilus, and Sitana can only autotomize and do not regenerate, whereas most of the other agamids listed above do both. If intervertebral autotomy usually evolves before the ability to regenerate, it will only become established in situations in which permanent loss of much of the tail does not carry heavy penalities, or in which loss can usefully be limited to a small portion of it. Such situations may be rather rare.

Circumstantial evidence suggests that the tail is often less important to rock-dwelling species than to those in other habitats (Section IV. F.), so tail loss may be survivable by the former even without regeneration. In long-tailed, frequently bipedal agamids in which autotomy is usually restricted to the distal part of the organ, such modest loss may have only a mimical effect on the balancing function of the tail. This suggests that analogous pipedal forms, such as the basiliscine iguanids, should also have distal autotomy, but they do not. Perhaps this reflects the way in which the ability to shed the tail has been lost in groups such as iguanids that were originally intravertebral autotomizes. Both in ontogeny and phylogeny, the fracture planes in the vertebrae fuse from the caudal tip forwards (Etheridge, 1967). Consequently, if selection favours retaining the greater part of the tail for balancing purposes in bipedal forms, the whole organ must become nonautotomous.

Among snakes, intervetebral autotomy is reported in Pliocercus, Rhadiena decorata, Scaphiodontophis, and Sibynophis. It is uncertain what factors may have caused its development in these few forms, all of which are characterized by very long tails. The rarity of the condition suggests that it has arisen only in rather special circumstances, but what these were is still a matter for conjecture. In Scaphigdontophis, the tail may be used to flush prey out of litter but also shows high incidence of damage (Henderson, 1984).

Restriction of Autotomy to the Caudal Base

Restriction of autotomy planes to the vertebrae of caudal base occurs quite widely in geckos (Arnold, 1984), including the eublepharine Hoplodactylus, the diplodactylines

Carphodactylus, Diplodactylus stenodactylus, D. strophurus, D. vittatus, Heteropholis, Naultinus, Nephrurus laevis, Oedura castelnaui, O. marmoratus, Phyllurus, and Gymnodactylus, and the gekkonines Colopus, Kaokogecko, Geckonia, Pachydactylus mariouensis, Falmatogecko, Paroedura, Stenodactylus offinis, S. doriae, S. grandicers, S. lemocosymbotes, S. petrii S. slevini, S. sthenodactylus, S. yemenicus. Tetraredupis fasciata, and Uroplatus. In others, autotomy planes may occur more widely but autotomy is still most usually basal, for example among the eublepharines Aeluroscolabotes, Eublepharis, and Hemitheconyx and the gekkonines Chondrodactylus, Crossobamon eversmanni, Phyllodactylus mermoratus (Daniels, 1985a). Pristurus carteri P. callaris, P. ornithocarhalus, and Ptenopus. Basal restriction of autotomy planes also occurs in some species of Anolis, but in some cases autotomy can still occur more distally (Cox, 1969). Because of this, discussion will be limited to geckos.

In most lizards, the tail is autotomous throughout its length distal to the pygal vertebrae, and breakage often takes place just in front of the place at which the tail is seized, or contacted, by a predator. This economy of autotomy" minimizes costs for a lizard breaking away from the grip of a pursuer. Nevertheless, some species always shed the whole will although it often contains substantial amounts of stored fat (for example in Phyllodactylus marmoratus; Daniels, 1983). However, braking away from a predator is only likely to be successful if the lizard has the speed to evade further pursuit or if a refuge is close.

Relatively slow species, particularly ones living in habitats with few secure shelters, probably rely on a different strategy in which the tail is used to deflect the attention of predators from the more vulnerable head and body. However, sacrifice of the smallest portion of the tail possible may not be sufficient to retain the attention of a predator and restriction of autotomy to the caudal base could be a means of ensuring that a large enough "bait" is provided to deflect an attacker. The distribution of rasal autotomy tends to support such an interpretation. Many of the geckos listed above are indeed relatively slow, slower than relatives with functioning autotomy planes throughout the tail (Arnold, 1984). Thus, basally autotomizing Pristurus and Stenodactylus are slower than other members of these genera. Further, many basal autotomizers occupy open arid country, in which refuges are likely to be sparse; most other basal autotomizers climb in bushy vegetation in which the same may be true.

Ontogenetic Loss of Autotomy

Juveniles of many species of lizards possess well developed autotomy mechanisms and reduce, or more often, obliterate them by maturity. Examples are Iguana, Cyclura spp. Ctenosaura acanthura, Sauromalus, Basiliscus, Enyalioides, most members of the Anolis latifrons series, Trachydosaurus, Tiliqua rugosus, Tiliqua scincoides, and some Tupinambis teguixin.

Presumably the change results from a shift in the balance of costs and benefits either the former rising or the latter falling with increase in body size, or both. The benefits of autotomy are probably greater in young animals, because of their more obvious vulnerability compared with adults (Section IV. G). Not surprisingly, ontogenetic loss of autotomy is commonest in quite large species.

(*e*) Short-term Changes Within the Individual

The strength of individual fracture planes determines the maximum force necessary to produce physical breakage of a particular part of the tail; but much smaller stimuli may initiate autotomy, and the threshold for this often varies considerably within individual lizards.

Such variation has undergone little formal investigation but a number of observations indicate that it exists.

(1) Conscious animals typically shed their tails more readily than unconscious ones (see for example Slotopolsky, 1922).

(2) Ease of autotomy may vary with temperature, for example in Uta stansburiana (Brattstrom, 1965), Podarcis sicula (Quattrini, 1952), Stenodactylus (Werner, 1964). Gehyra variegata (Bustard, 1968), and Phyllodactylus marmoratus (Daniels, 1984). In most cases, autotomy occurs most easily at high temperatures, but in Gehyra it takes place readily at high or at low temperatures.

(3) Captive lizards, including many lacertids, that become tame will tolerate a degree of handling of the tail that would produce immediate autotomy in wild animals.

(4) In some species, such as the lacertids Podarcis sicula and Lacerta vivipara, the tail is shed quite readily if grasped as the lizard runs away. However, if the tail is pulled or twisted while the lizard is held by the body the tail can be detached only with difficulty.

(5) Ease of autotomy may change with physiological state; thus, starved Phyllodactylus marmoratus some the tail less readily than well fed ones (Daniels, 1984).

(6) Scincella (Dial and Fitzpatrick 1983), Colenyx (Congdon et al., 1974), and Eumeces (Vitt and Cooper, 1986) do not autotomize their tails immediately upon being grasped by a predator. This may allow the lizard the opportunity to escape with an intact tail if the predator's grasp is poor.

These examples suggest that tail shedding threshold is modified in the individual by neurological means in response to external circumstances. If the mechanism is optimally adjusted by natural selection, autotomy should only occur easily situations for which likely benefits exceed likely costs. The likely costs of losing a given portion of tail are probably quite constant in the short term (although they may very seasonally, for instance if large amounts of fat are stored in the tail at certain times of the year). Likely benefits on other hand change very rapidly. The benefits of autotomizing the tail are probably low until a predator is at close quarters.

They will rire further if the predator actually makes contact with the tail and wonld be expected to be higher still if the predator is of a kind that is efficient at capturing lizards.

At least some of the known individual variation in autotomy threshold appears to be adaptive and Brattstrom (1965) and Butard (1968) provide functional hypotheses for temperature-correlated change in Uta and Gehyra, respectively. In the case of a lizard held by the body, tail shedding is probably of little advantage as it will not aid escape. In this situation, a high autotomy threshold would make retention of the tail more probable; if the lizard subsequently breaks free, it could still utilize caudal autotomy in any further pursuit.

Interspecific Variation in the Readiness With Which the Tail is Shed

Not only should the complete loss of autotomy mechanisms be determined by the balance of costs and benefits but so should relative readiness with which the tail is shed by different autotomizing species. Tails should be relinquished most easily in those species with the lowest costs of loss, in terms of future reproduction. If the loss of the tail is comparatively inexpensive the lizard can "afford" not to take high risks and autotomy can be brought into operation even in circumstances in which the changes of capture are not especially high. Conversely, if loss of the tail is likely to be very costly, autotomy should be delayed until the risk of capture without autotomy approaches certainty. For example, if loss of the tail only slightly reduces future reproduction on average autotomy should be initiated whenever the risks of being caught and killed by the pursuing predator are also low. However, if chances of failing to reproduce after tail loss are high, autotomy should be delayed until the probability of being killed by the predator rises to this level. Animals with low-cost tails should therefore autotomize earlier in response to a particular kind of attack and at lower levels of tactile stimulation.

That such interspecific differences in autotomy threshold exist is apparent to anyone who has tried to catch diverse species of lizards. Some species shed the tail if it is only lightly touched during pursuit, whereas others only break in the organ is grasped very firmly. Among European lacertids, tails appear to break most readily in rock climbing species, such as Lacerta oxycephala, and, in Arabian geckos, fragility seems to be greater in climbing forms than in ground dwellers. Such assessments are, of course, subjective as we lack satisfactory comparative measurements of tail fragility. This is largely because of the difficulty of delivering calibrated stimuli to lizards under natural conditions. Moreover, habituation and restraint may alter autotomy thresholds of captive animals, perhaps differentially among species. In addition, applied stimuli should mimic the kinds of contacts made by predators; the attachment of weights to the tail (Quattrini, 1952; Brattstorm, 1965; Danices, 1984), though easy to apply and measure, are unlike anything that lizards usually encounter.

Brightly Coloured Tails

Many lizards have their tails colored in a way that contrasts with that of the rest of the body. This can enhance the effectiveness of caudal autotomy (Section III.H), although a number of alternative functions are possible (Copper and Vitt, 1985. Markings, such as bold transverse bars, may be disruptive, breaking up the outline of the tail. Caudal coloring might be used in intraspecific signalling or contribute to other antipredator devices (Section III.F, III.G). None of the available explanations of conspicuous caudal markings necessarily rules out the others. Conspicuous tail markings occur in some nonautotomous species, for instance some Varanus and Phrynocephalus, although not as commonly as in autotomous ones. In many cases, the main function is probably not to direct attention to the tail. However, this explanation cannot be completely excluded, as diversion of attack from more vital areas to the tail may be advantageous although the tail cannot be shed. This usage is suggested for snakes with bright, nonautotomic tails (Greene, 1973).

Regularities in the occurrence of bright caudal coloring have been surveyed in lizard faunas of Arabia and in those of Europe west of the U.S.S.R. (Arnold, 1984); the following remarks are largely based on these surveys. In nearly all cases in which conspicuously coloured tails occur, the caudal colouring is substantially exposed in the living lizard. It typically occurs on the dorsal caudal surface and is most intense here; although a few species have bright colouring on the underside of the tail so coloured. In species tnat practice economy of autotomy, caudal colouring is more intense distally; it presummably directs attention away from the more vulnerable body and towards. the caudal tip, so that the amount of tail lost will be minimal In forms that shed the whole tail, colouring tends to be more uniformly distributed.

Permanently exposed bright caudal coloring carries the risk of attracting the attention of predators that would not have otherwise detectcd the lizard. At first sight, a better strategy would be to have conspicuous coloring concealed, for instance beneath the tail, and only exposed once a predator is at close quaters. This, however, would require the approach of the predator to be detected sufficiently early to allow time for the caudal coloring to be deployed. Most lizards, particularly active foragers, may not be able afford the high degree of vigilance involved, especially at times during which their attention is largely devoted to huntting. The risk inherent in permanently exposed bright colouring, namely that of sometimes attracting a predator that otherwise might have ignored the lizard, may be offset by the advantage that the attention of the predator will be constantly directed toward the tail.

Conspicuous caudal coloring is better developed in young than in adults, and often disappears entirely before maturity. This may reflect the greater vulnerability of juveniles to predation. Being small, they can be tackled by a greater range of predators. Frequently, they are built more delicately than equivalent-sized adult lizards and are likely to be slower. Newly hatched or dispersing juveniles are unfamiliar with their immediate surrounding and cannot flee to refuges or otherwise evade capture with the speed and facility provided by detailed knowledge of a home range. The loss of the colour at maturity may partly relate to the disastrous effect of loss of tail lipids on reproduction of females (Vitt and Cooper, 1986).

Conspicuous caudal making is typically associated with open situation lacking much vegetation and the shadows it throws. Lizards living in open, uniform habitats are quite easily seen; this may increase the need for diverting attention from head to the tail. In Arabia, ground dwelling form open sand gravel substrates have contrastingly coloured tails, as do specses climbing on continuous open surfaces, such as rock faces. In contrast, bright tails are absent in species inhabiting litter, dense vegetation and rocky surfaces with good plant cover. Most European lizard habitats are relatively enclosed and bright caudal color here is uncommon and largely confined to rock-dwelling forms. In Arabia and Europe, most species with conspicuous tails are active, widely-ranging foragers. These place limited reliance on crypsis and may be more vulnerable to predators than are sit-and-wait species. Whether bright caudal colouring and activity are really correlated is difficult to confirm in these areas, as autotomizing passive foragers are uncommon. In the New World, conspicuous caudal coloring is more frequent in actively hunting groups, such as teiids and skinks, than among sit-and-wait iguanids; however, the above association with type of habitat is unsupported (Vitt and Cooper, 1986).

Nocturnal species (predominantly geckos) with conspicuous tails nearly always have the tail patterned with light and dark areas, often forming transverse bands. In contrast, most diurnal forms have bright colours of which blue and blue-green are by far the most common. This difference correlates with the usual visual capacity of noctuanal and diurnal predators. Night operating hunters have poor colour vision and probably the objects most conspicuous to them are ones in which light and dark areas are juxtaposed; on the other hand, many diurnal hunters such as birds can distinguish colours.

The predomidance of blue and blue-green colouring in diurnal lizards may be adaptive because these colours are conspicuous from a short distance but much less arresting frcm far away compared to reds and yellows. This may have the advantage of making the tail stand out at close quarters without attracting more distant predators.

(h) **Regeneration of the Tail**

Most lizards that autotomize the tail regenerate the shed portion. In many cases, the greater part is replaced quite rapidly. For instance, Lacerta dugesii takes only 12 weeks to regenerate 90% of the length of the portion originally lost (Bryant and Bellairs, 1976), and the tail of Coleonyx brevis may regenerate in 4—5 weeks (Dial, 1978). In constrast, a number of forms reproduce the tail very slowly, if at all. This is the case in snakes and some agamids; autotomizing amphisbaenians do not regenerate (Gans, 1978). Rate and extent of caudal regeneration differs not only among taxa but also among ontogenetic stages and sexes of the same species and even in the same individual in different situations.

Except, perhaps, in some forms that have completely lost the ability to regenerate, rate and extent of caudal replacement should be determined by the balance between present costs and benefits. Benefits of regeneration include the restitution of the tail as an antipredator device and all the other things listed as costs of autotomy in Section III.F. Costs of regeneration, on the other hand, involve the energetic and material expense of regenerating the tail and that of maintaining it subsequently. As we have seen, these costs may possibly be met by increased food intake or diversion of energy from other functions, such as somatic growth or reproductive effort (Section III.F). The optimal balance between these factors varies with circumstances and, in many cases, there is considerable compromise between them. Thus, young Ulta stansburiana may curtail general growth while the tail regenerates (Ballinger and Tinkle, 1979). In constrast, juveniles of Coleonyx variegatus, Eumeces skiltonianus and E. gilberti devote a smaller proportion of their disposable energy to regeneration than do adults. In female Coleonyx brevis, the rate of regeneration is generally less than in males, apparently because of the need to use energy during vitellogenesis; regeneration of the tail is show until vitellogeness has been completed (Dial and Fitzpatrick, 1983).

The rate of regeneration might correlate with the history of species. Long-lived, slowly maturing forms with short annual breeding seasons are thought to give precedence to reproduction and to regenerate slowly, whereas short-lived species that have several breeding episodes each year will reproduce the tail rapidly (Vitt *et al.*, 1977). Elgaria multicarinata conforms to the first type, and Coleonyx approaches the second (Vitt et al., 1977; Dial and Fitzpatrick, 1981). More data are needed to test this supposed regularity. Mabuya heathi, which has a very short annual breeding season, regenerates the tail very quickly (Vitt, 1981).

Slow regeneration rates appear to be most frequent in elongate, usually legless lepidosaurians that burrow or live cryptically close to the earth-vegetation interface or under objects. Thus, Anguis fragilis typically produces a regenerate less than 5 mm long in 14 weeks (Bryant and Bellairs, 1976) and the skink Ophomorus streeti may regenerate even more slowly (Rathor, 1971). Similarly, a captive Anniella pulchra developed a regenerate of only 4.1 mm in 11 months (Miller, 1944). Detailed information on regeneration rates is not available for other lizards with this general mode of life history. Only small regenerates found in museum material of many such species, suggesting that caudal replacement is very incomplete, or at least extremely slow. This appears to be true of the anguids, Ophisaurus and Ophiodes and such skinks as Acontias, Melanoseps, and Ophiomorus species, in addition to O. streeti, Parachalcides, Scelotes, and Sphenops. As previously noted, caudal regeneration is lacking in all of the autotomizing amphisbaenians and in the few colubrid snakes that are known to shed portions of the tail.

Benefits of regeneration may not be high in these squamates for few of the potential advantages of possesing a tail seem likely to be important to them. The tail may contribute to locomotion, but in limbless forms moving among herbage or loose soil, additional length beyond a critical level may not increase velocity. Also, a long tail may be less important as an antipredator device in these largely cryptic forms. Nevertheless, in at least some of them, it seems to allow breaking way from predators and may do so more than once, in spite of poor regenerative capacity. In Anguis, for instance, although many animals have damaged tails, these are rarely broken close to the vent. Often most of the organ is retained, suggesting that losses are often small and that the tail may be able to function in a number of predator encounters, even without intervening regrowth. Costs of regeneration may be considerable. Possibly they are important in dictating very restricted replacement in slowly-growing forms such as Anuis.

4

Parental Care

Prolonged care of eggs or offspring is a common feature of vertebrate reproduction. Parental care takes diverse forms, and presumably has evolved independently many times (*e.g.*, Gross and Shine, 1981). Although parental care is exhibited by only a minority of reptilian species, these include a diverse array of lizards, snakes and crocodilians. The present chapter (a) reviews the occurrence and nature of reptilian parental care and (b) poses and tests hypotheses on the selective forces responsible for the evolution of such behaviour. The only previous comprehensive reviews of reptilian parental care are those of Noble and Manson (1933) on squamates and of Cott (1971) on crocodilians.

I define "parental care" as any form of postovipositional parental behavior; that is, any action of the parent after oviposition or parturition, that increases the chances of survival of the offspring. This definition differs from the concept of "parental investment" (Trivers, 1972), which (a) includes investment prior to oviposition, and (b) is restricted to behaviour that decreases the probable survivourship or future fecundity of the parent. This latter condition is difficult to verify empirically and hard to apply to certain cases. For example, if the "costs" of a female guarding her eggs are independent of the number of eggs guarded, the addition of more eggs would not increase the mother's "parental investment"; however, her behaviour towards those additional eggs clearly would qualify as "parental care."

Parental care may be divided conveniently into two major categories: care of the eggs and care of the young. Care of the eggs is widespread but uncommon in reptiles, and generally consists of the parent remaining with the eggs for some time after oviposition. This behaviour usually is termed "egg-brooding" or "egg-guarding." However, although protection against predators is probably the major advantage conferred by the parent's presence (see below), actual defense against predators has been witnessed only rarely. Hence, the terms "nest-guarding" and "egg-guarding" should be applied only to species in which active nest defense has been observed. Cases in which the parent remains with the eggs, but without documentation of nest defense, may be termed "nest attendance" or egg-attendance" (Tryon, 1979). Finally, the term "egg-brooding" may be restrieted to species in which the female facilitates incubation by raising nest temperatures above ambient (Campbell and Quinn, 1975). I prefer this terminology to the proposal (Groves, 1982) that the term "egg-brooding" should be reserved for species that actively manipulate their eggs.

THE OCCURRENCE OF PARENTAL CARE

(a) Taxa Reported to Show Parental Care

1. *Testudines*

Parental care is rare or absent in turtles; apparently the only record is that of Hodsdon and Pearson (1943), who described maternal behaviour in the Bahamian emydid Pseudemys malonei. According to these authors (and see Oliver, 1955), the female turtles returned to their nest sites immediately before hatching and dug away hard-packed soil above the nest-chambers so that the hatchlings could emerge successfully. Fresh-water turtles have attracted considerable ecological study since the time of Hodsdon and Pearson's observations, and if maternal behaviour of the type described above is a general phenomenon, it is surprising that it has not been noted by other workers.

2. *Crocodilians*

In contrast to the turtles, parental care is common (possibly ubiquitous) among crocodilians. Descriptions of crocodilian nest defense may be found in the works of Pliny and Aristotle and more recent scholars have documented parental behaviour in virtually every crocodilian species studied in detail.

Although the frequency of nest attendance and active nest defense varies interspecifically, and even intraspecifically the broad outlines of parental care are similar in all crocodilians for which data are available. The pattern is quite different from that described in any squamate reptile. The female parent remains in the vicinity of the nest after laying, or returns to it at intervals. Potential predators on the eggs may or may not be attacked; species that the not egg predators are ignored. The female opens the nest at the time of hatching, and carries full-term eggs and newly-hatched young to the water (Pooley, 1977). The young remain in a group for days or weeks and are actively defended by the adults. If attacked, even older juveniles will give "distress cals" that will initiate aggressive behavior from adult crocodilians toward the attacker (*e.g.*, Hunt, 1974, 1975). Parental behaviours, such as nest-guarding and nest-opening, are probably universal among present-day crocodilians (Greer, 1971; Kushlan and Simon, 1981).

3. *Squamates*

Postoviposional parental behaviour by squamate reptiles can take many forms of which the simplest is concealment of eggs immediately after oviposmon. Oviparous squamates generally lay eggs in places with appropriate conditions of temperature and humidity (Muth, 1980) and away from the attention of egg predators (see Packard and Packard, this volume). Several species of lizards manipulate their eggs after oviposition, presumably to place them into suitable microhabitats. This behaviour has been reported in lacertids (Hilzheimer, *1910),* iguanids (Gordon, 1980), Smith et al., 1972; Tokarz and Jones, 1979; Jones and Guillette, 1982) scincids (Noble and Mason, 1933), agamids (Asana, 1941), and gekkonids (Nettman and Rykena, 1979; Phyllurus platurus, Shine, pers, obs.). Presumably such behavior is widespread among reptiles.

More common or at least longer-term, parental care is shown by many species. Among the squamates, at least 103 species have been reported to have one or both parents remain with the eggs after oviposition (Table 4.1). These species include a taxonomically diverse array of lizards and snakes. Unfortunately, some of the cases listed in Table II are poorly documented; at least 15 of the 103 examples depend upon single unservations of adults found near eggs. Undoubtedly, some of the cases do not indicate parental care, but may be coincidental or reflect incipient cophagy or discovery of a female with a recently oviposited clutch. The force of this objection is diminished by the consistency with which the adult is reported to be coiled around the eggs rather than merely close to them, but nonetheless these reports should be regarded with caution. I once found a small snake (Unechis gouldi) only a few centimeters from four eggs under a piece of tin; the obvious interence would have bee parental care, except that the snake species is vivparous, and the eggs were those of an agamid lizard.

Records of parental care for the Varanidae are particularly difficult to evaluate. No well-documented case is available, but at least three anecdotal reports are suggestive. For example Cogger (1967-60) notes that "on several occasions when adult goannas have been disturbed at termite mounds, newborn young have been found emerging. This suggests that the female may remain near the nest until the young are ready to hatch at which time she makes a new tunnel to release the young". In each case, the species involved was Varanus varius (H. Cogger, pers. comm.). On two separate occasions, a captive female Varanus mitchelli was found with the tail coiled around the eggs (G. Gow, pers. Comm.). Parental care is suggested by the behaviour of a female Varanus salvator that remained at the nest site for several days after oviposition

TABLE 4.1

Squamate Taxa in Which the Female Parent Has Been Reported to Remain With the Eggs After Oviposition

Family	*Species*	*Authority*
(*1*)	(*2*)	(*3*)
Lizards		
Anguidae	Diploglossus delasagra	Barbour and Ramsden, 1919; Greer, 1967
	D. bilobatus	Taylor, 1956; Fitch, 1970
	Gerrhonotus liocephalus	Greene and Dail, 1966; Tinkle and Gibbons 1977
	Elgaria multicarinatus	Langerwerf, 1981
	Opeisaurus apodus	Langerweri, 1981
	O. attenuatus	Fitch, 1970
	O. gracilis	Daniel, 1983
	O. harti	Smith, 1935
	O. ventralis	Noble and Mason, 1933 Vinegar, 1968; Mount, 1975
Iguanidae	Amblyrhynchus cristatus :	Carpenter, 1966; Eibl-Eibesfeldt, 1966; Trillmich, 1979
	Brachylophus fasciatus	Cogger, 1974; Gibbons and Watkins, 1982
	B. Vitiensis	Gibbons and Watkins, 1981
	Conolophus pallidus	Christian and Tracy, 1982
	C. subcristatus	Werner, 1982
	Cyclura carinata	Iverson, 1979
	C. stejnegeri (=C. cornuta)	Wiewandt, 1977, 1979

(1)	*(2)*	*(3)*
	C. cyclura	Carey, 1975
	C. nubila	Shaw, 1954
	Iguana iguana	Rand, 1968; Alvarez del Toro, 1972; Rand and Rand, 1976; Wiewandt, 1982
Scincidae	Emoia cyanura (?)	Fitch, 1970
	Eumeces anthracinus	Smith, 1946; Dowling, 1950; Collins, 1974
	E. brevilineatus	Werler, 1951
	E. callicephalus	Campbell and Simmons, 1961
	E. chinensis	Wang, 1966
	E. egregius	Mount, 1963
	E. elegans	Mell, 1929; Hikida, 1981
Scincidae	E. fasciatus	Ruthven, 1911; Dunn, 1920; Blanchard, 1922; Smith, 1882; Brimley, 1904; Ditmars, 1904; Brady, 1927; Burt, 1928; Corrington, 1929; Klots, 1930; Cagle, 1940; Fitch, 1954; Strecker 1908; Allard, 1909; Hurter 1911; McCauley, 1939, 1945; Bishop, 1926; Evans and Roecker, 1951; Minton, 1972; Cooper *et al.*, 1983
	E. inexpectatus	Duellman and Schwartz 1958
	E. laticeps	Smith, 1946; Martof, 1956; Vitt and Cooper, 1985
	E. latiscutatus	Hikida, 1981

(*1*)	(*2*)	(*3*)
	E. multivirgatus	Gehlbach, 1965; Van Devender and Van Devender, 1975
	E. obsoletus	Smith, 1946; Evans, 1959; Fitch, 1955, 1970; Hall, 1971
	E. okadae	Hikida, 1975; Hasegawa, 1984
	E. oshimensis	Toyama, 1975
	E. quadrilineatus	Mell, 1929
	E. "pekinesis" (=xanthi)	Mell, 1929
	E. septentrionalis	Breckenridge, 1943; Smith, 1946
	E. skiltonianus	Tanner, 1943, 1957; Smith, 1935
	E. stimsoni (?)	Taylor, 1935
	E. tetragrammus	Behler and King, 1979
	E. xanthi	Schmidt. 1927; Tavlor, 1935
	Neoseps reynoldsi	Telford, 1959
Teiidae	Tupinambis teguixin	Krieg, 1925
Varanidae	Varanus mitchelli (?)	G. Gow, pers. comm.
	V. salvator (?)	Biswas and Kar, 1981
	V. varius (?)	Cogger, 1967
Snakes		
Boidae	Aspidites melanocephalus	Boos, 1979; Barker, 1981; Charles et al., 1985
	Chondropython viridis	Kratzer, 1962; Mackay, 1973; Walsh, 1979; Switak, 1975
	Liasis albertisii	Ross, 1978
	L. amethystinus	Pope 1961; Boos, 1979; Charles et al., 1985
	L. boa	Barker, 1981

(1)	(2)	(3)
	L. childreni	Ross, 1973; Dunn, 1979a; Barker, 1981
	L. fuscus	Kinghorn, 1956; Ross, 1978; Boss, 19'9; Shine, pers. obs.
	L. mackloti	Ross and Larman, 1977; Barker, 1981
	Morelia bredli	Gow, 1981
	M. spilotes	Cogger and Holmes, 1960; Boos, 1979; Harlow and Grigg, 1983; Charles *et al.*, 1983
	Python curtus	Noble, 1935; Pope 1961; Vinegar *et al.*, 1970
	P. moluris	Lamarre-Picquot, 1835, 1842; Valenciennes, 1841; Forbes, 1881; Abercromby, 1983; Wall, 1921; Stemmler-Morath, 1956; Pope, 1961; Hutchinson et al., 1966; Yadav, 1967; Vinegar et al., 1970; Wagner, 1973, 1986; Van Mieroo and Barnard, 1976, 1978. Acharjyo and Misra, 1976; Corborn, 1975; Foekema, 1975
	P. regius	Logan, 1973; Pitmen, 1938; Walsh, 1979; Boos, 1979; Schivre, 1972; Van Mierop and Bessette, 1981

(1)	(2)	(3)
Boidae	P. reticulatus	Wray, 1862; Wall 1926; Leader, 1944; Pope, 1961; Taylor, 1965; Vinegar et. al., 1970; Honeggar, 1970; La Panouse and Pellier 1973; Hediger, 1968; Muller, 1970.
	P. sebae	Sclater, 1862 Fitz-Simons, 1930; Benedict, 1932; Pitman, 1938; Broadley, 1959; Dowling, 1960; Pope, 1961; Schutte, 1970; Vinegar *et al.*, 1970; Dunn, 1979b; Schmidt, 1973
	P. timoriensis	Murphy *et al.*, 1981; Barker, 1981
Colubridae	Amphiema mairii	J. Bredl, pers. comm.
	A. stolata	Wall, 1921; Mell, 1929
	Cemophora coccinca	Ditmars, 1942
	Elaphe climacophora	Fukada, 1965
	E. guttata	D. Kent, pers. comm.
	E. obsoleta	Oliver, 1955; Medsger, 1919
	E. quarivirgata	Fukada, 1965
	Farancia abacura	Meade, 1940; Goldstein, 1941; Riemer, 1957; Ashton and Ashton, 1981
	F. erytrogramma (?)	Ashton and Ashton, 1981
	Heterodon platyrhinos (?)	Hahn, 1969

(1)	(2)	(3)
	Lampropeltis trianguum	Babcock, 1929; Ditmars, 1942; Minton and Minton, 1973; G. Marsec, pers. comm.
	Lycodon aulicus	Smith, 1943
	L. striatus (?)	Wall, 1921
	Natrix natrix	Gallwey, 1932; Smith, 1973
	Opheodrys vernalis (?)	Blanchard, 1933
	Opisthotropis latouchi (?)	Pope. 1935
	Pituophis melanolecus (?)	Carl, 1944
Colubridae	Psammophylax rhombeatus	FitzSimons, 1962; Le Roux, 1964; Bourquin, 1970, Visser, 1971; Broadley, 1977
	P. tritaeniatus (?)	Sweeney, 1981; but see Broadley, 1977
	P. variabilis	Broadley, 1977
	Ptyas korros	Mell, 1929
	P. mucosus	Wall, 1907, 1921; Mell, 1929; Pope, 1935
	Rhabdophis subminiatus	Mell, 1929
	Xenochrophis piscator	Mell, 1929; Abercromby, 1913
Elapidae	Bungarus caernleus	Smith, 1943; Wall, 1921
	B. candidus	Mell, 1929
	B. ceylonicus	Green, 1905, Wall 1921
	B. fasciatus	Mell, 1929, Pope, 1935; Smith, 1943
	Callophis maculiceps (?)	Frith, 1977

(1)	*(2)*	*(3)*
	Demansia papuensis (?)	Parker, 1982
	Laticauda colubrina	Smedley, 1931; Neill, 1964; Taylor, 1965; but see Dunson, 1975
	Micrurus corallinus	Mole, 1924
	M. fulvius	Campbell, 1973
	Naja naja	Wall, 192.; Mell, 1929; Smith, 1937; Kopstein, 1938; Smith, 1943; Tweedie, 1957; Petzold, 1968; Campbell and Quinn, 1975
	N. melanoleuca	Try on, 1979
	Ophiophagus hannah	W. Smith, 1935; H Smith, 1936; Must ll, 1936; M. Smith. 1936; Mustill, 1936; M. Smith, 1943; Oliver, 1956; Leakey, 1969; Shaw and Shebbeare, 1931
	Pseudechis butleri (?)	Fitzgerald and Mengden, 1987
	Pseudonaja textilis (?)	Fleay, 1943; R. Wells, pers. comm; J. Edwards, pers. comm.
Leptotyphlopidae	Leptotyphlops dulcis	Hibbard, 1964
Typhlopidae	Rhintyphlops cacus	Bogert, 1949; but see Erasmus and Brand 1983.

(*1*)	(*2*)	(*3*)
	Rhamphotyphlops braminus (?)	Mell, 1929
Viperidae	Causus rhombeatus (?)	Sweeney, 1961; Broadley, 1983
	Calloselsma rhodostoma	Smith, 1915, 1943; Tweedie, 1957; Leakey 1969
	Lachesis muta	Mole, 1924
	Trimeresurus monticola	Leigh, 1910; Prope, 1935

A question mark indicates that the data are anecdotal and possibly unreliable.

(Biswas and Kar, 1981). A less reliable acconnt of possible parental care in varanids, probably a case of misinterpreted egg predation, has been reported by Barney (1936).

Parental care in squamates may be divided into four major categories, the first three of which deal with care of the eggs rather than the offspring:

(1) female buries eggs and defends nest site briefly against conspecific females;

(2) female coils around eggs, defends them against predators and warms them by shivering thermogenesis;

(3) female remains with eggs after oviposition and may defend them against predation or pathogens;

(4) female aids newly born or newly hatched young.

Nest-site defense against conspecific female is widespread in iguanine lizards (Table 4.1) but has not been reported in any other squamate group. In some cases, the nest site is defended

TABLE 4.2

Shivering Thermogenesis in Brooding Female pythons : Disagreements in the Literature

Species	*Shivering thermogenesis* Occurs	*Shivering thermogenesis* Does not occur
Aspidites melanocephalus	Boos, 1979; Charles *et al*., 1983	Murphy *et al*., 1981
Liasis amethystinus	Boos, 1979	Charles *et al*., 1983
L. mackloti	Charles *et al*., 1985	Ross and Larman, 1977
Morelia spilotes	Harlow and Grigg, 1983	Charles *et al.* 1935
Python curtus	Vinegar *et al*., 1970	Nbole, 1935
P. regius	Pitman, 1974; Logan, 1973	Van Mierop and Bessette, 1981
P. reticulatus	Hediger, 1968; Muller, 1970	Honegger, 1970; La Panouse and Pellier, 1973; Pitman, 1974; Vinegar *et al*., 1970
P. sebae	Benedict, 1932; Noble, 1935	Vinegar *et al*., 1970; Pitmann, 1974

even prior to oviposition (Wiewandt, 1982). The intensity of nest site defense varies considerably among species, and even among geographic areas within a species. Some of this variability may be attributed to the physical condition of the females: only animala in good condition can engage in vigorous nest defense (Wiewandt, 1982; references in Table 4.2).

Egg brooding (shivering thermogenesis) has been recorded only in pythons (Table 4.2). The female coils tightly around the cluch, so that the eggs are completely hidden. In at least some species, rhythmic muscular contractions of the mothers' body produce sufficient heat to maintain a relatively high and constant temperature in the egg mass. This phenomenon has been described in Aspidites melanocephalus, Chondropython viridis, Liasis amethystinus, L. mackloti, Morelia spilotes, Python curtus, P. reticulatus, and P. sebae (references in Table 4.2), and may be universal within the pythons.

Certainly the pythons known to show shivering thermogensis cover a broad taxonomic range, including small as well as large species, and exclusively topical as well as temperate-zone taxa. Although several authors (especially Vinegar *et al.*, 1970) have stressed that only some python species utilize shivering thermogenesis, the evidence for interspecific differences in such behaviour is weak. Commonly, studies of the same species by different workers are contradictory with respect to whether or not shivering thermogenesis is exhibited (Table 4.2). These disagreements are due to several factors, including—

(a) the type of evidence used (*e.g.* temperature measurement versus presence of muscular twitching),

(b) the temperature-dependence of shivering thermogenesis (if the room is warm, no shivering is observed),

(c) the difficulties in interpreting slight temperature differences between the egg mass and ambient, because of the great thermal inertia of these large reptiles (*e.g.*, Cogger and Holmes, 1960), and

(d) the low sample sizes (often only one or two animals). Some python species may not utilize metabolic heat production during brooding but convincing data are currently unavailable.

The maintenance of high temperatures during incubation undoubtedly quickens embryonic development (*e.g.*, Joshi, 1967) and also may increase embryonic survivourship. Python embryos develop normally only at high temperatures (approximately 30°C), and are much more stenothermic in this regard than embryos of most other squamates (Joshi, 1967; Vinegar, 1973; Harlow and Grigg, 1984). One hypothesis is that maternal brooding is an adaptation to this embryonic thermal sensitivity (Vinegar, 1973). However, distinguishing cause and effect is difficult in such a correlation. Equally plausibly, the embryonic sensitivity may have eveolved because of maternal brooding rater than vice versa. If the eggs are always kept warm by the brooding female, selection to maintain embryonic tolerance of low temperature would be weak or absent. An additional hypothesis is that maternal thermogenesis allow pythons to reproduce successfully under colder climatic conditions than would otherwise be possible (Harlow and Grigg, 1984).

The apparent restriction of shivering thermogenesis to phtons might be a function of their massively developed lateral musculature. Presumably, most repties would lack the physiological ability for sustained muscular contraction and hence, significant metabolic heat production. The substantial increase in metabolic rate of brooding female pythons (Hutchinson *et al* , 1966; Van Mierop and Barnard, 1976, 1978; Harlow and Grigg, 1984) means that maternal care in these snakes cofers a massive energetic cost. Curiously, this group has not followed the boines in evolving viviparity, a strategy allowing egg temperatures to be raised by maternal behavioural (rather than physiological) thermoregulation. One reason may be that an egg-retaining female python could not subdue prey by constriction, because of the risk of damaging the eggs. An egg-brooding female would not face this restriction.

The most widespread form of squamate parental care consists of the female remaining with the eggs after oviposition. Her presence presumably serves some function, which has rarely been demonstrated. Functions documented in some species include defense against predators, and regulation of egg temperature or moisture levels (discussed further in Section III.A). Many of the reported cases of egg-attending (Table II) are likely to be invalid inferences, because they are based on single observations of adults with eggs. Nonetheless, the consistency with which this behaviour has been reported in particular species cleary indicates that egg-attendance is widespread in squamates (Table II).

Parental care of offspring rather than of egg is remarkably rare in reptiles, perhaps because the mother and neonates are rarely in close proximity. One exception involves egg-attending (or guarding or brooding) species in which the female often is present at hatching. In at least one such species (Eumeces obsoletus), a female was observed to help the young hatch, and later "groom" the offspring (Evans, 1959). Even more remarkable is the observation (unfortunately uncorroborated) of newly-hatched Python molurus returning to their empty eggshells at night, with the mother than coiling around them and heating them by shivering thermogensis (Doflein, 1932; cited in La Panoese and Pellier, 1973). An aborigine from Mudginberri Station (Alligator Rivers region, tropical Australia) suggested that femal water pythons (Liasis fuscus) "took the babies down to the water and taught them how to swin" (R. Lambeck, pers, comn.).

If the scarcity of posthatching parental care in squamates is due to the lack of physical association between mother and young in most oviparous forms, this problem is overcome in viviparous squamates, which comprise approximately one-fifth of all species of lizards and snakes (shine, 1985). Observations of viviparous females at parturition confirm that mothers in some squamate species may stimulate the young to emerge from their membranous sacs by pushing with the snout, or may lizards (Cowles, 1944; Miller, 1954), scincid lizards (Niekisch, 1975; Reboucas-Spieker and Vanzolini, 1978; Shine, pers. obs. for Sphenomorphis quoyii), and boid snakes (Groves, 1981).

However, mothers apparently do not open the fetal membrances at birth in viviparous iguanid, diploglossid; or lacertid lizards (Reboucas-Spieker and Vanzolini, 1978). In some colubrid and boid snakes, the mother ingests dead young and birth debris after the living young have emerged from their membranes (Neill and Allen, 1962; Rieppel, 1970; Groves, 1981; Holmstrom and Behler, 1981). A similar phenomenon occurs in oviparous reptiles, the mothers of which ingest spoiled eggs *e.g.*, crocodilians: Kushlan and Simon, 1981; squamates:

Groves, 1982). Such behaviour has been interpreted as parental care, because—

(a) it may prevent fungal infection spreading to other eggs, and

(b) it may prevent odors attracting predators (Groves, 1982). Alternatively, rotting eggs may have a different odor than healthy eggs, and the females may mistake them for food items (R.B. Huey, pers. comn.).

One further form of squamate parental care, which has received great publicity, is the alleged ability of female snakes to swallow their young for protection when danger threatens. The first published reference to this behaviour comes from the Eygptians at about 2500 B.C. (Speck, 1923), whereas the first reference in the English language is in Spenser's "The Faerie Queene" (1950). The story has been applied to many species, especially to viviparous forms, and often to species that have been observed in detail by other workers who have failed to report the same phenomenon (*e.g.*, Bothrops neuwiidi: Schupp, 1913; Melgarejo, 1977). The extensive literature on the subject has been summarized and evaluated multiple times (Noble, 1921; Ortenburger, 1930; Schmidt, 1929; Angel, 1950; Rose, 1962; Klauber, 1972). The consensus is that no satisfactory evidence of the phenomenon has ever been produced. The stories seem unlikely to be true because—

(a) the behaviour has never been reported in captivity or from field observations of scientisis;

(b) undigested young have never been discovered in stomachs of snakes dissected for dietary studies;

(c) offspring are probably unable to survive for significant periods inside the stomach of the mother (*e..g.*, Klauber, 1972). Nevertheless, the occurrence of mouth-brooding and transport of young in cichlid fishes and crocodilians as well as the recent discovery of gastric brooding in an Australian frog (Corben *et al.*, 1973), suggest that gastric protection of young in snakes is not impossible. Nonetheless, reliable data are totally lacking.

Other popular tales concerning reptilian parental care include the mother snakes nourishing the young inside her stomach (Rivers, 1874; Stanley, 1897; Burroughs, 1985, the mother rattlesnake crooning (by rattling mildly) to her newborn young (Crites, 1952), and the mother rattlesnake caring for her young until they are well-grown, and then finding suitable home sites for them (Meek, 1946). Charming as these stories are, their validity seems doubtful.

Although the present review is concerned only with living reptiles, parental care also may have been shown in a variety fo groups that are now extinct (*e.g.*, Coombs, 1982; Horner, 1984).

Taxonomic Biases

1. *General*

The incidence of reptilian parental care is not distributed randomly among different taxonomic groups. This taxonomic bias is evident at the levels of the order, suborder, family, and genus.

2. *Comparisons Among Orders*

The three orders of living reptilies are only distantly related to each other, having been evolving as separate lineages for almost 300 million years (*e.g.*, Romer, 1966). This long divergence has resulted in great differences among the living orders in morphology and, as shown by the present review in the frequency and form of parental care. Parental behaviour appears universal in crocodilians (and in their relatives, the birds) but seems to be lacking in testudines. Parental care is seen in many squamate taxa, but has been described in only about 2% of all oviparous squamate species (7% of oviparous genera). Of course, many species have yet to be examined in this regard.

3. *Comparisons between Suborders*

Among the squamate suborders, parental care is unknown in amphisbaenians (possibly because of a paucity of field data), rare in lizards (recorded in 41 of approximately 3000 oviparous species, or 1.3%), and more common in snakes (47 of 1700 species, or 2.8%). If each species is treated as an independent data point (see Ridley, 1983, for a critique of this assumption), the proportion of oviparous species recorded to show parental care is significantly lower in lizards than in snakes (2×2 table, 1 d.f., $\chi^2=18.5$, $p<0.001$). However, this may simply reflect a greater tendency for herpetologists to keep and observe snakes rather than lizards.

4. *Comparison Among Families*

Among the squamates, the percentage of oviparous species reported to show clutch attendance ranges from zero in many families (Agamidae, Chamaeleonide, Cordylidae, Dibamidae, Gekkonidae, Helodermatidae, Lacertidae, and Pygopodidate) up to virtually 100% in the oviparous Boidae. The variation in percentages among families is too great to be attributed to chance alone (combining families to ensure adequate sample sizes; 7×2 table, 6 d.f., $\chi^2(=103.2$, $p<0.0001)$.

5. *Comparisons Among Genera*

The 103 squamate species showing parental care are distributed among 13 genera of lizards and 33 of snakes. If parental care was randomly distributed with respect to taxonomy, the proportion of genera containg more than one nest-attending species should be low. In reality, many instances of squamate parental care are found in congeneric species; for example, half of all the examples in lizards (20 of 41 cases) belong to the single scincid genus Eumeces. Parental care is probably ubiquitous in this genus, as well as in others such as Ophisaurs and Python. Such data ars clearly sufficient to reject the null hypothesis of taxonomically random occurrence of parental care. Instead, parental care, although rare overall, is very common within certain taxonomic groups.

A strong taxanomic bias is evident not only in the frequency of parental care, but also in its form. Simple clutch attendance is usually the only form of parental care, but different behaviour are exhibited by iguanine lizards (eggs buried, and nests defended against conspecifics), boid snakes (eggs warmed by shivering thermogenesis), and crocodilians (complex biparental behaviour before and after hatching.)

Number of Independent Evolutionary Origins

The tendency for parental care to be widespread within a single phylogenetic lineage suggests by parsimony that in such cases it has evolved only once, early in the group's history, and has been retained during subsequent speciation (Ridley, 1983). The extreme case of the alternative hypothesis is that all species showing parental care have independently evolved this behaviour. Although the strong taxonomic bias in parental care argues against this latter hypothesis, it is difficult to dismiss in specific cases, particularly should only a few distantly-related species show the trait. This could result from either .

(a) lack of data on other related species which show parental care although it has not been recorded

(b) an evolutionary loss of parental behaviour in these related forms,

(c) separate evolutionary origins of parental care in each species showing this trata.

A typical example of such ambiguity occurs in ihe colubrid snake genus Elamphe, two species of which show material care for several days after laying (E. climacopiera, E. quadrivirgata), whereas a close relative (E. conspicillata) does not Fukada, 1256). Nest attendance also occurs in a distantly related North America congener (Table II).

In the following analysis, I have adopted a conservative approach by considering any cases of parental care in the same subfamily as representing a single evolutionary origin, except in the cases of the speciose snake subfamilies Colubrinae and Elapinae. In these groups, available phylogenetic reconstructions suggest multiple origins of parental care, based on low probability of relatedness among the species involved.

1. *Colubrine*

Omitting the doubtful record for Opheodrys (Table 4.1), all of North American colubrids reported to show parental care (Elaphe, Cemophera, Lampropeltis, and Pituophis) probably are derived from a single adaptive radiation from Old World Elaphe (Underwood, 1967). Hence, parental care may parsimoniouly be regarded as having a single evolutionary origin in these snakes. However, the other colubrine texa showing parental care (Farancia, Lycodon, Ptyas) are only distantly related to each other and to Elaphe (Underwood, 1967), suggesting that egg-attending has evolved independently several times in colubrids.

2. *Natricinae*

The natricine snakes (family Colubridae) pose a similar problem. Parental care has been reported in five genera (six species), which appear to be relatively closely related (Malnate, 1960; Underwood, 1967). Hence, these cases will be treated as a single evolutionary origin of parental care. This decision may be overly conservative because parental care has been reported from only one or two species within each genus; if it were a primitive character for the entire natricine radiation, one might expect to see more occurrences of the trait. The high probability that parental behaviour has been overlooked in many taxa respresent a problem with enumerating the number of independent evolutionary origins (as in Ridley, 1983) by application of a specific method. Hence it seems most parsimonious to court this as a single example of the evolution of parental care.

This analysis suggests that parental care has evolved independently remarkably few times in extant reptiles ; the number may be as low as 21 (Table 4.2). Of these origins, only

TABLE 4.2

The Incidence of Parental Care in Reptiles at the Level of Family, Subfamily, Genus and Species With Estimatcs of the Probable Number of Independent Evolutionary Origins of Parental Care

Family	*Subfamily*	*Number of taxa in which parental core recorded*		*Probable number of evolutionary origins*
		Genera	*Species*	
(*1*)	(*2*)	(*3*)	(*4*)	(*5*)
Squamata				
Anguidae	Aniguinae	1	5	1 (Aniguinae, Diploglossinae and Gerrhonotinae together)
	Diploglossinae	1	2	
	Gerrhonotinae	1	2	
Iguanidae	Iguaninae	4	10	1
Scincidae	Lygosominae	1	1	doubtful record
	Scincinae	2	21	1
Teriidae	Teiinae	1	1	1
Varanidae	Varaninae	1	3	1

(1)	(2)	(3)	(4)	(5)
Boidae	Pythoninae	5	16	1
Colubridae	Bioginae	1	3	1
	Colubrinae	8	13	4
	Natricinae	5	6	>1
	Xenodontinae	1	1	doubtful record
Elapidae	Elapinae	6	11	>4
	Laticaudinae	1	1	1
Leptotyphlopidae	Leptotyphlopinae	1	1	1
Typhlopidae	Typhlepinae	1	2	1
Viperidae	Crotalinae	4	4	1
Testudines				
Emydidae	Emydinae	1	1	doubtful record
Crocodylia				
Crocodylidae	Alligatorinae	4	6 }	
	Crocodylinae	2	10 }	1
Gavialidae	Gavialinae	1	1 }	

5 have occurred among the lizards compared to 15 among the snakes. The ratio of evolutionary origins to the number of present-day oviparous taxa in the suborder thus is much higher (5 in 3000 versus 15 in 1700) in the latter group.

3. *Elapinae*

The six elapine genera recorded to show parental care are only distantly related to each other and hence are likely to represent independent evolutionary origins of the trait. The records for Australasian species are questionable and may be omitted, but this still leaves four probable origins within the elapines.

Male, Female, and Biparental Care

Parental care in reptiles is performed almost always by the female (Table 4.1), sometimes in conjunction with the male. Even in crocodilians, the only group in which male parental care has been reported commonly, the male plays a much less significant role than does the female in nest-guarding or transport of young. However, males may be important in subsequent defense of hatchings from predators.

Records of male parental care in squamate reptiles are often anecdotal or poorly documented. Egg-attendance by a male Diploglossus delasagra were reported by Barbour and Ramsden (1919), but they did not note how the parent's gender was determined. Egg-attendence has also been reported for a female of this species (Taylor, 1956), and for several related forms ; Barbour and Ramsden's (1919) animal may have incorrectly sexed.

Biparental care has been reported in four species of snakes : the Asian elapids Bungarus ceylonicus, Ophiophagus hannah and Naja naja, and the north American colubrid Elaphe obsoleta. The record for Bungarus (Green, 1905) is suspect; two adults of unspecified sex were observed curled up in a hollow with eggs and hatchlings. Other records of parental care in this genus involve only the female remaining with the eggs. The record for Ophiophagus rests upon Wall's statement that 'during the incubation period there is no doubt at any rate in some instances, that the male is in close attendance on his mate" (1924:193). This statement is repeated by Tweedie (1957); but detailed studies in capacity (Oliver, 1956), and collection of adults from 15 nests in the wild (Leakey, 1969) provide strong evidence that only the female remains with the eggs.

The record of biparental care in Naja naja is based upon a captive pair; the female usually stayed witd the eggs but "the male, which was always near the mound, (took) the place of the female during the three hours that she left the eggs to drink and feed. Both reptiles were particularly vicious during the incubations period" (Jennison, 1931; repeated by Smith, 1937; Tweedie, 1957). Deraniyagala (1955) reported that both male and female of a captive pair of Naja naja attended the nest, but that only the female resumed this behaviour after disturbance. A recent detailed study of reproduction in captive Naja naja did not report any assistance by the male to the egg-attending female (Campbell and Quinn, 1975). A statement by Jennison (1931) that the female Naja comes obediently to the whistle of the keeper, also seems inconsistent with observations of captive snakes. Hence, I am skeptical of the reliability of the reported biparental care.

The remaining case of parental care is in Elaphe obsoleta at Jacob's Creek, Pennsylvania (Medsger, 1919). The eggs were buried in an old sawdust pile, and the two "parent" snakes remained on, in, or near the pile for at least 3 weeks. At one point, "Mr. Medsger secured a fork, and at a depth of twelve inches dug up 44 eggs of the pilot snake. The male snake was coiled around the eggs" (Medsger, 1919: 28). Mr. Medsger's tallk on this subject was illustrated by photographs, so there seems no reason to doubt his observations. However, his means of sexing the adult snakes may have been in error. The reported clutch size is very high; usual clutches for this species are 7—12 eggs (Fitch, 1970), and communal oviposition has been recorded (Lynch, 1966). Hence, the clutch was probably a communal one, and the two snakes observed by Medsger may both have been females. Further observations on reproduction of this relatively common snake would be of interest.

HYPOTHESES ON THE EVOLUTION OF PARENTAL CARE

(a) Cost-Benefit Models

1. *General*

The selective forces responsible for the evolution of reptilian parental care may be investigated by an analysis of the possible "costs" and "benefits" of such behaviour. Thus, the fitness (lifetime reproductive success) of a hypothetical reptile showing parental care is compared to that of an otherwise identical animal not showing care, and the ecological conditions or species characteristics conferring a higher fitness to the former individal are considered.

As described in theoretical models for the evolution of viviparity (Shine, 1985), the primary "costs" of parental care are likely to the decrements in the food intake, survivorship, or subsequent fecundity of the parent. Remaining with the eggs may increase vulnerability of the parent to predators, depending on the site chosen and on whether the eggs have an odor detectable to predators. Remaining with the eggs certainly is likely to reduce the feeding opportunities of the parents. This in turn may prevent the accumulation of energy required for production of a second clutch. In males, restriction to the nest site and guarding of young may reduce opportunities for further copulation.

The "benefit" of parental care presumably is to increase offspring fitness, either by increasing survivourship of embryos and hatchlings or by accelerating embryogensis so that hatching occurs at a favourable time. The increase in offspring survivorship could result from parental protection against many potential sources of mortality (*e.g.*, predation, dessication, flooding, and fungal attack). These factors are considered in mare detail below.

2. *Benefits of Parental Care*

Table 4.3 summarizes published hypotheses and available evidence on the functions (benefits) of repitian parental care. Several of these presumed benefits may be valid, and indeed any single case of parental care may well have evolved for a variety of reasons. The common observations that females are particularly aggressive during the defense of nests (Table 4.3) suggests that deterring predators is a major function of parental care. This phenomenon is particularly striking in cases in which the individual reptile, or the species to which it belongs, is generally nonaggressive (*e.g.*, Bungarus, Laticauda, Naja, Python regius; Table 4.3). However, many nest-attending females are not aggressive: females of several pythons and crocodilians, as well as those of Ophisaurus apodus. O. ventralis, Naja melanoleuca, and Ophiophagus hannah, have all been reported to be relatively quiescent (in some cases, almost comatose) while attending eggs (Barker, 1981; Cott, 1971; Langerwerf, 1981; Noble and Mason, 1933; Tryon, 1979; Oliver, 1956). Nonetheless, it cannot be claimed unambiguously that any of these changes in maternal behaviour (*e.g.*, increased or decreased levels of aggression) are adaptations to parental care. The behaviour of the plarents could alternativsly be interpreted as a direct response to a changed thermal environment endocrine modifications associated with oviposition, the physiological stress of oviposition, or some other factor. This does not mean that the behaviour of the parent fails to protect the eggs, but merely that it may be a direct consequence of the physiology of the parent rather than an adaption per se.

Although the studies cited in Table 4.3 show that females may actively defend their eggs against potential predators, the only work to compare predators rates on defended versus undefensed clutches is that of Metzen (1977), who studied 110 nests of Alligator in the Okefenokee swamp. Most nests (96) were not defended by females, and almost all of these nests (92, or 96%) were destroyed by predators, especially bears. However, in 14 actively-defended nests, predation was rare (four nests destroyed, or 26%).

TABLE 4.3

"Benefits" of Parental Care: Hypotheses and Evidence

Benefit	*Species*	*Evidence*	*Authority*
(1)	*(2)*	*(3)*	*(4)*
1. Deter potential egg predators	Elgaria multicarinatns	—Aggressive to humans	Langerwerf, 1981
	Eumeces egregius	—Aggressive to potential predators	Mount, 1963
	E. fasciatus	—Aggressive to potential predators	Noble and Mason, 1933
	E. laticeps	—Aggressive to potential predators	Noble and Mason, 1933
	Tupinambis leguixin	—Aggressive to humans	Kreig, 1925
	Brachylophus fasciatus	—Aggressive to humans	Gibbons and Walkings, 1982
	Liasis albertisii	—Aggressive to humans	Ross and Larman, 1977
	Morelia spilotes	—Aggressive to humans	Charles *et al.*, 1983
	Python regius	—Aggressive to humans	Van Mierop and Bessette, 1981
	Natrix natrix	—Aggressive to humans	Gallwey, 1932
	Bungarus candidus	—Aggressive to humans	Mell, 1929

(1)	(2)	(3)	(4)
	Laticauda colubrina	—Aggressive to humans	Smedley, 1931
	Naja naja	—Aggressive to humans	Jennison, 1931; Campbell and Quinn, 1975
	Alligator mississippiensis	—Aggressive to potential predators, humans	Mellhenny, 1934; Dietz and Jackson, 1979
		—Presence reduces egg predation	Metzen 1977
	Crocodylus acutus	—Aggressive to potential predators	Dugan *et al*., 1981
	C. niloticus	—Aggressive to potential predators	Modha, 1962;Cott, 1971
	C. palustris	—Aggressive to potential predators, humans	Dharmarkumarsinhji, 1947; Anon, 1982
	C. porosus	—Aggressive to potential predators, humans	Cott, 1971; Biswas, 1977; Acharjyo and Mishra, 1981; Bustard and Choudhury, 1980; Bustard and Kar, 1981; Bustard and Maharana, 1982
	Gavialis gangeticus	—Aggressive to humans	Bustard and Singh, 1981

(1)	*(2)*	*(3)*	*(4)*
2. Prevent disturbance to nest by othe ovipositing females	All iguanine species listed to Table	—Attacks conspecific if they attempt to excavate nest	References in Table
	Alligator mississippiensis	—Stops turtles digging into nest	Dietz and Jackson, 1979
3. Provide additional heat to speed embryonic development	Probably all pythons	—Produces heat by "shivering"	References in Tables
	Eumeces fasciatus	—Basks, returns to eggs	Noble and Mason, 1933; but see Fitch, 1954
	Python regius	—Basks, returns to eggs	Van Mierop and Bessette, 1981
	Morelia spilotes	—Basks, returns to eggs	Harlow and Grigg, 1984; Charles *et al.*, 1995
	Elaphe obsolecta	Parents bask, return to eggs	Medsger, 1919
4. Regulate egg temperatures	Ophisaurus ventralis	—Adjusts depth to which eggs are buried	Vinegar, 1968
		—Adjusts depth to which eggs are buried	Fitch, 1954
	Crocodylus porosus	—Lies no nest on sunny day	Deraniyagala, 1939

(1)	(2)	(3)	(4)
5. Prevent eggs drowning	Eumeces fasciatus	—Moves eggs from flooded burrows	Fitch, 1951
	Opisthotropis latouchi	Nest site is vulnerable to flooding	Fitch, 1970
6. Provide moisture for eggs	Eumeces fasciatus	—Urinates on nest	Fitch, 1954
	Crocodylus palustris	—Urinates on rest	Whitaker and Whitaker, 1977b
	C. niloticus	Guarded nest is damp	Pooley, 1976
	C. porosus	Guarded nest is damp	Webb *et al* , 1977
7. Eat or remove spoilt eggs, to deter fungal attack or discovery by predators	Eumeces fascialus	—Eats spoilt eggs	McCauley, 1945; Gravers, 1982
	Gerrhonotus liocephalus	—Eats eggs	Tinkle and Gibbons, 1977
	Ophisaurus attenuatus	—Eats eggs	Fitch, 1970
8. Open nest to enable hatehlings to emerge	Pseudemys malonei	—Opens nest at hatching time	Hodsdon and Pearson, 1973
	Varanus varius	—In vicinity of nest a, hatching time	Cogger, 1967
	Probably all crocodilians	—Opens nest	References in Table

(1)	(2)	(3)	(4)
9. Keep eggs covered and hidden	Iguana iguana	—Returns to nest daily, makes repairs	Alvarez del Toro, 1972
	Ophiophagus hanah	—Recovers exposed eggs	Oliver, 1956
10. Deter predators of hatchlings	Many crocodilians	Adults respond to alarm call	References in Table
11. Turn eggs to prevent growth of mould	Eumeces fasciatus	—Turns eggs, unattended eggs rot	Fitch, 1954

An alternative threat to eggs may come from their being dug up by other nesting females. This has been reported to occur in sea turtles (Carr, 1967) and presumably is the selective force favoring nest-site defense in iguanine lizards (Table 4.3). Suitable nesting areas are rare in some habitats occupied by these lizards, so that disturbance of an earlier clutch by an oviposing female is frequent. In these species, nests are defended only until the end of the egg-laying season A similar situation occurs in Alligator. Turtles attempting to nest on their nest mounds may disturb their eggs and nest-attending female Alligator attack them under such circumstances (Dietz and Jackson, 1979).

The only other major function of reptilian parental care is likely to be its effect on the thermal environment of the eggs (Packard and Packard, this volume). Egg temperatures may be modified by—

(a) the parent adjusting the depth at which the eggs are laid beneath the soil surface (Fitch, 1954; Vinegar, 1968),

(b) the parent basking and then returning to the eggs (Medsger, 1919; Noble and Mason, 1933, but see Fitch, 1954; also Harlow and Grigg, 1984, for Morelia);

(c) the parent producing heat metabolically to warm the eggs. The last of these alternatives is employed commonly by boid snakes; shivering thermogensis has been repoated in at least nine species (see above).

Apart from benefits accuring from defense against predators or disturbance, and from thermoregulation, other functions of parental care have only been hypothesize rather than demonstrated (Table 4.3). Most seem intuitively reasonable, but are likely to be important only in a restricted number of cases. Other hypotheses seem invalid; for example, Coborn's (1975) explanations for "shivering" in breeding pythons are that this behaviour—

(1) promotes circulation of air around the eggs,

TABLE 4.4

"Costs" of Parental Care: Hypotheses ane Evidence

Cost	*Species*	*Evidence*	*Authority*
(1)	*(2)*	*(3)*	*(4)*
1. Reduction of food intake	Eumeces egregius	—Remains with eggs constantly	Mount, 1963
	E fasciatus	—Remains with eggs constantly	Fitch, 1954
	E. obsoletus	—Remains with eggs constantly	Fitch, 1955
	E. okadae	—Remains with eggs constantly	Hasegawa, 1984
	E. septentrionalis	—Remains with eggs constantly	Breckenridge, 1943
	Ophisaurus attenuatus	Empty stomachs in egg-attending—	Fitch, 1970
	Ophiophagus hannah	Empty stomachs in nest-attending—	Leakey, 1969
	Morelia spilotes	Captive—refuse food	Shine, 1980
	Python molurus	Captive—refuse food	Taylor, 1965; Van Mierop and Barnard, 1978
	P. reticulatus	Captive—refuse food	La Panouse and Pellier, 1973

(1)	(2)	(3)	(4)
	P. regius	Captive—refuse food	Boos, 1979; Ross, 1978; Van Mierop and Bessette, 1981
	P. sebae	Captive—refuse food	FitzSimons, 1930
	Aspidites melanocephalus	Captive—refuse food	Charles *et al*., 1985
	Liasis amethystinus	Captive—refuse food	Charles *et al*., 1985
	Crocodylus niloticus	Empty stomachs in nest-attending—	Pitman, 1930; Pooley, 1977; Cott, 1971
	C. porosus	Empty stomaches in nest-attending—	Deraniyagala, 1939; Choudhury and Bustard 1979; Bustard and Maharana, 1982
2. Reduction of subsequent fecundity	Python reticulatus	—Breeds less often than annually	Fitch 1970
	Morelia spilotes	—Breeds less often than annually	Harlow and Grigg, 1984
	Eumeces obsoletus	—Breeds less often than annually	Fitch, 1970
	Alligator messissippiensis	—Breeds less often than annually	Neil, 1971

(2) increases maternal circulation,

(3) is a maternal reaction to the presence of eggs between the coils. Undoubtedly, the posthatching parental care of crocodilians protects offspring from predatory attacks (*e.g.*, Hunt, 1974, 1975). The parent—offspring bond in crocodilians also may enable the parents to "teach" specific behaviour patterns to their offspring, but no data are available to test this hypothesis.

3. *Costs of parental Care*

By staying with the eggs after oviposition, the adult reptile may forego opportunities for feeding. This in turn may reduce energy intake to the point that production of a subsequent clutch of eggs is delayed. Alternatively, a lower growth rate may depress subsequent fecundity. In either case, the reproductive success of the animal is reduced. A more direct "cost" of parental care may be a decrement in survivourship, due to increased exposure to predators or other hazards. This might be likely if parental care occurs in a habitat different from that normally occupied by adults, or if predators can detect eggs more easily than they detect adults. The high metabolic expenditure of brooding pythons, which raise metabolic rate to warm the eggs, is another clear "cost" of parental care.

A reduced food intak, often a total cessation of feeding, has been documented in several egg-attending species both in the field and in captivity (Table 4.4). Observations of captive specimens indicate that inanition during this period is due to a specific disinclination to feed rather than a lack of encounters with suitable prey. Apart from the cases cited in Table 4.4, many other reports of nest attendance refer to the constant presence of the female with the eggs. Presumably, food intake is reduced in most or all of these species. Egg-attending females often are reported as appearing emaciated by the time that parental care terminates (*e.g.*, Mell, 1929). Interspecific differences may occur in fidelity to the nest; females of some, species regularly leave the eggs and forage, whereas those of other species never do so (Mell, 1929). However, specific data on this point are lacking. Females remaining with eggs potentially could obtain considerable quantities of food by consuming potential egg-predators attracted to the nest (Tinkle and Gibbons, 1977). Again, data are lacking.

A reduced food intake probably is common, but not universal, in egg-attending females (*e.g.*, Ross, 1978; Boos, 1979). The same situation occurs with other reproductive modes in reptiles; for example, food intake is reduced during gestation in many, but not all, viviparous species (Shine, 1980).

Without direct experimental manipulation (such as removal of clutches), longer-term costs of parental care are difficult to assess. The hypothesized delay in subsequent reproduction is supported by long intervals (greater than 1 year) among successve clutches in at least four egg-attending species of reptiles (Table 4.4). Indeed, a strong correlation has been demonstrated both for reptiles and amphibians, among low frequencies of reproduction and "accessory costs", such as potential care, viviparity, and long breeding migrations (Bull and Shine, 1979). Low reproductive frequencies are common in taxa that do have such "costs." However, this correlation may not reflect the fact that parental care confers a high cost per se, but rather that the cost is relatively independent of fecundity. For example, the reduction in food intake of a brooding female python probably is independent of the number of eggs she is brooding. Under these circumstances, calculations suggest that infrequent reproduction, in each case with the production of a large clutch, will be the optimal life-history strategy (Bull and Shine, 1979). A recent study of Eumeces okadae (Hasegawa, 1984) confirmed that postovipositional weight loss of egg-attending females was not correlated with their fecundity.

Although the egg-attending species with low frequencies of reproduction (Table 4.4) provide circumstantial evidence in support of "costs" of parental care, the opposite extreme also occurs; many egg-attending species, especially those from tropical areas, have been reported or suggested to produce more than a single clutch per year. These include Gerrhonotus liocephalus, Elgaria multicarinatus, Tupinambis teguixin, Rhabdophis subminiatus, Xenochrophis piscator, Cemophora coccinea, Lycodon aulicus, Pituophis melanoleucus, Ptyas korros, and P. mucosus (data and references in Fitch, 1970). However, these records may be based on females that have lost or abandoned their first clutch (Fitch, 1970).

4. *Conclusion*

Overall, the available data are consistent with hypothesized "benefits" and "costs" of parental care, but are generally circumstantial in nature. The only direct demonstration of the effectiveness of parental care in increasing egg survivourship comes from a study of Alligator (Metzen, 1977). Good quantitative estimates of costs of parental care are lacking, These inadequacies in the available data are regrettable, because studies to measure the relevant variables are entirely feasible. Experimental manipulations could provide measurements both of the benefits of parental care (what is the survivorship of attended and unattended eggs?) and its associated costs (what are the differences in food intake, growth rate, and in the time of subsequent reproduction, between those females that are attending eggs and females of which the clutches of eggs have been removed?). Sample manipulations might involve:

(a) removal of par or all of the clutch of an egg-attending female.

(b) removal of the female from the clutch. Studies on parental care in salamanders (*e.g.*, Tilley 1973; Huheey and Brandon, 1975; Forester, 1979a.

(c) frogs (*e.g.*, Taiger et al., 1984), and insects (*e.g.*, Tallamy and Denno, 1982) have progressed much further in this respect, than have those on reptiles.

Intermediate Stages in the Evolution of Parental Care

The vast majority of reptilian species do not show any form of parental care, but a small minority have been reported to remain with the eggs after oviposition. In most cases, the parent is believed to remain with the clutch for the entire duration of incubation, and only to leave after the eggs have hatched. This situation raises the question of intermediate stages in the evolution of parental care. Presumably, a species without egg-attending behavior does not give rise by a single mutation to individuals that remain with eggs throughout development. Instead, the two more likely scenarios are as follows.

1. The female remains with the eggs for only a short time after oviposition and then leaves. This could occur if the female was "exhausted" by oviposition. Given the chance association between parent and eggs, selection could then favor "parental" behaviour (*e.g.*, defense against predators) and ultimately prolonged nest-attendance (Noble, 1931). The feasibility of short-term egg-attendance as an intermediate stage towards prolonged parental care is supported by records of short-term egg attendance in several squamate species. Unfortunately, many case cases of this phenomenon have been reported in captive animals, even in species known to attend eggs throughout incubation in the wild state (*e.g.*, FitzSimons, 1930; Boos, 1979). This suggests that nest desertion in captivity may be an artifact arising from human disturbance. Variation in the duration of nest defense in iguanine lizards may reflect maternal condition; emaciated females defend nests only briefly, then leave to feed (Wiewandt, 1982).

2. The female leaves the eggs after oviposition, but returns to them regularly because they are located under a favored refuge. This intermittent prorinity of parent and eggs provides the opportunity for selection to favor "parental" behaviour, and perhaps leads eventually to more persistent nest attendance. We do not know how commonly this phenomenon occurs in nature, although Mell (1929) suggests that there is a continuum among species from intermittent to constant egg attenders.

Both pathways may well have been involved in the evolution of parental care, as is suggested by the apparent occurrence of putative "intermediate stages" in modern reptiles. However, such strategies in modern reptiles may be secondary modifications of prolonged nert attendance, and intermediates in the evolution of this trait.

Factors Increasing the Benefits of Parental Care

1. *Prediction*

If the primary benefit of parental care is an increase in egg survivorship, then parental care should evolve most readily in situations in which it has a large effect on survival of the eggs. This may occur in species or environments in which the attending parents are unusually successful at reducing moraliiy rates of eggs, either because of the abilittes of the parents, or the high susceptibility of the unprotected eggs.

2. *Exposed Verus Hidden Nest Sites*

Eggs that are laid under superficial cover on the surface of the ground, rather than being deeply buried, may be particularly vulnerable to predation. Hence, parental care might evolve more readily in species that do not bury their eggs. (Noble and Mason, 1933). This hypothesis has been used to explain the presence of parental care in Tupinambis teguixin, and its absence in the synonymous T. nigropunctatus, which oviposits inside termite mounds (Noble and Mason, 1933; but see Goeldi, 1902 and Riley *et. al.*, 1985, who report that T. teguixin oviposits in termitaria). However, the argument can be generalized to predict that parental care should be more common in snakes, which typically do not bury their eggs (probably because of the lack of limbs with which to excavate a nest-hole) then in lizards, which typically dig nest-holes. For the same reason, parental care should be more common in limbless (nonburrowing) lizards than in species with fully-developed limbs.

Data presented earlier support the prediction that parental care is more common in snakes than in lizards (3% of oviparous species versus 1%) and has evolved more often in the former group (15 origins in 11 families versus 5 origin in 15 families). Limblessness has evolved many times among lizards, but is shown by only a small proportion of taxa (Gans, 1975). Hence, the parental care of the limbless Ophisaurus offers further, although weak, support for the hypothesis that limblessness favors the evolution of parental care.

3. *Ability of Parents to Defend Eggs*

If a major benefit of parental care is the repulsion of potential egg-predators, then parental care should evolve most often in species in which the parent is capable of deterring predators. This most likely to be true of large and of venomous species. No such trend is evident among the lizards, except for Tupinambis and the tentative records for Varanus.

However, in snakes there is a strong bias such parental care in large and venomous species. The only major taxon in which parental care is common (probably universal) is the Pythoninae, and this group contains the largest oviparous snakes of the world. Parental care also occurs in 4 of the 8 oviparous genera of the Viperidae (50%) and 7 of the 17 oviparous genera of the Elapidae (41%). In contrast, the much larger (>200 genera) Colubridae, which consists primarily of nonvenomous (or less venomous) species, show parental care less commonly (15 genera, or 7.5%). Several of the colubrids reported to show parental care are unusually large (*e.g.*, Elaphe Farancia, Ptyas) or belong to the minority of venomous forms within the family (*e.g.*, Psammophylax, Rhabdophis). The prevalence of parental care crocodilians—most of which are large and formidable—also is consistent with the prediction that this behaviour should be shown most frequently by groups in which parents are well able to defend their eggs. Interestingly, many taxa of marine invertebrates show the opposite trend; parental care is most common in the smaller species of marine invertebrates (Strathmann and Strathmann, 1982). Nonetheless, extended parental care may also be common in highly venomous forms (*e.g.*, Hapalochlaena, Chironex; Sutherland, 1983).

2. *Limited Availability of Nest Sites*

If nest sites are scarce relative to the number of nesting females, older nests may be excavated and destroyed by new arrivals. This situation has been described in sea turtles and island populations of iguanine lizards. Active defense of nest sites by postovipositional females is common in iguanines. Intraspecific variations in intensity of iguanine nest defense are correlated with the degree to which suitable nest sites are available, and the ease of digging burrows. The selective advantage of parental care in this situation is increased by the relatively synchronous nesting of the population; thus even brief nest defense is effective.

Although these arguments are consistent with the behaviour seen in iguanines, they are unlikely to be of general importance because scarcity of suitable nesting sites may be a rare phenomenon. The only parallels to iguanine nest defense may be in alligators, in which the nest-guarding female prevents ovipositing turtles from digging into the nest mound, and tuatars, in which resting femals defend nesting sites against other females.

(b) **Facturs Reducing the Costs of Parental Care**

1. *Prediction*

If the major "costs" of parental care are a reduction in food intake, probable survivourship, or subsequent fecundity of the reproducing female, then parental care would be expected to evolve most often in species and environments in which such costs were minor or insignificant.

2. *Selection for "Risky" Life-History Strategies*

Life-history theory predicts that small, short-lived species are more likely to pursue "risky" reproductive strategies than are large, long-lived ones. Although parental care was suggested to be "risky," a test on fishes revealed no clear trend for more parental care in smaller species, possibly because the smaller species were not capable of deterring egg-predators. Similarly, a compilation of data on 49 species of lizards showed a trend reverse to that predicted; parental care was more common in late-maturing species than in those that matured early. This contradiction has been explained by suggesting that paretal care is not as "risky" as it appears; "lizards that practice it usually remain with their eggs in a well-hidden site ih which they may be less exposed to dangers than female that does not tend her eggs". Indeed, following a general discussion of the evolution of lizard life histories, Tinkle predicted that parental care will generally be found in long-lived iteroparous species, especially those with a short, annual breeding season.

To the degree that large body size is correlated with late attainment of maturity the trend for parental care in large reptiles (see previous section) is consistent with Tinkle's (1969) prediction. However, it is consistent also with the hypothesis that parental care by small species is ineffective against predators, or the alternative hypothesis that in small species, parents themselves are too vulnerable to predation to permit the evolution of parental care Certainly, the data on reptile are inconsistent with the prediction that parental care will be most common in small, short-lived species.

3. *Frequency of Reproduction*

Egg-attending reduces food intake of the reproducing female (Table 4.4), and thus may delay the time at which a subsequent clutch can be produced. Hence, parental care is likely to evolve in species that produce only a single clutch of eggs per year. The costs of remaining with the eggs may be independent of the number of eggs guarded, whereas the benefits increase with the number of eggs. Hence, parental care might be more likely to evolve in species the produce large but infrequent clutches rather than in those that produce small and frequent clutches. Both hypotheses predict that parental care should be most common in species that produce clutches only once a year or less often. Data on lizards support the prediction, although there are some puzzling cases of multiple-clutching, egg-atting species. The correlation between parental care and extremely low frequencies of reproduction also is consistent with this prediction, but is open to the other interpretations discussed easlier.

Because the annual production of multiple clutches is common only in the tropics, parental care should typify temperate rather than tropical species. The available data are biased by the concentration of scientific study in the temperate zone, but even so, the prediction is refuted; parental care is widespred in tropical reptiles. The reptilian subfamilies in which parental care is most common are all primarily tropical groups. Overall, the predicted bias toward temperate-zone species is conspicuously lacking.

4. *Suitability of Habitat for Clutch Attendance*

Parental care might be unlikely to evolve whenever eggs are laid in a habitat different from that usually occupied by the adult. In such a situation, the parent may be unusually susceptible to predation or physiological stress. This hypothesis suggests that parental care should be rare in aquatic or arboreal species. It is an obvious explanation for the lack of testudinian parental care, but unconvincing because

(a) many turtles are terrestrial;

(b) many other aquatic reptiles show parental care.

5. *Harsh and Unpredictable Environment*

Prolonged attendance on the clutch might be most likely to be favored in environments in which resources for the adult, at about the time of egg deposition become increasingly scarce or unpredictable so that searching for such resources becomes a high-risk endeavour. Under these circumstances, parental care might strongly benefit egg survivourship while conferring only a minor cost on the opportunities of the adult for future reproduction. This hypothesis predicts that parental care will most often be found in environments in which resources for adults are limited at the time of egg deposition. The prediction is difficult to test, but seems inconsistent with the strong taxonomic bias in the distribution of reptilian parental care. If specific environmental variables are important, one might expect to see parental care restricted to particular habitat types rather than to all species within a given taxon. However, available data are insufficient to convincingly refute the prediction.

6. *Brief Incubation Periods*

Parental care might be associated with brief incubation periods for three reasons:

1. The "costs" of remaining with the eggs depend upon the incubation period: if the eggs hatch soon after oviposition, the female is burdened only briefly by egg attendance.

2. Natural selection may favour prolonged oviducal retention of eggs in species with parental care, because the costs of this retention may be no higher than the costs of parental care. In contrast; species without parental care are less likely to evolve egg retention; in this case, the physical burdening of the gravid female imposes too high a cost on survivourship and subsequent fecundity.

3. Parental care and prolonged uterine retention of eggs are both examples of increased maternal investment and may be favoured by the same selective forces. Hence. they are likely to occur in the same species and the same environments.

A recent review of incubation periods showed "at least an indication that egg-guarding species may have slightly shorter development times than those that do not guard eggs". More detailed analysis of the data from Tinkle and Gibbons confirms that the men incubation period of egg-attending species is lower than that of nonattending speeies but the variances are so high that the difference fails to reach statistical significance. One confounding variable in this test, however, is the trend for parental care in larger species (see earlier); large species tend to have larger eggs, which in turn have longer incubation periods.

An alternative test involves the examination of the embryonic stage of development at oviposition in egg-attending and non-egg-attending species. "Visible embryonic development at oviposition" has been reported more commonly in species with parental care (Shine and Bull, 1979). However, a more recent study, which used objective criteria to stage embryonic development, found that relatively prolonged retention of eggs is the rule rather than the exception in oviparous squamates (Shine, 1983a). No evidence exists to show that uterine retention is more prolonged among egg-attending species (Shine, 1983a).

A recent anatysis of parental care in salamanders argues that parental care should evolve in species with long incubation periods; that is, it predicts the reverse of the above-discussed prediction. Nussbaum notes that salamanders with parental care tend to have large eggs, and that such eggs take a long time to develop. From these data, he argues that parental care has evolved to reduce the otherwise high rate of mortality of these slowly-developing embryos. An alternative interpretation of the same data is that natural selection has favoured an increase in egg size in species with parental care, because the offspring may thereby be kept for longer in a low-risk situation (protected eggs) rather than as unprotected free-living juveniles. The association of parental care with large offspring is less clear in reptiles than in many other animal taxa, possibly because of the concentration of most reptilian embryonic mortality to a short postovipositional period.

Which Sex Should Show Parental Care ?

One consistent feature of reptilian care parental is that involvement by the male is nonexistent (squamates) or relatively minor (crocodilians). In contrast, male parental care is common in amphibians, fishes, and birds. The evolutionary basis for sex differences in the tendency to show parental care has been the subject of several recent discussions. Two hypotheses for the selection of parental care have received wide attention:

1. Selection against male parental care, because internal fertilization results in a delay between insemination and oviposition, making the paternity of any given clutch uncertain (unlike the situation in most external fertilization). This low reliability of paternity may select against male parental care in species with internal fertilization including reptiles.
2. The unlikelihood of selection for male parental care in species with internal fertilization, because the time delay between insemination and oviposition means that a male and his offspring may never be in close proximity.

Both of these hypotheses correctly predict that reptilian care parental should be performed by females rather than males. Although the reptilian data therefore do not permit a test between the two hypotheses, the latter (parent-offspring promixity) model seems more accurately to predict the distribution of parental care in telecosts and amphibians (Gross and Shine, 1981). The paternity hypothesis also has been criticized on the grounds of faulty logic (Western *et al.*, 1980).

An alternative approach is to consider the effects of reproductive activities on the parents. If the female is "exhausted" by oviposition, she may likely to stay with the eggs until she recovers, protoadapting the species for female parental care (Noble, 1931). However, exactly the opposition prediction is made by Maynard Smith (1977), because the females exhausted, she has a greater need than the male to recommence feeding as possible. Thus, Maynard Smith predicts that this situation should favor the evolution of male parental care. The reptilian pattern is consistent with the prediction of Noble rather than with that of Maynard Smith, but does not provide strong support the former hypothesis, because the predominance of maternal care is consistent with several alternative theories.

A related question is why biparental care is virtually unknown in squamates, and probably rare even in crocodilians. This result is consistent with a general trend for biparental care to be less common in ectotherms than in endothermic vertebrates. Biparental care may be most likely to evolve when the form of care is such that two parents are much more effective than one; for example, feeding or guarding mobile young rather them merely guarding eggs. This hypothesis predicts that biparental care in reptiles should be restricted to crocodiliane (the only group to guard offspring after hatching), and that the contribution of the male should commence only after the eggs hatch. This prediction is strongly supported by data.

Why is Parental Care Rare in Reptiles ?

An analysis of parental care at the familial level reveals major differences among reptiles, teleost fishes, and amphibians in the frequency and form of parental care they exhibit. The proportion of families containing care-giving species is highest among amphibians, in terms of parental care by either sex, or overall. Reptiles differ strongly from the other two groups in their low frequency of male parental care. This probably reflects the lack of externally-fertilizing reptilian species. Although the overall proportion of families showing parental care is similar in fishes andreptiles, the direct comparison is misleading. Most fish species produce pelagic eggs, so that parental care is impossible. The proportion of species showing parental care would be much higher in demersal-spaw-ning ielecosts than in reptiles. This difference may be attributable ;

(a) the high incidence of external fertilization in teleosts, protoadapting for male parental care.

(b) the brief incubation period of telecost eggs, requiring only a short duration of parental care.

TABLE 4.5

The Distribution of Parental Care and Viviparity Among Families of Teleost Fishes, Amphibians, and Reptiles

(1)	*Teleosts* (2)	*Amphibians* (3)	*Reptiles* (4)
Number of families for which data available	182	35	43
Number with male parental care (and proportion with male parental care)	63 (.35)	16 (.46)	1 (.02)
Number with female parental care (and proportion)	28 (.15)	17 (.49)	13 (.30)
Number with parental care by either parent (and proportion)	68 (.37)	23 (.66)	13 (.43)

(*1*)	(*2*)	(*3*)	(*4*)
Number with viviparity (and proportion)	11 (.06)	6 (.17)	18 (.42)
Number with parental care and/or viviparity (and proportion)	77 (.42)	25 (.71)	24 (.56)

Data for telecosts and amphibiants from Gross and Shine (1981), and references therein.

Although parental care is protionally less common among reptiles than among other ectothermic vertebrates, an alternative form of increased parental investment—viviparity—is strikingly more common in reptiles. This may reflect two reptilian features: internal fertilization and behavioural thermoregulation. The former protoadapts the species to viviparity whereas the latter enables rates of embryonic development to be accelerated greatly by uterine retention of eggs. Because parental core of eggs may serve as a protoadaption to viviparity, the incidence of reptilian parental care may be lowered by a trend for viviparity to evolve in care-giving species.

5

Ecology of Reptilian Eggs and Embryos

Oviparity is the prevalent mode of reproduction among reptiles (Shine, 1985). Females of all extant turtles and crocodilians and of most lepidosaurians lay relatively large eggs containing all of the nutrients needed to produce viable offspring. These eggs usually are deposited in subterranean nests or in other sheltered sites, and embryonic development then proceeds under physical conditions dictated largely by the location of the site, the type of nest, and prevailing weather. In this chapter, we shall summarize information on reptilian eggs and nests, on physical conditions inside these nests, and on effects of the physical environment on eggs and embryos. Our focus will be on temperature, water, and respitatory gases, because these variables are likely to be the primary determinants of metabolism, growth, and survival by embryos developing in nature. We shall also evaluate some of the methods used in recent research on reptilian eggs, and shall try to identify some of the more important topics for future study.

TYPES OF EGGS

A. Testudines

Many species of turtles, including all cheloniids, chelydrids, and dremochelyids and most pelomedusids and emydids, produce eggs with flexible shells. However, morphology of flexible eggshells varies considerably among species. Eggs of sea turtles of the family Cheloniidae have a shell comprised of a relatively thick, fibrous membrane overlain by a calcareous layer formed of numerous small, often poorly defined, shell units. Eggshells of chelydrids and emydids are similar, except that shell units are larger and more clearly defined and the shell membrane accounts for a smaller proportion of the total thickness of the eggshell. Large gaps between adjacent shell units (or between groups of shell units) provide for the exchange of respiratory gases and water between embryos and the nest environment. The crystalline material in the eggshell usually is aragonite, but traces of calcite may form at the nucleation sites for shell units (Solomon and Baird, 1976; but see Silyn-Roberts and Sharp,

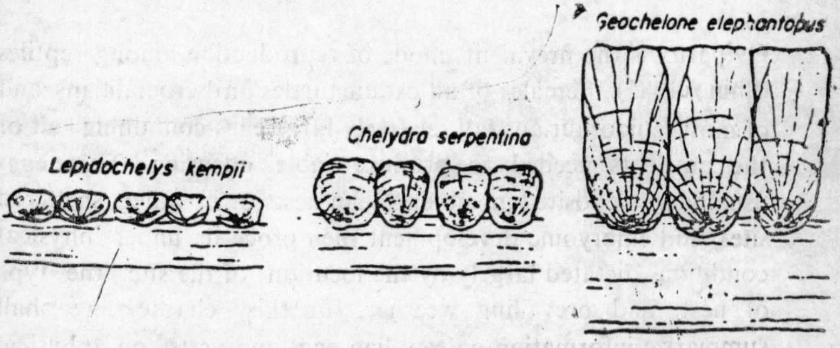

Fig. 5.1. Schematic illustrations of radial sections through shells of turtle eggs, showinig the relative importance of membranous and calcareous layers and the size and configuration of shell units. All three eggshells are drawn to the same scole with the outside on top. Note the large gaps between shell units of the eggshell of Lepidochelys kempii (Cheloniidae) and the smaller gaps between shell units of the eggshell of Chelydra serpentina (Chelydridae). Occasional pores (not figured) pass between adjacent shell units of the eggshell of Geochelone elephantopus (Testudinidae).

1985). Calcite may also comprise a portion of the calcareous layer of eggshells produced by animals reared in captivity (Baird and Solomon, 1979), wnich raises the possibility that characteristics of eggshells may be modified by the diet on which females are maintained.

Other species of turtles, including carettochelids, chelids, dermatemyids, kinosternids, testudinids, trionychids, and some pelomedusids and emydids, lay eggs having rigid shells (Ewert, 1979, 1985). The shell of such eggs is comprised of a relatively thin shell membrane (or two) overlain by a calcareous layer (Ewert, 1979; M.J. and G.C. Packard, 1979; Young, 1950). The calcareous layer is formed of crystals of aragonite arranged into discrete shell units that abut tightly, thereby causing the layer to be rigid and noncompliant (Figure 5.1; also Erben, 1970; Erben *et al.*, 1979; Ewert, 1985; Ewert *et al.*, 1984; G.C. and M.J. Packard, 1980; M.J. and G.C. Packard, 1979; M.J. Packard *et al.*, 1982c, 1984a; Thompson, 1983, Silyn-Roberts and Sharp, 1985; Woodall, 1984). The crystalline layer is interrupted occasionally by distinct pores that provide for an exchange of respiratory gases and water between the contained embryo and the nest environment. These pores are more concentrated at the equator of elliptical eggs than at the poles (Ewert *et al.*, 1984; Thompson, 1983; Woodall, 1984); their distribution in spherical eggs has not been determined.

The ultrastructure of rigid eggshells appears at first to be very different from that of flexible eggshells (M.J. Packard *et al.*, 1982c). However, eggs of Kinosternon flavescens at intermediate stages of calcification have a flexible shell that is indistinguishable from shells of oviposited eggs of chelydrids and emydids; that is, the calcareous layer has numerous small, widely spaced shell units (M.J. Packard *et al.*, 1984b). By the time of oviposition, however, the calcareous layer is a rigid structure characterized by elongate shell units that abut tightly (M. J. Packard *et al.*, 1984a, b). Thus, the differences in morphology of flexible and rigid eggshells may arise simply in consequence of different amounts of time that crystallization proceeds inside oviducts of females. This question deserves further attention, because control of the duration of calcification appears to be implicated in the evolution of interspecific differences in shell structure.

Turtle embryos are in the gastrula (*i.e.*, in a presomite stage of development) at the time of oviposition, regardless of how long eggs may have been retained within the maternal oviducts prior to oviposition (*e.g.*, Crastz, 1982; Cunningham, 1922; Decker, 1967; Domantay, 1968; Ewert, 1979, 1985; Lynn and von Brand, 1945; Mahmoud *et al.*, 1973; Risley, 1933, 1944; Yntema, 1968). This consistency contrasts mnrkedly with the variation in developmental stage of oviposition in squamates (Saint Girons, 1985; Shine, 1985). The cause for the apparent arrest in development at this stage has not been determined (Ewert, 1985), but hypoxia may be a factor (Risler, 1944).

Eggs of turtles have a thick layer of albumen at oviposition (Agassiz, 1857; Ewert, 1979; Mitsukuri, 1890, 1891; Morris *et al.*, 1983; G.C. Packard *et al.*, 1981a, 1983), and this layer holds much of the water that the female parent vests in her eggs for support of embryogenesis (Morris *et al.*, 1983; G.C. Packard *et al.*, 1981a, 1983). Most of the water passes from the albumen into the vitelline sac during the first 1-2 weeks of incubation, however, so the albumen layer is not prominent in eggs that have undergone much development (Agassiz, 1857; Ewert, 1979; Mitsukuri, 1890, 1891; Morris *et al.*, 1983; G.C. Packard *et al.*, 1981a, 1983).

Turtle eggs have a large yolk at oviposition, but the yolk becomes even larger during the first 1-2 weeks of development as water flows inward from the albumen (Morris *et al.*, 1983; G.C. Packard *et al.*, 1981a, 1983). Indeed, by the end of this period, the yolk occupies almost the entire interior of viable eggs. This compartment contains all of the lipid and most (or all) of the protein required to sustain metabolism and growth of embryos (Congdon *et al.*, 1983c; Karashima, 1929; Koga *et al.*, 1978; von Brand and Lynn, 1947). The yolk also supplies part of the calcium used by embryos for osteogenesis (Bustard *et al.*, 1969; M.J. and G.C. Packard, 1986; M.J. Packard *et al.*, 1984d: Simkiss, 1962, 1967) and is the proximate source fo water used by embryos (Morris *et al.*, 1983; G.C. Packard *et al.*, 1983).

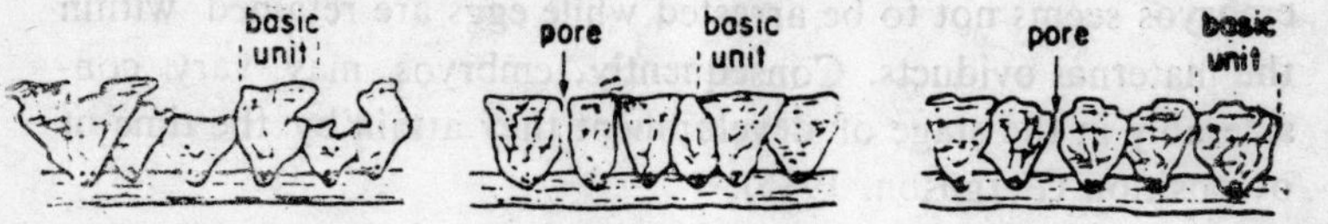

Fig. 5.2. Schematic illustrations of radial sections through shells of crocodilian eggs, showing the relative importance of membranous and calcareous layers and the size and configuration of shell units. All three eggshells are drawn to the same scale (which differs from that characterizing turtle eggshells in Figure 5.1). Note the pores between shell units in eggshells of Crocodylus niloticus and Alligator mississippiesis, and the erosion craters (Ferguson, 1881a), b) on the external surface of eggshells of Melanosuchus niger and Alligator mississppiensis.

B. Crocodylia

All crocodilians by elliptical eggs having rigid, calcareous shells (Ferguson, 1985; Guggisberg, 1972; McIlhenny, 1935). The shell is formed of two fibrous membranes overlain by a relatively thick layer of calcium carbonate (Bigalke, 1931; Clarke, 1888, 1891; Deraniyagala, 1936, 1939, Ferguson, 1982; Reese, 1915; Voeltzkow, 1892). The calcareous layer is comprised of crystals of calcite (Erben, 1970; Ferguson, 1982; Jenkins, 1975; Schmidt and Schonwetter, 1943; Silyn-Roberts and Sharp, 1985; Solomon and Reid, 1983) organized into large, wedge-shaped shell units that abut tightly (Figure 5.2), thereby conferring rigidity to the eggshell (Deraniyagala), 1936, 1939; Erben, 1970; Ferguson, 1982: Schmidt and Schonwetter, 1943). The crystalline layer occasionally is interrupted by pores that provide for an exchange of water and respiratory gases between the enclosed embryo and the nest environment. These pores tend to be more numerous around the equator of the egg than at either of the poles (Ferguson, 1982).

Embryonic crocodilians in freshly laid eggs commonly have 16-18 somites, and therefore are somewhat more advanced at oviposition than are embryonic turtles (Ferguson, 1982, 1985; Webb *et al.*, 1988b). However, development of crocodilian embryos seems not to be arrested while eggs are retained within the maternal oviducts. Consequently, embryos may vary considerably in the stage of development they attain by the time of oviposition (Ferguson, 1985).

Crocodilian eggs contain a distinct layer of albumen at oviposition, and this albumen surrounds a large yolk (Bigalke, 1931; Clarke, 1888, 1891; Deraniyagala, 1936, 1939; Ferguson, 1982; Reese, 1915). The albumen probably contains much of the water required by the developing embryo, because the volume of albumen declines appreciably during the course of incubation (Ferguson, 1982). Additional water is contained within the yolk, which seemingly is the source also for all of the lipid and most (or all) of the protein required to sustain the embryo. Although a portion of the calcium required for skeletal ossification apparently is moblized from the yolk, most of it comes from the calcareous layer of the eggshell (Jenkins, 1975).

C. Lepidosauria

Most oviparous lepidosaurians lay eggs with thin, flexible shells. However, both flexibility and morphology of eggshells vary considerably among and within species, so the descriptive terms that are commonly used to characterize lepidosaurian eggs (*e.g.*, "parchment-like" and "flexible-shelled") actually encompass a wide range of morphological states.

The eggshell in some lepidosaurians is comprised of a fibrous membrane that seemingly is devoid of calcareous deposits (Figure 5.3; also Andrews and Sexton, 1981; DeSalle *et al.*, 1984; Krampitz *et al.*, 1973; M. J. Packard *et al.*, 1982c; Sexton *et al.*, 1979, Trauth and Fagerberg, 1984). In other species, small amounts of calcium carbonate are found as isolated crystals among fibers of the shell membrane (Kriesten, 1975) or as larger, but widely separated, deposits on the outer surface of the membrane (M.J. Packard *et al.*, 1982c). Finally, in still other species, a layer of calcareous material forms the outer surface of the eggshell (Figure 5.3). Sometimes this calcareous layer is formed of numerous thin, flat plaques (or spheres) attached to the surface of the shell membrane (Guillette and Jones, 1985; Lillywhite and Ackerman, 1984; M.J. Packard *et al.*, 1982a, c; Solomon and Reid, 1983), but in other instances the calcareous material is formed into columns that penetrate deeply into the membrane (M.J. Packard *et al.*, 1982b), The calcium carbonate usually exists as crystals of calcite (M.J. Packard *et al.*, 1982a, b), but crystals of aragonite are interspersed among the crystals of calcite in eggshells of Python regius (Solomon and Reid, 1983). However, variation occurs within some species both in the type of crystal forming the calcareous layer (Solomon and Reid, 1983) and in morphology of the shell itself (M.J. Packard *et al.*, 1982a), and eggshells of a single species have occasionally been assigned to different categories by different investigators (*e.g.*, Giersberg, 1922; Krampitz *et al.*, 1973).

Lizards of the family Dibamidae are said to produce eggs with rigid, brittle shells (Boulenger, 1912), but the original observation has never been repeated. On the other hand, gekkonid lizards in the subfamilies Gekkoniaae and Sphaerodactylinae are well known to produce eggs with hard, calcareous shells (Bustard, 1968; Werner, 1972). Crystals of calcite in the calcareous layer are arranged into jagged blocks that abut tightly and interlock (Erben and Newesely, 1972; Kramptiz *et al.*, 1972; M.J. Packard *et al.*, 1982c; Schmidt, 1943), thereby rendering the layer rigid and noncompliant. The proximal portion of these calcareous blocks invests fibers at the outer surface of the shell membrane and anchors the two layers together (M.J. Packard *et al,*, 1982c). Pores have not been observed in the calcareous layer; but all of the studies were preliminary, and pores could easily have been overlooked.

Embryos in freshly laid eggs of most lepidosaurians have attained stage 27-29 (or its equivalent) in the development series of Dufaure and Hubert (1961; see Shine, 1983a), and therefore have 23-30 pairs of somties. In Opheodrys vernails and Sceloporus aeneus, however, eggs are retained for extended periods within oviducts of females, and embryonic development proceeds well beyond stage 29. Oviposited eggs of these species contain

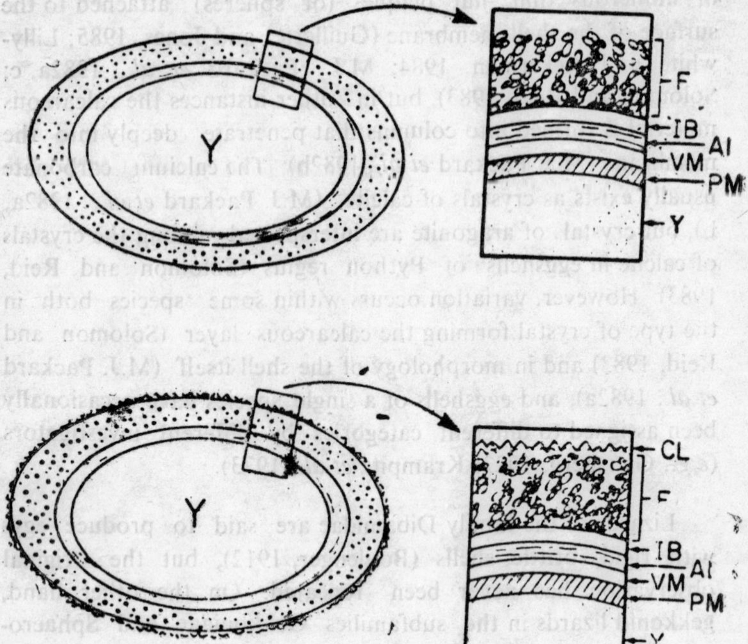

Fig. 5.3. Schematic illustrations ot radial sections through shells of lepidosaurian eggs, showing the differencc between eggs having no calcareous material in the eggshell and those having a distinct calcareous layer. Insets show enlarged sections passing from the outer surface of eggshell into the yolk. None of the layers is drawn to scale. CL, calcareous layer; F., fibres of shell membrane; IB, inner boundary of shell membrane; Al, albumen; VM, viteline membrane; PM, plasma membrane; Y, Yolk.

embryos of substantial size, and incubation to hatching may be completed within a matter of days. On the other hand, eggs of Chamaeleo lateralis contain embryos in a presomite stage of development and those of Sphenodon punctatus apparently contain embryos that are in gastrulation. Considerable variation therefore exists in the stage of development attained by lepidosaurian embryos by the time of oviposition, but these embryos usually are more advanced than are those of turtles and crocodilians in freshly laid eggs.

Lepidosaurian eggs usually lack a distinct layer of albumen at oviposition but eggs of Conolophus subcristatus are reported to have an albumen accounting for 28 per cent of their initial mass. However, albumen was separated from yolk of the Conolophus eggs by low-speed centrifugation, so the fraction refereed to as albumen may simply have been water forced out of the yolk by ultrafiltration.

The albumen testudinian and crocodilian eggs is an important reservoir of water, so she apparent absence of an albumen layer in lepidosaurian eggs has been taken to indicate that these eggs contain insufficient water at oviposition to support embryos until hatching. Indeed, water represents only 63 per cent of the initial mass in eggs of Amphibolurus barbatus ahd 59 per cent in eggs of Coluber constrictor. However, 74 per cent of the initial mass of eggs is water in Pituphis melanleucus. Comparable values for eggs of turtles fall between 68-72 per cent.

Many authors have reported the presence of a large layer of albumen in squamate eggs that have undergone substantial incubation. Hower, the fraction they identified as albumen may actually have been material from the allantoic sac. The allantoic sac of embryonic Sphenodon punctatus becomes greatly enlarged during incubation, owing to the accumulation of a clear, semigelatinous fluid resembling the albumen of an avian egg. Whenever a swollen egg ruptures, thus fluid is released onto the surface of the eggshell where it could easily be mistaken for albumen.

The yolk occupies most of the interior of lepidosaurian eggs at oviposition, and presumably provides embryos with almost all of the nutrients required to support development (Body, 1985; Crodzinski, 1949; also Bellairs, 1959; Stewart and Castillo, 1984; J. Thompson, 1981). All of the water vested in eggs for support of embryogenesis is in this compartment (M.J. Packward *et al.*, 1984c, 1985), and most (or all) of the calcium required for skeletal ossification also is stored there (Jenkins and Simkiss, 1968; M.J. Packard *et al.*, 1984c, 1985).

TYPES OF NESTS

A. Testudines

Most species of turtles dig a flask shaped hole in the ground in which eggs are deposited (*e.g.*, Carr, 195?; Ernest and Barbour, 1972). In some instances, excavation of the nest chamber is preceded by construction of a "body pit," thereby removing surface debris and loose soil that might otherwise sift into the forming cavity (*e.g.*, Alho and Padua, 1982; Carr, 1952; Hendrickson, 1958; Richmond, 1945). Among species producing small clutches, all of the eggs rest on the floor of the cavity and therefore have some contact with the substrate. Among species producing large clutches, however, many eggs are supported by points of contact with other eggs and consequently make no contact with the substrate (Ackerman, 1977; Hendrickson, 1958; G.C. Packard *et al.*, 1981c). After all of the eggs have been laid, the female kicks durt into the neck of the chamber. Some of this dirt may sift downward among the eggs, but air space between the eggs and at the top of the cavity usually are not obliterated completely (Ackerman, 1977; Allard, 1935; Burger, 1976: Cagle, 1937, 1950; Hendrickson, 1958; G.C. Packard *et al.*, 1981c). These air spaces persist for the duration of incubation in nests constructed in stable soils (Breitrnbach *et al.*, 1984), but may be obliterated by infiltration of soil into nests located in less stable substrates (Hotaling *et al.*, 1985). The infiltration of soll or sand into nests sometimes leads to increased mortality among embryos (Simon, 1975), so persistence of air spaces may be more important to suceessful incubation than generally is realized.

Some species of the families Chelidae and Kinosternidae depart from tne pattern of nest construction cearacterizing most turtles. For example, Strenotherus minar and sternotherus odouratus may simply back up to a fallen log and deposit their eggs beneath the edge; if females fail to scrape leaf litter over these "nests," the eggs remain partially exposed to the elements (Cagle, 1937; Carr, 1952). Eggs of Phynops gibbus frequently are deposited on the surface of the ground or with only a thin covering of soil (Mittermeier *et al.*, 1978). In contrast, Kinosternon flavescens from regions of sandy soil in the central U.S. burrow through the ground prior to ovaposition, and eggs are simply deposited in the burrow system constructed by each female (J.B. Iverson, pers. comm.).

B. Crocodylia

Crocodilians construct mound nests or hole nests (Campbell, 1972; Ferguson, 1985; Greer, 1970). Mound nests, such as those characterizing Alligator mississippicnsis, Caiman crocodilus, and Crocoidylus porosus, are located in piles of mud and/or vegetation scraped together by gravid females (Deraniyagala, 1936, 1939; McIlhenny, 1935; Station and Dixon. J977; Webb *et el.*, 1983c). A nest chamber is dug in the top of each mound, and eggs are deposited inside. Eggs at the bottom and periphery of the pile contact walls of the cavity, but those toward the center of the mass do not. The female closes the cavity, after all eggs have been laid and may then crawl across the top several times, thereby compressing the loose mud and vegetation. Spaces between eggs occasionally are filled as the nest is closed (Clarke, 1888), but pockets of air usually persist within the cavity for the duration of incubation (McIlhenny, 1935).

Other crocodilians, like Crocodylus johnsoni, Crocodylus niloticus, and Gavialis gangeticus, dig a hole obliquely downward into a sand, gravel, or dirt bank (Modha, 1967; Subba Rao and Bustard, 1979; Webb *et al.*, 1983a). The narrow tunnel leads to a spacious egg-chamber at the bottom (Whitaker and Whitaker, 1984). As eggs are laid, they simply roll to the bottom and form a pile. Eggs in the middle of the pile make no contact with the substrate, whereas those at the periphery have some contact with soil. The female subsequently closes the tunnel with sand or dirt. Some soil and debris occasionally is packed between eggs (Pooley, 1969), but air spaces usually persist inside the egg chamber (Modha, 1967, Ogden, 1978), presumably for the duration of incubation.

C. Lepidosauria

Such a diverse array of sites is used for oviposition by different species of lepidasaurians that generalization about nests of these teptiles is virtually impossible. Developing embryos of lepidosaurians encounter diverse ecological and physiological problems. This diversity represents both a challenge and an obstacle: A challenge, in that a great deal of new research is needed to describe nests in the detail required for understanding these problems; an obstacle, in that consequent variation in physiological ecology of embryos may defy simple generalization.

Because snakes lack limbs, most species apparently are unable to contruct nests specifically for their eggs, and rely instead on pre-existing sites that are suitable for oviposition (see Burger and Zappalorti, 1986, for an exception). What constitutes a suitable site varies somewhat with the size of the species in question. For example, small snakes, such as the colubrid Diadophis punctatus, frequently deposit eggs in small crevices and ant galleries inside rotting logs and stumps (Blanchard, 1936). Eggs of larger species, like the colubrids Coluber constructor, Masticophis taeniatus, and Pituophis melanoleucus, often are deposited in the subterranean burrows of rodents (Eitch, 1963; Parker and Brown, 1972, 1980). The largest of snakes, such as Python sebar, usually oviposit in large cavities beneath objects on the ground or in abandoned burrows of hardvarks and anteaters. In the absence of such sites, pythons may lay eggs on the surface of the ground in dense grass or in pockets of dead leaves (FitzSimons, 1962; Pope, 1961). Detailed descriptions are not available for nests of any snake, but the lower aspect of some for all) eggs in a clutch probably contacts are subtrate, whereas the upper surface of the eggs probably is exposed to air inside the nest cavity. The proportion of the surface contacting the subtrate has important implications for temperature of eggs and for exchanges of water with the environment, so detailed descriptions of egg chambers are badly needed.

One species or another of saurian uses almost every imaginable place to lay its eggs. Limbless lizards, such as Ophisaurus attentuatus and Pygopus lepidopodus, probably use the same kinds of sites for oviposition as are used by snakes of similar size (Vinegar, 1968; Wells and Husband, 1979). Small skinks, anoles, and other lizards lay their eggs in cavities in rotting logs or beneath surface objects (Cagle, 1940; Fitch, 1954; Magnusson and Lima, 1984; Rand, 1967; Whitaker, 1968), whereas others oviposit on the surface of the ground beneath a protective layer of leaf litter (Rand, 1967: Sexton *et al.*, 1964). Some geckos lay their eggs beneath fallen limbs and debris (Bustard, 1969a), but others conceal their eggs beneath loose bark on standing trees or in clumps of dead leaves collecting in crotches of these trees (Sexton *et al.*, 1964; Sexton and Turner, 1971). Many species in tropical regions, including some of the larger varanids, oviposit insite termite mounds (Cowles, 1928; Riley *et al.*. 1985). In every ease, the site selected for oviposition is a primary determinant of the physical connitions to eggs will be exposed, but detailed descriptions of such sites are wanting for most of these saurians.

A number of species of saurians, including many agamids, chamaeleonids, iguanids, lacertids, and teiids, construct nests specifically for their eggs. A tunnel is excavated at an angle into the soil, and an enlarged chamber usually is located at the end (Bartholomew, 1966; Blair, 1960; Brown, 1956; Cooper, 1958; Hudson, 1977; Johnson, 1979; Menzies, 1958; Shaw, 1960; Subba Rao and Rajabi, 1972; Van Devender and Howard, 1973). After the female lays her eggs, the ncst is closed. Females of some species apparently pack loose dirt into the nest chamber during closure, thereby oblitsrating air spaces among the eggs (Johnson, 1965; Rand and Dugan, 1983), whereas females of other species do not (Brown, 1956; Carpenter, 1966; Guillettee and Lara Gongora, 1986; Iverson, 1979; Muth, 1977; Shaw, 1960). The presence or absence of air spaces inside-nests has important implications concerning exchanges of heat, water, and gases between developing embryos and their surroundings. Also, obliterating these air spaces is known to increase embryonic mortality in some species (Muth, 1977). Thus, it is important to determine wherher filling the cavity occurs normally in the species for which it has been reported (see Rend and Dugan, 1983) or whether the published reports are of atypical nests (see Brown, 1956).

TEMPERATURE

A. Factors Affecting Temperature of Eggs

Several factors affect the temperature of reptilian eggs and embryos incubating in natural nests, as can be illustrated with a simple energy balance equation for a single egg:

$$q_S = q_M + q_{CD} + q_{CV} + q_E + q_R$$

where q_S is heat storage (or loss), q_M is heat released by the embryo in intermediary metabolism, q_{CD} is heat gained (or lost) by conduction from the substrate, q_{CV} is heat gained (or lost) by convention from air inside the nest cavity, q_E is heat gained (or lost) by changes in state of water, and q_R is heat gained (or lost) by thermal radiation (see Gates, 1980); Tracy, 1982). whenever the input of heat from various sources equals the output, storage of heat (q_S) is zero, and temperature of the egg does not change. Whenever the input of heat does not balance the output, however, heat must be stored in the egg or lost from it, and, temperature of the egg therefore must increase or decrease, respectively. A fuller appreciation of the several processes affecting energy exchange and of their relative importance as determinants of egg temperature can now be gained by considering each of the terms in order.

B. Temperature in Nests

1. Determinants of Nest Temperature

Numerous factors influence the heat exchange and temperature of reptilian eggs Although air temperature inside the nest is only one of these factors, it is especially important because it can be measured with relative ease and because it usually yields a reasonoble approvimation to the temperature of the eggs themselves. Thus, an understanding of the factors affecting temperature in nests is fundamental to any discussion of temperature relations of reptilian eggs.

The diurnal cycle in net radiation at the surface of bare ground is the primary determinant of temperature in the upper 50 cm of the soil profile (Marshall and Holmes, 1979: Taylor and Ashcroft, 1972). Shortwave radiation reaching the ground during daylight hours heats the surface layer and leads to a subsequent flow of heat downward (Carson and Moses, 1963). Conversely, emission of longwave radiation at night cools the surface layer and causes heat to flow upward through the soil (Carson and Moses, 1963). These cycles in heating/cooling and in heat flow are manifested as daily cycles in soil temperature also Buchan, 1982; Carson and Mosses, 1963; Parton, 1984; Wierenga a*t al.*, 1969). The amplitude of the temperature cycles decreases with increasing depth, and maxima and minima occur progressively later in the day as distance from the surface increases.

Our interest in temperature cycles in the upper 50 cm of the soil centres on the fact that this in the zone in which subterranean nests of most turtles and of many crocodilians and lepidosaurians are located. With relatively few exceptions (see below), the primary determinant of temperature inside these nests in the daily cycle in heating/cooling of the surface of the soil and attendant cycles in transport of heat through upper reaches of the profile. This view is supported by measurements showing that temperatures at the top, middle, and bottom of hole nests located in the open exhibit the same phase-shift and damping of diel cycles with increasing depth that are expected to occur in the soil at large also Burger, 1976; G.C. Packard *et al.*, 1985b). Indeed, the daily cycle in net radiation probably is the primary factor also influencing temperature inside the mound nests of crocodilians and inside the nests of lepidosaurians located in logs or beneath surface objects (Magnusson; 1979; Staton and Dixon, 1977; Webb *et al.*, 1983a).

Several factors modify the impact of diel cycles in heat

transport through upper layers of the soil and thereby affect temperatures at the level of nests. For example, nests located deep in the ground tend to have lower temperatures than shallower nests at the same site (Wilhoft *et al.*, 1983), simply because of the resistance imposed by the soil itself to transport of heat from the warm surface to deeper layers in the profile. Shaded nests generally are cooler than nests located at similar depths in the open (Alho *et al.*, 1985; Bull, 1985; G.C. Packard *et al.*, 1985b; Webb *et al.*, 1983a), because shading reduces insolation and heating of the surface of the soil. Moreover, temperatures are lower during periods of overcast and rainfall than during periods of fair weather (Burger, 1976; G.C. Packard *et al.*, 1985b; Magnusson, 1979), owing to reduced insolation and to absorption of heat by infiltrating water. Finally, subtle factors like slope and aspect of the nest site influence the absorption of solar radiation (Mahrer, 1982; Mahrer and Avissar, 1985); and texture, composition and water content of the soil influence heat transport (Marshall and Holmes, 1979; Taylor and Ashcroft, 1972).

Several notable exceptions to these generalizations occur. For example, female sea turtles (families Cheloniidae and Dermochelyidae) deposit large numbers of relatively large eggs in a chamber deep beneath the surface of the sand. Late in incubation, embryos in these eggs generate more heat than is transported away from the cavity, and temperature inside the chamber therefore increases by 5—7°C above that of the adjacent sand (Bustard, 1972; Carr and Hirth, 1961; Hendrickson, 1958; Raj, 1976). If nests are deep and thus unaffected by daily temperature waves, concentric gradients in temperature may be created, with eggs at the center being exposed to higher temperatures than eggs at the periphery (Bustard, 1972).

Metabolic heat generated by embryos of the turtle Malaclemps terrapin also causes temperature in nests to rise late in incubation by 2—7°C over temperuture in adjacent soil (Burger, 1976). The magnitude of the observed increase is surprising, as the nests were shallow and each nest contained only eight eggs. Clutches this small seem to be unlikely sources for sufficient heat to elicit increases in temperature similar to those occurring in the much larger nests of sea turtles. Nevertheless, differences in temperature could not be detected between empty nests and adjacent soil, nor between nests containing freshly laid eggs and the surrounding substrate see production of metabolic heat is a viable explanation for the runings.

Metabolic heat also may influence temperature in nests of some crocodilians, because temperature at the bottom of hole nests occasionally is higher than that as the top (Modha, 1967). However, the higher temperatures as the bomam occur at a time of day when deeper layers of the soil might be extected to be warmer than shallower layers in consequence of the difference in phase of temperature cycles occurring at these levels. Thus, the question of significant metabolic heating in hole nests or crocodilians remains open.

Considerable attention has focused on the possible release of heat by fermenting vegetamonsterming walls of mound nets of many crocodilians (Mcllhenny, 1935 Reese, 1915). Although heat of fermentation clearly affects nest temperature in some instances (Magnusson, 1979; Webb *et al.*, 1977), it may not occur in more than a small proportion of nests (Magnusson 1970 Webb *et al.*, 1977) and therefore may not be a phenomenon of general importance (Magnusson *et al.*, 1985; Staton and Dixon, 1977).

Finally, we must meneon the interesting (case of Paleosuchus trigonatus, a South American crocodilian that nests beneath a closed canopy of trees in tropical jungle in which input of heat from direct solar radiation is precluded. The mound nests of this crocodilian usually are built against, or on top of, the mounds of a termite. Metabolic heat generated by the termites helps to mainum nest temperature several degrees above ambient and within the range conducive to embryonic development (Magnusson *et al.*, 1985). The implicit assumption is that exploitation of termite mounds as nests sites has enabled Palaeosuchus to occupy a habiiat from which it would otherwise be excluded by ambient conditions unfavourable to reproduction.

2. *Measurements of Nest Temperature*

Temperature has been measured inside reptilian nests on numerous occasions, but most of the resulting data are of limited value in understanding the thermal environment to which reptilian eggs are exposed. In most instances temperature was measured once in each of several nests at unspecified times during the day, and the values then were averaged. Unfortunately, values obtained in this way are not representative of temperatures to which eggs are exposed over the course of incubation and they usually are not indicative even of an extreme. A few measurements taken with maximum/minimum thermometers provide reliable information on extremes, but again do not characterize conditions to which eggs are exposed for the greater part of incubation. Thus, the database for nest temperatures is flawed, and we have confidence in only the most general of trends and tendencies.

Average temperatures usually are 26—32°C inside nests of most testudinians (*e.g.*, Burger, 1976; Bustard, 1972; Carr and Hirth, 1961; Goode and Russell, 1968; Hendrickson, 1958; Limpus *et al.*, 1983a; MacFerland *et al.*, 1974; McGehee, 1979; Pieau, 1982; Thompson, 1983; Vogt, 1980; Wilhoft *et al.*, 1983), crocodilians (*e.g.*, Chabreck, 1973; Deitz and Hines, 1980; Lutz and Dunbar-Cooper, 1984; Mngnusson, 1979; Magnusson *et al.*, 1985; Modha, 1967; Staton and Dixon, 1977; Webb *et al.*, 1977, 1983a; Whitaker and Whitaker, 1984), and lepidosaurians (*e.g.*, Bartholomew, 1966; Magnusson and Lima, 1984; Rand, 1972; Vaz-Ferreira *et al.*, 1970). Higher values have been reported for nests of thermophilic teiids (Brown, 1956) and iguanids (Johnson, 1965), and lower temperatures have been measured in nests of relatively cryophilic skinks (Shine, 1983b) and geckos (Bustard, 1969a). Nests of montane lizards may also have relatively low temperatures (Guillette and Lara Gongora, 1986).

Superficial nests constructed in shaded areas tend to track cycles in ambient temperature relatively closely (Shine, 1983b). However, nests in the open must be located beneath a protective layer of soil or vegetation to prevent the eggs from being exposed to lethal temperature. In such instances, cycles in temperature still occur inside nests, but the mean and amplitude from cycles in ambient temperarure. Eggs incubating in the large mound nests of crocodilians seldom are exposed to temperatures fluctuating by more than 2—4°C over the course of a day (Chabreck, 1973; Magnusson, 1979; Station and Dixon, 1977; Webb *et al.*, 1977), and a similar range characterizes the deep hole-nests of other crocodilians (Modha, 1967), some turtles (Alho and Padua, 1982; Carr and Hirth, 1961; Hendrickson, 1958; MacFarland *et al.*, 1974; McGehee, 1979), and a few lepidosaurians (Bartholomew, 1966; Rand, 1972). However, the daily cycle in temperature in the shallow nests of a few crocodilians, many turtles, and numerous lepidosaurians commonly spans 6—10°C also Bnrger, 1976; Bustard, 1969a; Ewert, 1979, Guillette and Lara Gongora, 1986, Magnusson and Lima, 1984; Moll and Legler, 1971; G.C. Packard *et al.*, 1985b; Pieau, 1982; Sexton and Claypool, 1978; Thompson, 1983; Vogt, 1980; Whitaker and Whitaker, 1984; Wilhoft *el al.*, 1983).

Temperature in superficial nests of the snake Opheodrys vernalis may exceed 40°C for part of the day (Sexton and Claypool, 1978), and temperatures may briefly reach this level also in shallow nests of the turtles Chrysemys picta and Pseudemys scripta (Ewert, 1979; Moll and Legler, 1971; Schwarzkopi and Books, 1985). Most surprising, however, is the report that temperatures in nests of the turtle Podocnemis expansa may not fall below 37°C for upwards of 9 days (Alho and Padua, 1982); prolongsd exposure to temperatures this high would be lethal to embryos of most species.

Eggs located at the top of nests of crocodilians and many turtles are exposed to diel cycles in temperature having higher means and greater emplitudes than are eggs at the bottom of the same nests also Ferguson and Joanen, 1982, 1983; Modha, 1967; G.C. Packard *et al.*, 1985b; Thompson, 1983; Webb *et al.*, 1983a; Wilhoft *et al.*, 1983). Maxima and minima also occur somewhat earlier at the top of the egg mass than at the bottom also Burger, 1976; G.C. Packard *et al.*, 1985b; Webb *et al.*, 1983a; Wilhoft *et al.*, 1983). As was indicated earlier, such cycles in temperature occur normally in the upper layers of the soil, so eggs apparently are being exposed to temperature regimes appropriate for that depth in the soil profile.

C. Physiological Effects of Temperature

1. *Limits of Tolerance*

Several studies have characterized the thermal tolerances of embryonic reptiles developing at constant temperatures in the laboratory. The general finding is that embryos cannot survive exposure to as great a range of temperatures as can adults of the same species. Embryonic crocodilians usually survive to hatching when incubated at temperatures of 28—34°C, but mortality increases rapidly as temperatures go above or below this range (Bustard, 1971b; Ferguson and Joanen, 1982, 1983). Young of the green turtle (Chelonia mydas have high hatching success at 25—33°C (Bustard 1971a; Bustard and Greenham, 1968; Miller, 1982), and a similar range of tolerances characterizes developing young of Caretta caretta (McGehee, 1979; Miller, 1982; Yntema and Mrosovsky, 1980) and Chelonia depressa (Miller, 1982). The range of thermal tolerance of embryonic Chelydra serpentina (Yntema, 1968, 1978), Pseudemys scripta (Moll and Legler, 1971), Terrapene carolina (Dodge *et al.*, 1978), and Terrapene ornata (Legler, 1960) is similar extent to that of sea turtles, but is displaced downward by approximately 2°C.

Data for lepidosaurians indicate that there may be a relation between limits of tolerance and local climate. Embryos of Sceloporus undulatus, a lizard from temperate regions of North America, have a range of thermal tolerance extending from

25°C to nearly 35°C (Sexton and Marion, 1974), whereas those of Dipsosaurus dorsalis, an iguanid inhabiting deserts of the American Southwest, survive exposure to 28—38°C (Muth, 1980). Furthermore, young of the tropical lizard Iguana iguana, which normally develop in nests that are well buffered against changes in temperature (Rand, 1972), will not complete development at temperatures departing by more than 2°C from an optimum at 30°C (Licht and Moberly, 1965). Finally, the lower limit of thermal tolerance for embryos of the snake Python molurus is approximately 28°C (Vinegar, 1973), suggesting adaptation of embryos to the relatively high temperatures that are likely to be encountered during incubation of this tropical species (Shine, this volume).

Causes of death in embryos developing at temperatures exceeding the limits of tolerance have not determined. However, the high incidence of developmental anomalies among young developing at temperatures towards the extremes of the range of tolerance indicates a general disruption of highly integrated physiological systems (Bustard, 1969b; Fox, 1948; Fox *et al.*, 1961; Gutzke and Packard, 1987a; Gutzke, 1984a: Osgood, 1978; Vinegar, 1973, 1974; Webb *et al.*, 1983a, c; Zweifel, 1980).

Data gathered in studies performed at constant temperatures are valuable in establishing the range of conditions that embryos can withstand for extended periods, but are of limited relevance to development in nature because of the diel cycles in temperature to which eggs usually are exposed. With the possible exception of those crocodilians, sea turtles, and others having nests that are buffered against changes in temperature, reptilian embryos probably are exposed periodically to temperatures exceeding the specified levels of tolerance, but they nonetheless develop normally. For example, embryos of the skinks Eumeces fasciatus and Eumeses obsoletus probably could not survive prolonged expocure to temperatures exceeding 40°C (see Muth, 1980; Sexton and Marion, 1974). Nevertheless, young can withstand temperature of 42°C for 4 hours (Fitch and Fitch; 1967) and 43°C for 30 min (Fitch, 1964). Thus, duration of exposure must be considered in any discussion of limits of thermal tolerance.

Embryos of numerous reptiles from temperate regions can withstand exposure for extended periods to low temperatures of 4—12°C (*e.g.*, Cunningham, 1922; Fitch, 1954; Legler, 1960; Maderson and Bellairs, 1962), but die at temperatures below this range (Fitch, 1964; Fitch and Fitch, 1967). Older embryos may be better able to tolerate such exposure than younger ones (Holder and Bellairs, 1962). During these extended periods at low temperature, metabolism of young apparently is arrested almost completely. Development resumes when embryos are subsequently returned to favourable thermal environments, and the only apparent effect of the sojourn at low temperature is prolongation of incubation by a period equivalent to the duration of the cold exposure (Legler, 1960).

2. *Metabolism*

Because reptilian embryos are ectotherms, their metabolism is expected to vary as a positive function of temperature (G.C. Packard *et al.*, 1977). This prediction has been tested only once, however, in a study of oxygen consumption by advance embryos of the snake Opheodrys vernalis (Zarrow and Pomerat, 1937). Four eggs from a single clutch found beneath a log were brought to the laboratory and studies at constant temperatures spanning the range 12—30°C. Metabolism of the embryos varied with temperature in the predicted manner, with a higher Q_{10} between 12—24°C than between 25—30°C.

A substantial body of indirect evidence also supports the concept that metabolism of embryonic reptiles is positively correlated with temperature. Embryos of the Python molurus grow more rapidly at high temperatures than they do at low temperatures (Vinegar, 1973), and developing young of numerous squamates (Clark, 1970; Dmi'el, 1967; Gutzke and Packard, 1987a; Mendelssohn, 1963; 1965, Muth, 1980; Platt, 1969; Sexton and Marion, 1974; Zweife, 1980) and turtles (Dodge *et al.*, 1978; Gutzke, 1984a; Legler, 1960; Miller, 1982, 1985; Miller and Limpus, 1981; Mrosovsky and Yntema, 1980; G.C. Packard *et al.*, 1987; Yntema, 1968, 1978) complete incubation sooner in warm environments than they do in cooler ones, Rapid growth and early completion of development at high temperatures presumably are reflections of higher rates of metabolism under these conditions than prevail at lower temperatures.

Another effect of temperature is manifested as variation in size of carcasses and residual yolk of hatchlings. Painted turtles (Chrysemys picta) and bullsnakes (Pituophis melanoleucus) hatching from eggs incubated at low (22°C) and intermediate (27°C) temperatures have larger bodies, but smaller residual yolks, than young emerging from eggs held at high (32°C) temperatures (Gutzke, 1984a; Gutzke and Packard, 1987a). Similar findings have been reported for snapping turtles (Chelydra serpentina) incubated at 26.0°C, 28.5°C, and 31.0°C (G.C. Packard *et al.*, 1987). Thus, embryos developing at high temperatures seemingly have higher rates of metabolism than those at low temperatures, but their metabolic efficiency—defined here as the efficiency of converting energy reserves into tissues—is reduced.

In contrast to these species, alligators (Alligator mississippiensis) hatching from eggs incubated at high temperatures (34°C) are reported to have smaller residual yolks than animals emerging from eggs incubated at lower temperatures (30°C) (Ferguson and Joanen, 1982, 1983). The study was not based on hatchling alligators, however, but on embryos that had undergone 60 days of development irrespective of incubation temperature (Ferguson and Joanen, 1982, 1983). Embryos sampled at 34°C probably were nearer to hatching than those sampled at 30°C, so findings for these two groups of embryos may not be representative of hatchlings. This study needs to be repeated to resolve the question of residual yolk size.

All of the aforementioned investigations were performed at constant temperatures. We look forward to learning whether the metabolic response of an embryo to a particular temperature is the same under transient and steady-state conditions.

3. Acid/Base Balance

Acid base physiology has been studied extensively in adult reptiles (Howell and Rahn, 1976), but not as yet in embryos. We nevertheless introduce the subject here to illustrate the pervasive effects of temperature on the physiology of reptilian embryos, and to gain perspective concerning problems of physiological control in animals exposed to sublethal fluctuations in temperature.

Consider the blood plasma of a reptilian embryo. The acid/base status of this liquid can be characterized using nine simultaneous equations for chemical equilibria (see Stewart, 1981):

1. $K'_W = [H^+][OH^-]$
2. $K_A [HA] = [H^+][A^-]$
3. $[A_{TOTAL}] = [HA[+ [A^-]$
4. $[CO_2] = S_{CO_2} \times P_{CO_2}$
5. $K'_{CO_2} [CO_2] = [H_2CO_3]$
6. $K_{H_2CO_3} [H_2CO_3 = [H^+][HCO_3^-]$
7. $K_{HCO_3} [HCO_3^-] = [H^+][HCO_3^{-2}]$
8. $[CO_2] + [H_2CO_3] + [HCO_3^-] + [CO_3^-{}_2] = CO_{2TOLAL}]$
9. $[SID] + [H^+] - [OH^-] - [A^-] - [HCO_3^-] - [CO_3^{-2}] = 0$

where K_W is the ion product for water, $[H^+]$ is the concentration of hydrogen ions (equivalents per litre), $[OH^-]$ is the concentration of hydroxyl ions (equivalents per litre), K_A is the dissociation constant for weak acid, [HA] is the concentration of weak acid in undissociated form (equivalents per litre), $[A^-]$ is the concentration of anion of the weak acid (equivalents per litre). $[A_{TOTAL}]$ is the total concentration of weak acid (equivalents per litre), $[CO_2]$ is the concentration of carbon dioxide in solution in the plasma (equivalents per litre), S_{CO_2} is the solubility coefficient for carbon dioxide in plasma (equivalents per litre per kPa), P_{CO_2} is the partial pressure of carbon dioxide (kPa), K'_{CO_2} is the equilibrium constant for hydration of carbon dioxide to form carbonic acid, $]H_2CO_3]$ is the concentration of carbonic acid (equivalents per liter), $K_{H_2CO_3}$ is the dissociation constant for carbonic acid, $[HCO_3^-]$ is the concentration of bicarbonate ions (equivalents per litre), K_{HCO_3} is the dissociation constant for bicarbonate, $[CO_3^{-2}]$ is the concentration of carbonate ions (equivalent per litre), $[CO_{2TOTAL}]$ is the total concentration of carbon dioxide (equivalents per litre), and [SID] is the strong ion difference (equivalents per litre). The strong ion difference is, effectively the sum of $[Na^+]$ and $[Ca^{-2}]$ miuus $[Cl^-]$.

Equation (1) is a formal statement of the dissociation of pure water into hydrogen and hydroxyl ions; equation (2) characterizes the dissociation of weak acid into its component ions; equation (3) characterizes conservation of mass for the weak acid; equation (4) quantifies the amount of dissolved carbon dioxide in the liquid; equation (5) is the equilibrium reaction for hydration of carbon dioxide to form carbonic acid; equation (6) describes the dissocation of carbonic acid into hydrogen and bicarbonate ions; equation (7) describes the dissociation of bicarbonate into hydrogen and carbonate ions; equation (8) characterizes the conservation of mass for carbon dioxide; ar d cquation (9) satisfies the requirement for electrical neutrality of the solution (Stewart, 1981).

Three fundamentally different classes of variables appear in these several equations: independent physical variables independent physiological variables, and dependent physiological variable (Stewart, 1981). Independent physical variables are those that are governed by laws of nature and over which developing embryos have no control; K_W, K_A, K'_{CO_2}, $K_{H_2CO_3}$, K_{HCO_3}, and S_{CO_2} are examples. Independent physiological variables, on the other hand, are those potentially under the direct physiological control of developing embryos, via organs such as the kidneys, intestine, or chorioallantois; we include [SID], $[A_{TOTAL}]$, and

P_{CO_2} in this category. Finally, dependent physiological variables are those that change passively in response to changes in both of the preceding classes of variables and that can be regulated only indirectly; $[H^+]$, $[OH^-]$, [HA], $[A^-]$, $[CO_2]$, $[H_2CO_3]$, $[HCO_3^-]$, $[CO_3^{-2}]$, and $[CO_{2TOTAL}]$ are examples.

Because all of the independent physical variables change with temperature (Albers, 1970), fluctutations in temperature of reptilian eggs presumably could elicit shifts in the several chemical equilibria characterizing blood plasma of embryos, thereby bringing about major changes in two of the dependent physiological variables. However, changes in two of the dependent physiological variables, namely [HA] and $[A^-]$, could seriously impair the capacity for embryos to exercise control over numerous metabolic functions.

A variety of proteins together comprise the weak acid found in plasma (Stewart, 1981). Some of these proteins are hormones some function in the transport of other molecules, and some are enzymes. In almost every instance, however, the specific function of a particular protein molecule is dependent on its conformation (*i.e.*, its tertiary and/or quaternary structure). Conformation of these proteins is affected by their net electrical charge, which is determined, in turn, by the degree of ionization of side-groups of the many amino acid residues; particularly the imidazole group of histidine (Reeves, 1977; Reeves and Rahn, 1979). To conserve conformation of all of these proteins and thereby to assure their continued role in a variety of metabolic activities, the fractional dissociation of the imidazolium must be held constant. However, the total concentration of protein, $[A_{TOTAL}]$, probably does not vary appreciably over the course of a single day (see Stewart, 1981), so control of fractional dissociation is equivalent to holding constant the concentrations of imidazolium in ionized ($[A^-]$) and unionized states ([HA]).

Among adult reptiles confronted with changes in body temperature, regulation of [HA] and $[A^-]$ is accomplished by altering the rate of ventilation relative to metabolic demand for oxygen (Howell and Rahn, 1976; Rceves. 1977). The P_{CO_2} of plasma is shifted upward at high temperatures and downward at low temperatures, thereby compensating for effects of temperature on the independent physical variables and holding [HA] and [A⁻] relatively constant (Reeves, 1977; Reeves and Rahn, 1979). If a similar mechanism for compensating temperature changes is manifested by embryos, it must entail shifts in blood flow through the chorioallantois and attendant changes in resistance of the membrane to gas exchange. Of course, control of [HA] and [A⁻] could be effected also by altering [SID] via the kidneys or the chorioallantois, but such mechanisms probably would be too slow to allow changes in acid/base status to keep pace with changes in temperature in the nest chamber.

This analysis could easily be extended to the intracellular compartment, where thermally-induced perturbations of acid/base status presumably would have further effects on metabolism (see Busa and Nuccitelli, 1984; Walsh and Moon, 1982; Yancey and Somero, 1978). Regardless of whether attention is focused on intracellular or extracellular compartments, however diel cycles in temperature inside nests of many reptiles clearly have the potential to disrupt integrated metabolic systems by virtue of effects on acid/base status of embryos. The degree to which embryos regulate acid/base status in the face of continuous changes in temperature, or tolerate disruptions of metabolism attending thermally-induced shifts in acid/base status, has yet to be evaluated.

4. Sexual Differentiation

All crocodilians, most turtles, and many lepidosaurians lack hetero-morphic sex chromosomes (Bull, 1980). Sexual differentiation by embryos in many of these animals is affected by the temperature of incubation (Bull, 1980). For example, incubation of turtle eggs at constant temperatures above 30°C usually elicits differentiation of phenotypic females, whereas exposure of their eggs to 24–28°C usually elicits differentiation of phenotypic males. Both males and females arise from eggs exposed to

28 – 30°C, but proportions vary from predominantly male near 28°C to predominantly female near 30°C. This general pattern of response has been reported for numerous cheloniids, chelydrids, dermochelyids, emydids, kinosternids, pelomedusids, and testudinids (Table 5.1). Exposure of eggs of some of these turtles to temperatures below 24°C again elicits differentiation of females, indicating that a second temperature "threshold" occasionally exists (Gutzke and Paukstis, 1984; Schwarzkopf and Brooks, 1985; Vogt *et al*,, 1982; Yntema, 1976, 1979, 1981). However, exposure of eggs of the other species to temperatures below 24°C continues to elicit differentiation of males (Pieau, 1978).

Sexual differentiation in chelids, trionychids, and some emydids (*e.g.*, Clemmys insculpta) is unaffected by temperature (Bull and Vogt, 1979; Bull *et al.*, 1985; Thompson, 1983; Vogt and Bull, 1982b), and presumably is controlled by the genotype. Genetic control is suspected to occur also in the few emydids and kinosternids (and chelids) possessing heteromorphic sex chromosomes (Bull *et al.*, 1974; Carr and Bickham, 1981; McBee *et al.*, 1985; Sites *et al.*, 1979).

Sexual differentiation by embryos of crocodilians and some lizards also is influenced by temperature. In contrast to the situation in turtles, however, the embryonic crocodilians and lizards usually differentiate as males at relatively high temperatures and as females at relatively low temperatures (Bull, 1980; Charnier, 1966; Ferguson and Joanen, 1982, 1983; Lang, 1985; Wagner, 1980). Embryos of Gekko japonicus and perhaps also of Crocodylus johnsoni are exceptional in showing patterns of sexual differentiation similar to those reported for many turtles, with females arising in eggs held at high or low temperatures and with males emerging from eggs exposed to intermodiate temperatures (Tokunaga, 1985; Webb and Smith, 1984; Webb *et al.*, 1983a).

Many lepidosaurians possess heteromorphic sex chromosomes (Bull, 1980; Gorman, 1973), and sexual differentiation in these forms presumably is under genetic control. Moreover, the

TABLE 5.1

Species of Turtles in Which Environmental Control of Sexual Differentiation has been Reported to Occur

Species	*Family*	*Reference*
1	*2*	*3*
Caretta caretta	Cheloniidae	Mrosovsky et al., 1984a; Yntema and Mrosovsky, 1980, 1982
Chelonia mydas	Cheloniidae	Miller and Limpus, 1981; Morreale et al., 1982; Mrosovsky, 1982; Mrosovsky et al., 1984b
Eretmochelys imbricata	Cheloniidae	Dalrymple et al., 1985
Lepidochelys olivacea	Cheloniidae	McCoy et al., 1983; Standora and Spotila, 1985
Chelydra serpentina	Chelydridae	G.C. Packard et al., 1987; Wilhoft et al., 1983; Yntema, 1976, 1979, 1981
Dermochelys coriacea	Dermochelyidae	Dutton et al., 1985; Mrosovsky et al., 1984b; Rimblot et al., 1985
Chrysemys picta	Emydidae	Bull and Vogt, 1979, 1981; Bull et at., 1982; Gutzke and Paukstis, 1983, 1984; Paukstis et al., 1984; Schwarzkopf and Brooks, 1985
Emydoidea blandingii	Emydidae	Gutzks and Packard, 1987b
Emys orbicularis	Emydidae	Pieau, 1971, 1972, 1975, 1976, 1978, 1982; Pieau and Dorizzi, 1981; Zaborski et al., 1982
Graptemys geographica	Emydidae	Bull and Vogt, 1979; Bull et al., 1982
Graptemys ouachitensis	Emydidae	Bull and Vogt, 1979, 1981; Bull et al., 1982

TABLE 5.1 (*Contd.*)

1	2	3
Graptemys pseudo-geographica	Emydidae	Bull and Vogt, 1979; Bull et al., 1982
Graptaemys pulchra	Emydidae	Bull et al,, 1982
Pseudemys scripta	Emydidae	Bull et al., 1982
Terrapene ornata	Emydidae	G.C. Packard et al., 1985a
Kinosternon flayescens	Kinosteruidae	Vogt et al., 1982
Sternotherus odouratus	Kinosternidae	Vogt et al., 1982
Podocnemis expansa	Pelomedusidae	Alho et al., 1985
Testudo graeca	Testudinidae	Pieau, 1971, 1972, 1975

process is under genetic control even in some of the species lacking distinct sex chromosomes (Muth and Bull, 1981; Raynaud and Pieau, 1972).

Patterns of sexual differentiation of embryonic turtles and crocodilians developing in natural nests generally are consistent with results of laboratory investigations, indicating that experiments performed at constant temperature are ecologically relevant. Nests of cheloniids, emydids, and pelomedusids located in warm, sunny sites give rise to a greater proportion of female hatchlings than do nests located in cool, shady settings (Alho *et al*,, 1985; Bull, 1985; Bull and Vogt, 1979, Morreale *et al.*, 1982; Vogt and Bull, 1984). Also, the proportion of females emerging from nests of Emys orbicularis is directly related to the amount of time that nest temperature is above 28.5°C (Pieau, 1982), whereas the proportion of males hatching from nests of Chrysemys picta seems to be related to the amount of the time eggs are exposed to temperatures between 20°C and 27.5°C (Schwarzkopf and Brooks, 1985). Hatchling Chelydra serpentina emerging from eggs in the warm, upper reaches of nests tend to be female, whereas those emerging from eggs in the cooler regions at the bottom of nests tend to be males (Wilhoft *et al.*, 1983). Finally, nests of alligators (Alligator mississippiensis) located on levees are much warmer than nests built in open marsh, and the former yield larger proportions of males than do the latter (Ferguson and Joanen, 1982, 1983).

The middle one-third of incubation is the period in which temperature exerts its control over sexual differentiation of turtle embryos (Bull and Vogt, 1981; Pieau and Dorizzi, 1981; Yntema, 1979: Yntema and Mrosovsky, 1982), but the period of sensitivity is somewhat earlier than this in crocodilian embryos (Ferguson and Joanen, 1982, 1983). However, crocodilian embryos are slightly more advanced at oviposition than are turtle embryos so the period of sensitivity may actually occur at equivalent stages in both groups. The period of sensitivity has not been established for any of the lizards in which enviranmental control of sexual differentiation occurs.

The exact way in which temperature intervenes in differentiation of the gonads is not clear (Raynaud and Pieau, 1985; Standora and Spotila, 1985; also Although none of the species exhibiting environmental control of sexual differentiation possesses heteromorphic sex chromosomes (Bull, 1980), genetic males and females nonetheless can be identified, at least in turtles, by the use of H-Y antigens (Engel *et al.*, 1981). Genetic factors are easily overridden by temperature (Zaborski *et al.*, 1982), however, leading to differentiation of the gonads into testes or ovaries. Secretion of steroid hormones presumably induces the subsequent differentiation of secondary sexual characters appropriate to males and females (Pieau *et al.*, 1982; Raynaud and Pieau, 1985).

WATER

A. Factors Affecting Water Balance of Eggs

Water presumably moves in both liquid and vapour phases down gradients in free energy in the soil-egg continuum (Corey and Klute, 1985; Taylor, 1965; Taylor and Stewart, 1960). Accordingly, any treatment of the factors affecting water transport in this continuum should be based on thermodynamic principles. Water potential is a measure of the energy required to overcome the forces of adhesion, cohesion, and ionic attraction that bind molecules of warer to particles of soil (Marshall and Holmes, 1972; Taylor and Ashcroft, 1972). We therefore adopt the concept of water potential for use in the present discussion.

The water balance of a reptilian egg can be described using an equation based on the principle of conservation of mass (see Ackerman *et al.*, 1985b; Tracy 1982; Tracy *et al.*, 1978):

$$m_S = m_M + m_L + m_V$$

where m_S is storage (or loss) of water inside the egg, m_M is water produced by the embryo in intermediary metabolism, m_L is uptake (or loss) of liquid from the environment, and m_V is uptake (or loss) of vapour from the surroundings. Whenever the uptake of water from the environment plus production of metabolic water balances water loss, there is no change in the mass of water available to support embryonic development. Whenever, the uptake of water plus metabolic water does not balance water loss, however, the mass of water must change. The terms in this equation for water storage (m_S in $mg \times s^{-1}$) and for production of metabolic water (m_M in $mg \times s^{-1}$) are straightforward and require no special explanation, but the other two terms can be expanded to reveal more explicity the factors affecting water transport between eggs and their environment.

Transport of liquid water can be described as

$$m_L = k_L \times A_S \times (\psi_T - \psi_E)$$

where k_L is a coeffieient for transport of liquid water across the eggshell ($mg \times s^{-1} \times cm^{-2} \times kPa^{-1}$), A_S is surface area of the eggshell contacting the substrate (cm^2), and $(\psi_S - \psi_E)$ is the difference in water potential between the soil and egg (kPa). All of the independent variable in this equation are subject to change during the course of incubation, thereby making transpost of liquid a very complex process. For instance, water potential of the soil may vary appreciably over the course of incubation in concert with cycles in wetting and drying. The effective water potential of the egg itself changes during development as different fluid compartments (with different solute potentials) come into proximity to the inside of the eggshell (Muth, 1981; G.C. Packard *et al.*, 1981a, 1983). Also, the surface area available for transport of liquid may change in consequence of dirt and debris infiltrating into air spaces in the nest chamber, and the transport coefficient may change in eggs with flexible shells owing to distension or contraction of fibres in the shell membrane.

The soil water potential of interest in education is that for soil contacting the surface of the eggshell. In most instances, this value will not be the same as that characterizing soil only a few centimeters away (Gutzke, 1984b; Tracy, 1982). Water generally cannot be transported through the substrate fast enough to keep pace with water exchanges at the surface of the eggshell. Consequently, the soil adjacent to the eggshell either accumulates water or is depleted of water, and therefore is at a different water potential than the soil at large. This phenomenon is familiar to plant physiologists, who have documented the major resistance to water transport imposed by soil adjacent to plant roots (*e.g.*, Legge, 1985).

Reptilian eggs also exchange vapour with their environment, as follows:

$$m_V = k_V \times A_A \times (p_E - p_A)$$

where k_V is the transport coefficient for vapour diffusing between the egg and surrounding air (including that in the soil) $(mg \times s^{-1} \times cm^{-2} \times kPa^{-1})$, A_A is the area of the eggshell across which vapour is transported (cm^2), and p_E and p_A are the vapour pressures of the egg and air, respectively (kPa). All of the independent variables in this equation also are subject to change during the course of incubation, so vapour transport is another eomplex process. For example, vapour pressure of both the eggs and surrounding air change with temperature, and therefore can be expected to change according to a daily cycle. Additionally, the transport coefficient may be higher at high ambient vapour pressures than at low ambient vapour pressures (Feder *et al.*, 1982; Lillywhite and Ackerman, 1984).

The present treatment of water balance differs in certain ways from earlier ones. A common assumption in prior research was that water produced by embryos in intermediary metabolism is not an important component of water balance (Ackerman, *et al.*, 1985b; G.C. Packard *et al.*, 1981a; Tracy, 1982; Tracy *et al.*, 1978). However, avian embryos produce sustantial quantities of water metabolically (Ar and Rahn, 1980), so it may be premature to assume that metabolic water is unimportant to reptilian embryos. Equation therefore includes a term for metabolic water.

Secondly, water exchange across reptilian eggshell has been attributed by some workers entirely to diffusion of vapour (Ackerman *et al.*, 1985b). However, the uptake of nonvolatile dyes by incubating turtle eggs indicates that transport of liquid also occurs (Thompson, 1987). We therefore include a term in equation for transport of liquid as well as one for transport of vapour. Transport of liquid may ultimately be shown to be a minor component of the water budget of an egg (Ackerman *et al.*, 1985b), but empirical evidence must be developed in support of this view.

Finally, transport of vapour across eggshells has been assumed by some workers to be continuously in an outward direction (Tracy, 1982; Tracy *et al.*, 1978). This assumption has recently been questioned on both theoretical (Ackerman *et al.*, 1985b) and empirical grounds (Morris *et al.*, 1983; G.C. Packard *et al.*, 1979a, 1980, 1981a), and seems no longer to be tenable. We subscribe to the view that vapour passes in either direction across eggshells, with the net direction of movement being established by gradients in vapour pressure.

B. Hydration of Nests

The flexible-shelled eggs of many testudinians and lepidosaurians have long been known to respond dramatically to variation in hydration of their environment (*e.g.*, Brimley, 1903; Coker, 1910; Cunningham and Huene, 1938; Cunningham and Hurwitz. 1936; Cunningham *et al.*, 1939; Dendy, 1899b). However despite the consequent awareness among herpetologists of the likely importance of water to developing young (G.C. Packard *et al.*, 1977), relatively few workers have measured moisture in reptilian nests. The few measurements fall into three categories based on methodology.

1. Several studies reported the mass of water per unit mass of sand or soil in the walls of subterranean nests of turtles and certain crocodilians (Bustard and Greenham, 1968; Coker, 1906; Lutz and Dunbar-Cooper, 1984; NcGehee, 1979; Webb *et al.*, 1983a), and one reported the water content of vegetation forming mound nests of another crocodilian (Chabreck, 1975). Although data such as these may reveal tendencies for walls of needs to collapse during construction (*e.g.*, Bustard and Greenham, 1968) or for gases to diffuse through substrates into nest cavities (*e.g.*, Lutz and Dunbar-Cooper, 1984), they are of little value in elucidating water relations *per se*. This point can be illustrated most readily by drawing an analogy between the water content (and water potential) of a substrate and its heat content (and temperature).

Heat content is a familiar measure for the total kinetic energy of all the molecules comprising the substrate in question. However, information on heat content alone cannot reveal whether heat will flow between this substrate and some reference system, or whether the two media are in a thermodynamic equilibrium. The direction of heat flow is determined by the average kinetic energy of molecules in the substrate relative to that of molecules in the reference system. Thus, temperature, which is a measure of average kinetic energy, is more useful than heat content in asseseing the tendency and direction for heat to move between the substrate and the reference system.

Measurements of gravimetric water-content are analogous to measurements of heat content, and measurements of water potential are analogous to measurements of temperature. Water content yields information on how much water actually is present in the substrate, but water potential yields information on the tendency for water to move isothermally between this substrate and some reference system. Thus, measurements of water potential are more useful than measurements of water content in elucidating the movements of water inside reptilian nests.

Water retention curves for three different kinds of soils illustrate this point further (Fig. 5.4). A sand substrate containing 4 per cent water is at equilibrium with a clay substrate containing 21 per cent water, because the water potential is the same in both media (*i.e.*,—200 kPa), Thus, there is no tendency for water to flow from the clay to the sand, even though much larger quautities of water are present in the former than in the latter. Conversely, a silty substrate containing 19 per cent water is not in equilibrium with a clay substrate at the same level of hydration, because water potential of the former is —200 kPa and that of the latter is —420 kPa. Water would flow from the silt to the clay despite the fact that water content of the two media is identical at the outset.

A surprising diversity of soil types is used for nesting by individual species of reptiles, even within a restricted geographic area (Lutz and Dunbar-Cooper, 1984; Stancyk and Ross, 1978; Thompson, 1983; Webb *et al.*, 1983a; Whitaker and Whitaker, 1984). The different water retention characteristics for different types of soil (Fig. 5.4) indicate clearly that comparisons among reptilian nests in different kinds of substrates should be undertaken with measurements of water potential, but not with measurements of gravimetric water content.

2. Hydration inside nests of the gecko Gehyra variegata and of the crocodilians Alligator mississippiensis and Caiman crocodilus was assessed by measuring relative humidity (RH) of air trapped inside the egg chamber (Bustard, 1969a; Joanen,

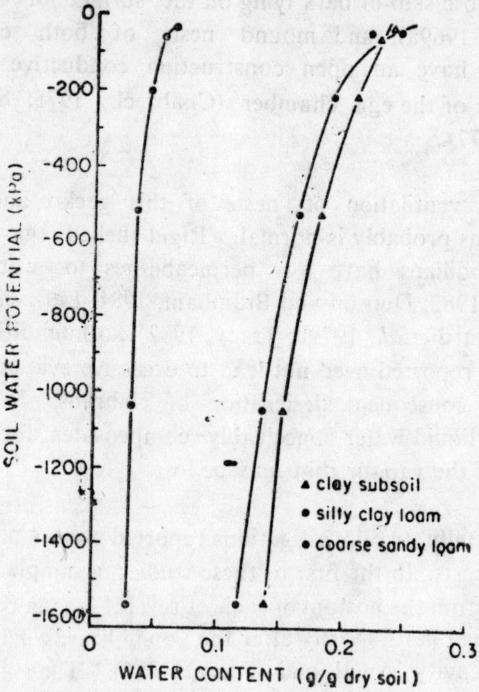

Fig. 5.4: Water retention curves for three substrates at a temperature of 15°C. The substrates vary in texture from a sandy loam (large particles) to a clay (fine particles).

1970; Staton and Dixon, 1977). Values of 55-80 per cent RH were recorded in one gecko nest, and 85-98 per cent RH in several nests of the crocodilians. Although relative humidity can be used in principle to infer water potential of substrates (Brown, 1970), the humidities in these nests would correspond to exceedingly dry conditions (Marshall and Holmes, 1979). Droplets of water nonetheless were observed in nests of one of the crocodilians (Joanen, 1970), so the nests probably were not as dry as the humidities would make it appear. We suspect that the nests were partially flushed with atmospheric air, thereby preventing vapour in air from coming into equilibrium with liquid in the substrate. This interpretation is consistent with location and construction of the nests. The gecko nest was under a loose slab of bark lying on the surface of the ground (Bustard, 1969a), and mound nests of both crocodilians commonly have an open construction conductive to partial ventilation of the egg chamber (Chabreck, 1975; Staton and Dixon, 1977).

Partial ventilation of nests of this gecko and the two crocodilians probably is normal. Rigid-shelled eggs of geckos and crocodilians have low permeabilities to water vapour (Dunson, 1982; Dunson and Bramham, 1981; Lutz *et al.*, 1980; G.C. Packard *et al.*, 1979b; Tracy, 1982), so humidities such as have been reported need not lead to excessive evaporation from eggs and consequent desiccation of embryos. Furthermore, uptake of liquid water conceivably compensates for at least a portion of the vapour that may be lost.

3. Finally, two investigations reported water potentials in reptilian nests. In the first of these studies, a sample of soil was removed from the bottom of each of several iguana (Conolophus subcristatus) nests shortly after the young lizards had emerged from the cavity (Snell and Tracy, 1985). The samples were sealed in glass vials and later were taken to the laboratory, where water potentials apparently were measured by dewpoint hygrometry (sensu Savage *et al.*, 1981). Values ranged from approximately—100 kPa in wet nests to less than —11,000 kPa in dry ones. Water potentials below —1,500 kPa generally lead to permanent wilting of plants (including many desert forms), so the data for iguanas indicate that their eggs may be exposed to a truly remarkable spectrum of conditions.

Collecting soil samples at the end of incubation and taking them into the laboratory for further analysis has the major advantage of being relatively simple. The important disadvantage of this method, however, is that a single measurement taken at the end of incubation affords only a rough index, at best, of the moisture conditions to which eggs were exposed throughout development. Also, the sampling-induced change in bulk density of the soil may reduce the accuracy of measurements

TABLE 5.2

Diel Cycle in Temperature, Water Potential, and Vapour Pressure Inside a Nest of a Snapping Turtle (Cheludra Serpentina)

Time	*Temperature (°C)*	*Water potential (kPa)*	*Relative humidity (%)*	*Saturation vapour pressure (kPa)*	*Nest vapour pressure (kPa)*
Top of nest					
0600	22.00	0	100.0	2.643	2.643
1200	24.25	0	100.0	3.028	3.028
1800	31.00	—560	99.6	4.492	4.474
2400	26.50	0	100.0	3.461	3.461
Middle of nest					
0600	12.75	—60	100.0	2.767	2.767
1200	22.50	—40	100.0	2.725	2.725
1800	28.75	0	100.0	3.947	3.947
2400	27.00	—60	100.0	3.565	3.565
Bottom of nest					
0600	23.25	0	100.0	2.851	2.851
1200	23.00	0	100.0	2.809	2.809
1800	26.50	0	100.0	3.461	3.461
2400	26.50	0	100.0	3.461	3.461

From G.C. Packard *et al.*, 1985b.

An important disadvantage of the psychrometric method is that it is highly labour-intensive. Even the calibration of psychrometers is extraordinarily time-consuming, because each probe should be individually calibrated against standard salt solutions before it is used in the field and again after all measurements have been completed (Brown and Bartos, 1982; Meyn and White, 1972; Savage *et al.*, 1981, 1983). A second disadvantage concerns the extraordinary sensitivity of psychrometers to the smell gradients in temperature that occur commonly in soil (Merrill and Rawlins, 1972; Wiebe and Brown, 1979; Wiebe *et al.*, 1977), Special care must be taken in placing probes in nests and in taking readings to assure that the measurements of water potential are reliable (Brown and Bartos, 1982).

C. Exchanges of Water by Eggs in Nature

Reptilian eggs incubating in natural nests seemingly experience a net uptake of water under some conditions and a net loss of water under others. The evidence supporting this statement admittedly if fragmentary and oftentimes circumstantial, but water exchanges undoubtedly occur.

For example, flexible-shelled eggs of a lizard and a turtle have been observed to swell during incubation and to become firmly wedged into their nest chambers (Cagle, 1950; Whitaker, 1968). Additionally, flexible-shelled eggs of turtles and rigid-shelled eggs of a crocodilian have been reported to weigh more after a period of incubation than they did to oviposition (Hendrickson, 1958; M.J. Packard *et al.*, 1982c: Staton and Dixon, 1977). Such findings are convincing evidence for net absorption of water by eggs incubating in nature.

The uptake of water by rigid-shelled eggs of turtles and crocodilians may even cause the shell to fracture under some circumstances (Goodwin and Marion, 1978; Joanen, 1970; McIlhenny, 1934, 1935: Plummer, 1976; Pooley, 1962; Webb *et al.*, 1977, 1983c), and associated swelling may cause edges of the cracks to draw apart to expose the underlying membrane. As long as the membrane remains intact, however, the embryo usually completes development successfully (Joanen, 1970; also Ferguson, 1982). Absorption of water to the point of fracturing the calcareous layer of the eggshell probably is the exception rather than the rule (G.C. Packard *et al.*, 1977), but may happen any time that free water stands in the nest chamber for an extended period. Water may collect in nests that are subject to inundation (Plummer, 1976) or that are sufficiently "tight" as to prevent rainwater from draining away (Legler, 1954).

Conversely, flexible-shelled eggs of turtles and rigid-shelled eggs of crocodilians have also been reported on other occasions to decline in mass during incubation (Ltuz and Dunbar-Cooper, 1984; Moore, 1953; M.J. Packard *et al.*, 1982c; Staton and Dixon, 1977), indicating that a net loss of water has occurred. The occasional formation of dents in flexible-shelled eggs of turtles (Cagle, 1937; Legler, 1954) and of air spaces inside rigid-shelled eggs of crocodilians is further evidence for loss of water from eggs incubating in natural nests (Ferguson, 1982; Webb *et al.*, 1983c).

Thus, eggs incubating in natural nests of a particular species probably vary widely in net water-exchange with the environment and consequently in overall water balance (M.J. Packard *et al.*, 1982c). We shall now consider the possible effects of such variation on developing embryos.

D. Physiological Effects of Water

1. *Methods of Study*

Most research into the importance of water to reptilian embryos has been performed in the laboratory, where the environment to which eggs are exposed is subject to a measure of control. Unfortunately, work reported before 1978 (and numerous accounts appearing more recently) suffered from inadequate characterization for hydration of the environment, and therefore is of limited value in elucidating either the patterns of water exchange by incubating eggs or the physiological effects of such exchange on developing embryos (G.C. Packard *et al.*, 1977). Substrates used in many of these earlier investigations seem to have been saturated with water —a condition that would have had a profound influence on patterns of water exchange by eggs and probably on physiology of developing young (G.C. Packard *et al.*, 1980). We therefore focus on those studies appearing since 1978 in which water potentials were reported.

Recent investigations of the water relations of reptilian eggs have made use of three different methodologies. Some workers have incubated eggs over salt solutions, thereby controlling water potential of the environment more precisely than would otherwise be possible (Muth, 1980, 1981). Other investigations have incubated eggs on substrates of moistened vermiculite, primarily because of the ease and convenience in working with this material (Black *et al.*, 1984; Gutzke, 1984a; Gutzke and Packard, 1986, 1987a; Morris *et al.*, 1983; G.C. and M.J. Packard, 1987; G.C. Packard *et al.*, 1979a, 1980, 1981a, b, c, 1982, 1983, 1984a, b, 1985a, 1987; M.J. and G.C. Packard, 1986; M.J. Packard *et el.*, 1980; Tracy, 1980; Tracy *et al.*, 1978). Still other investigators have used moistened soil or sand as a medium, thereby better simulating natural conditions for incubation (Andrews and Sexton, 1981; Thompson 1983; Tracy and Snell, 1985).

Recent research with a mechanistic model led to the prediction that patterns of net water-exchange by reptilian eggs will be profoundly affected by the type of substrate on which they rest (Ackerman *et al.*, 1985b). Simulations indicated that eggs resting on media of high thermal conductiviry (*e.g.*, sand) should have lower temperatures than eggs resting on media of low thermal conductivity (*e.g.*, vermiculite), primarily because of differences in the rate at which metabolic heat released by embryos is dissipated into the environment. The differences in temperature should, in turn, affect net exchanges of water between eggs and their surroundings. Accordingly, the use of different substrates for incubating eggs may have introduced a systematic bias into earlier investigations, thereby precluding comparison among many of the studies and among the species concerned.

However, the prediction of the model was not confirmed in a recent experiment, so the model needs to be reconsidered. The flexible-shelled eggs of snapping turtles (Chelydra serpentina) were incubated on sand and vermiculite at different temperatures and water potentials (G.C. Packard *et al.*, 1987). Net water-exchange by eggs was assessed, and so too were effects of the different substrates an hatching success and on characteristics of hatchlings. Although the type of substrate had a detectable influence on many (but not all) of these variables, this influence generally was minor. For instance, eggs resting on sand exhibited temporal patterns of net water-exchange that were virtually identical to those for eggs incubating on vermiculite at the same temperatures and water potentials. None of the conclusions from this investigation would have been affected by incubating eggs on just one medium or the other. Thus, valid comparisons apparently can be made even when eggs have been incubated on substrates having different physical characteristics.

Bias was much more likely to have been introduced into the earlier investigations by burying eggs in the substrate to different degrees. Some workers placed eggs on the surface so that only minimal contact was made between the eggshell and the medium (Muth, 1980, 1981). Others buried eggs partially (Andrews and Sexton, 1981; Gutzke, 1984a; Gutzke and Packard, 1986, 1987a; Morris *et al.*, 1983; G.C. and M.J. Packard, 1987; G.C. Packard *et al.*, 1979a, 1980, 1981a, b, c, 1982, 1983, 1984a, b, 1985a, 1987; M.J. and G.C. Packard, 1986; M.J. Packard *et al.*, 1980; Tracy, 1980; Tracy and Snell, 1985; Tracy *et al.*, 1978) or fully in the medium (Thompson, 1983). Both theoretical considerations (Ackerman *et al.*, 1985b; Tracy *et al.*, 1978) and empirical evidence Bustard, 1966; M.J. Packard *et al.*, 1980; Tracy and Snell, 1985) indicate that the pattern of net water-exchange characterizing an egg will be influenced by the amount of contact between the egg and the substrate. Thus, some of the reported variation among species (see below) probably reflects on different methods for incubating eggs.

2. *Limits of Tolerance*

An objective in most (but not all) studies of the water relations of reptilian eggs has been to define the hydric conditions necessary for successful completion of incubation. This goal has been elusive, however, because the water exchange between an egg and its environment is affected both by ambient temperature (and therefore egg temperature) and by water potential of the substrate (Gutzke, 1984a; Gutzke and Packard, 1987a; G.C. Packard *et al.*, 1987). Consequently, eggs incubating on a substrate a particular water potential may hatch at one temperature but not at another temperature within the range of thermal tolerance (G.C. Packard *et al.*, 1987). We shall return later to this matter of interaction of temperature and water potential. It suffices to say here that statements concerning effects of the hydric environment on hatching success apply only to the temperature at which the study was performed and not to other temperatures.

Flexible-shelled eggs of small lizard apparently must be deposited in settings conductive to absorption of substantial quantities of water during incubation. The mass of eggs must increase by 100-200 per cent over their mass at oviposition; lesser increases in mass are accompanied by high mortality of embryos (Andrews and Sexton, 1981; M.J. Packard *et al.*, 1980; Tracy, 1980). This requirement for water absorption is satisfied by substrates above —240 kPa (T_a = 35°C) for eggs of Callisaurus dranconoides, by substrates above —420 kPa (T_a= 28°C) for eggs of Sceloporus undulatus, and by substrates above-1480 kPa (T_a 23-28°C) for eggs of Anolis auratus and Anolis limifrons. We do not know whether eggs of Anolis are more tolerant of dry conditions than are eggs of Callisaurus and Sceloporus, or whether the observed survival of Anolis eggs at low water potentials resulted from their being incubated at lower temperatures than eggs of the other species. Embryos of Dipsosaurus dorsalis survive incubation on substrates at— 1000 kPa T_a = 36°C), but no information is available on their patterns of net water-exchange (Muth, 1980).

The flxible-shelled eggs of larger lizards and snakes will also absorb large amounts of water from suitable environments, but such uptake is not essential for the survival of embryos to hatching. Indeed, eggs of Conolophus, subcristatus, Caluber constricter, and Pituophts melanoleucus can sustain small (net) losses of water over the course of incubation without experiencing a reduction in hatching success (Gutzke and Packard, 1987a; G.C. and M.J. Packard, 1987; Tracy and Snell, 1985). The ability of these eggs to withstand losses of water probably is related to their relatively large size at oviposition and to the large reservoir of water that they therefore contain. In any event, eggs of Conolophus survive exposure to substrates a t—7500 kPa (T_a = ~30°C), whereas those of Goluber and Pituophus withstand exposure to substrates at—950 kPa (T_a= 27°C) and 1100 kPa (T_a = 22, 27, 32°C), respectively.

Flexible-shelled eggs of turtles may increase in mass during incubation owing to absorption of water from the environment, but this uptake of water usually is not essential for completion of development. Eggs of many species hatch successfully even after declining appreciably in mass (Bustard, 1971a). However, species differ in the amount of water than can be lost from eggs without ill effect. Eggs of Chelydra serpentina frequently hatch after declining in mass by 18 per cent over the course of incubation (Morris *et al.*, 1983), whereas eggs of Chrysemys picta seldom can withstand such a loss of water (G.C. Packard *et al.*, 1981a, 1983). The eggs of Chelydra are much larger than those of Chrysemys, and therefore contain more at oviposition. Thus, the larger eggs of Chelydra may differ from the smaller eggs of Chrysemys by simply being better buffered against the potentially lethal effects of water loss (see Gutzke and Packard, 1985).

Flexible-shelled eggs of both Chelydra serpentina and Chrysemys picta experience increased mortality at water potentials below—500 kPa (T_a = 26.0—31.0°C in Chelydra, T_a = 29.0°C in Chrysemys), but this mortality is more pronounced in the latter than in the former (G.C. Packard *et al.*, 1981a, 1983, 1987). The more rigid eggs of Emyioidea blandingii have high hatching success even when exposed to water potentials below —700 kPa (T_a = 29.0°C; G.C. Packard *et al.*, 1982).

Rigid eggs of turtles, and presumably of crocodilians and geckos, can successfully withstand exposure to a wide range of hydric conditions (G.C. Packard *et al.*, 1979a, 1981c; Thompson, 1983). The calcareous layer of the eggshell prevents the loss of large amounts of water to dry surroundings (Lutz *et al.*, 1980; G.C. Packard *et al.*, 1979a, b, 1981c; Thompson, 1983; Tracy, 1982), yet also opposes the uptake of water from wet environments (G.C. Packard *et al.*, 1979a, 1981c; Thompson,1983). Only the most severe of conditions are likely to elicit mortality of embryos developing in eggs such as these.

3. Metabolic Effects

Reptilian embryos in flexible-shelled eggs frequently hatch at a larger size when they are incubated in relatively wet environments than they do when they are incubated in relatively dry surroundings. The availability of water to embryos apparently is the determining factor, because the size of hatchlings is correlated with the net exchange of water experienced by their eggs.

Embryonic turtles take longer to develop to hatching in moist environments than they do in dry ones (Gutzke, 1984a; Morris *et al.*, 1983; G.C. Packard *et al.*, 1981a, 1983, 1984b, 1985a, 1987). Embryos developing in moist environments therefore have more time in which to grow before hatching than do those in dry settings. Thus, the larger size of turtles hatching in moist environments is due in part to their extended incubation. One hypothesis to account for such variation in length of incubation proposes that hatching is induced when water potential of the yolk (or some other compartment) declines to a threshold level, and that this threshold is reached sooner by embryos in dry settings than by those in wet ones. Embryos of some lizards and snakes also incubate longer in wet than in dry settings, so the hypothesis may apply to these animals as well.

Embryonic turtles in flexible-shelled eggs also have higher rates of metabolism in wet environments than they do in dry ones. Consequently, embryos in wet environments mobilize nutrients from the yolk more rapidly than do embryos in dry environments, and growth of embryos also is faster in wet conditions than in dry ones. An hypothesis to account for variation in metabolism and growth of embryos proposes that intermediary metabolism is inhibited by urea accumulating inside eggs. Because less water is available inside eggs in dry conditions to serve as a diluent for the urea, concentrations of this metabolite increase more rapidly in eggs in dry environments than in eggs in wet environments. Accordingly, inhibition of embryonic metabolism and growth is greater during the last one-third of incubation in dry environments than in wet ones. Growth rates of embryos have not been studied in those lizards and snakes in which size at hatching is influenced by hydration of the environment, so we do not know whether the present hypothesis has any broader application.

Embryonic turtles developing in rigid-shelled eggs seem not to be affected by variation in hydration or the environment, because netther oxygen consumpation of embryos not size of hatchlings is related to the wetness of the substrate on which eggs are incubated. However, this does not necessarily nean that embryos in rigid-shelled eggs are more tolerant of variation in the availability of water than are embryos in flexible-shelled eggs; it nay simply mean that embyos in rigid eggs are better insulated from the vagaries of the environment than are embryos of flexible eggs. The calcareous layer of rigid eggshells constraints movement of water between the interior and the surroundings, and rigid eggs in wet and dry environments therefore differ only slightly in net exchange of water. The resultant differences in availability of water to embryos developing in wet and dry conditions persumably are too small to elicit detectable differences in physiology of the young,

No information is available as yet on the possible influence of the hydric environment on duration of incubation, metabolism, or growth of embryonic crocodilians or of those geckos developing inside rigid-shelled eggs. We anticipate that their responses will be similar to those reported for turtle embroys developing in rigid, calcareous eggs.

4. Response of Eggs to Changing Hydration

In most studies of the water relations of reptilian embroys, individual eggs have been exposed to one, relatively constant water potential for the duration of incubation. However, one investigation considered the effect of exposing eggs to different water potentials at different times during development. In that study, flexible-shelled eggs of painted turtles (Chrysemys picta) were incubated in wet (—150 kPa) and dry—1100 kPa) environments at an ambient temperature of 28.5°C (Gutzke and Packard, 1980. Half of the eggs in each treatment were transferred to the other treatment at the end of the first trimester, and a similar shift was performed at the end of the second trimester. Thus, eight experimental groups were defined on the basis of hydric conditions to which eggs were exposes during each trimester. Some eggs were exposed to wet conditions for the first one-third of incubation and to dry conditions thereafter; other eggs were exposed to wet conditions only during the middle one-third to development; and so on.

Despite the fact that eggs on the dry substrate declined appreciably in mass daring the first trimester and that eggs on the wet medium simultaneously increased in mass (Figure 5.6), the condition of incubation during the first tremester had no apparent influence on survival of embryos or on size of hatchlings (Gutzke and Packard, 1986). This unexpected finding can be explained by compensatory exchanges of water occurring during the second trimeter. Eggs transferred from the wet substrate to the dry substrate at the end of the first trimester lost water more rapidly during the second trimester than did eggs exposed continuously to the dry environment (Figure 5.6). Consequently, the advantage held by eggs spending the first trimester in a wet environment was largely nullified by subsequent exposure to the dry setting. On the other hand, eggs transferred from the dry substrate to the wet substrate at the end of the first trimester absorbed water more rapidly during the recond trimester than did eggs exposed continuously to the wet conditions (Figure 5.6). Thus, the disadvantage of eggs spending the first trimester in a dry environment was largely

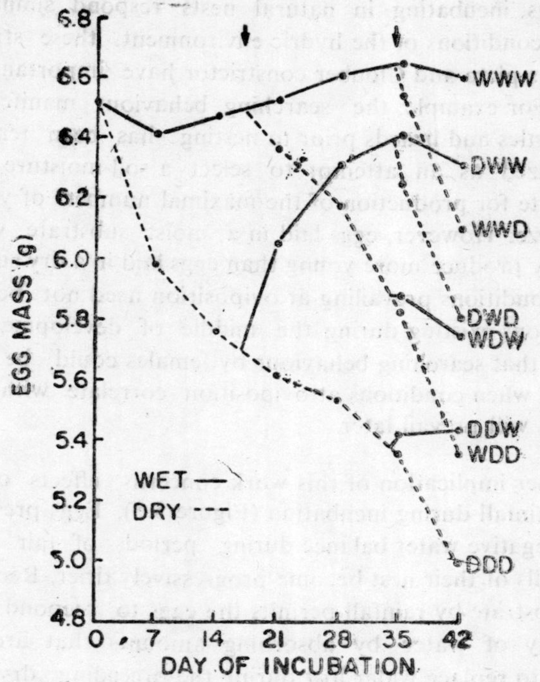

Fig. 5.6. Mass of eggs of painted turtles (Chrysemys picta) incubating on wet (—150 kPa) and dry (—1100 kPa) substrats at an ambient temperature of 28.5°C. The arrows mark the ends of the first and second trimesters of incubation, and identify times at which eggs were transferred from wet substrates to dry substrates and *vice versa*. The notation WWW indicates that eggs were exposed to the wet environment in all three trimesters; DWW indicates that eggs were exposed to the dry environment during thr first trimester and to the wet environment during the second and third trimesters; and so on.

overcome by subsequent exposure to a wet medium. Findings similar to these were reported over 32 years ago for eggs of Coluber constrictor, but were not widely accepted owing to uncertainty concerning the actual hyderation of substrates used for incubation (Clark, 1953a). No mechanistic explanation yet exists for this phenomenon of "compensatory water exchange.

If eggs incubating in natural nests respond similarly to changing conditions of the hydric environment, these studies of Chrysemys picta and Clouber constrictor have important implications. For example, the searching behaviour manifested by female turtles and lizards prior to nesting has been tentatively characterized as an attempt to select a soil-moisture regime appropriate for production of the maximal number of young of optimal size. However, eggs laid in a moist substrate will not necessarily produce more young than eggs laid in a dry substrate, because conditions prevailing at oviposition need not be similar to conditions existing during the middle of development. The only way that searching behaviour by females could be advantageous is when conditions at oviposition correlate with conditions that will prevail later.

Another implication of this work concerns effects of intermittent rainfall during incubation (Figure 5.5). Eggs presumably go into negative water balance during periods of fair weather as the walls of their nest become progressively drier. Recharging of the substrate by rainfall permits the eggs to reepond to the availability of water by absorbing amounts that are largely sufficient to replace water lost during the preceding dry period. In this way, both hatching success and size of hatchlings could be maximized.

5. *Sexual Differentiation*

One experiment has addressed the possibility that hydration of the environment influences the process of sexual differentiation in a species known already to respond to temperature. In this study, flexible-shelled eggs of the turtle Chrysemys picta were incubated on vermiculite at several different temperatures and water potentials (Gutzke and Paukstis 1983). All hatchlings emerging from eggs held at higher tempernture (30.5°C, 32.0°C) were females, irrespective of the wetness of the substrate. This finding is consistent with results of earlier research on this species (Bull and Vogt, 1979, 1981). At lower temperatures (16.5 C, 27.0°C), however, the expectation that all emergent young would be males (see Bull and Vogt, 1979, 1981) was realized only for eggs exposed to higher water potentials. Significant numbers of females emerged from eggs held on drier substrates.

Eggs in dry environments at a given ambient temperature presumably had slightly higher surface temperatures than eggs in wet settings (Figure 5.7), owing partly to the lower rates of dissipation of metabolic heat into dry substrates (Ackerman

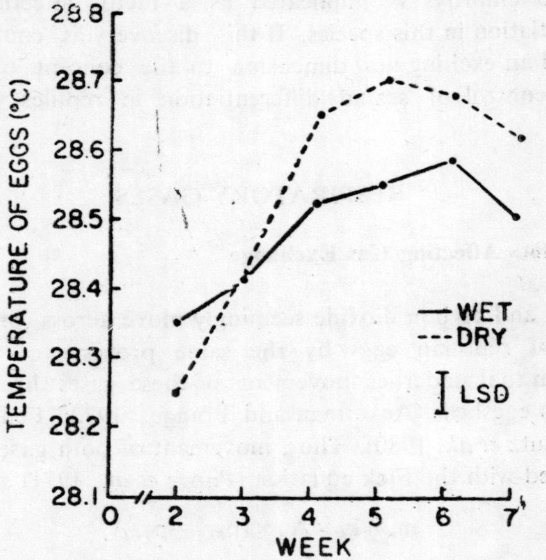

Fig. 5.7. Surface temperatures of eggs of painted turtles (Chrysemys picta) incubating on wet (—150 kPa) and dry (—1100 kPa) substrates of an ambient temerarure of 28.5°C. Means differing by the least significant difference (LSD) are significantly different at alpha=0.05. Note that surface temperatures were below ambient for the first 2 weeks and above ambient thereafter, but that the maximum difference between surface temperature and ambient temperature was only 0.2°C (on week 5). Also, the maximum difference in surface temperature between eggs on wet any dry substrates was only 0.15°C (again on week 5). Although eggs on the dry substrate had higher temperatures during the growth phase of development, metabolism and growth of embryos were lower in dry environments than in wet ones.

et al., 1985b) and partly to the smaller mass of eggs on dry media. However, the differences in surface temperature during the period of sexual differentiation probably were only 0.1-0.2°C (Figure 5.7), which is hardly sufficient to account for the formation of female turtles at ambient temperatures of both 26.5°C and 27.0°C. Because eggs in the different hydric environments experienced very different patterns of net water-exchange with their environment (see Gutzke, 1984a), the availability of water to embroys is implicated as a factor affecting sexual differentiation in this species. If this discovery is coufirmed, it will and an exciting new dimension to the concept of environ mental control of sexual differentiation in reptiles (see Bull, 1980).

RESPIRATORY GASES

A. Factors Affecting Gas Exchange

Oxygen and carbon dioxide seemingly move across (or through) shells of reptilian eggs by the same process of molecular diffusion that underlies movements of these gases through pores in avian eggshells (Ackerman and Prange, 1972; Feder *et al.*, 1982; Lutz *et al.*, 1980). Thus, movement of both gases can be described with the Fick equation (Piiper *et al.*, 1971) as

$$m_x = k_x \times A_A \times (p_{x,E} - p_{x,A})$$

where m*x* is the mass of gass "*x*" crossing the eggshell ($mg \times s^{-1}$), k_x is the transport coefficient for the gas in question ($mg \times s^{-1} \times cm^{2-} \times kPa^{-1}$), A_A is the area of the eggshell accross which exchange occurs (cm^2), and the term $p_{x \cdot E} - p_{x,A}$ is the difference in partial pressure of the gas across the eggshell (kPa). The rate of diffusion clearly is proportional to the area of eggshell exposed to air in the nest cavity (including that in spaces in the soil), as well as to the difference in partialpressure between the outer surface of the eggshell and the site of exchange inside the eggshell. Diffusion is proportional alsoto the transport coefficient, which is profoundly affected by whether the gas is diffusing through a medium of air or water.

Gases diffusing into rigid eggs of crocodilians and many turtles pass first through discrete pores in the calcareous layer and then between fibers forming the shell membrane(s) (Erben, 1970, Ewert, 1985; Ferguson. 1982; M.J. and G.C. Packard, 1979: M,J. Packard *et al.*, 1984a; Thompson, 1983; Woodall, 1984). The flexible-shelled eggs of lepidosaurians and many other turtles lack discrete pores in the ealcareous layer (if present), so gases merely pass through the large gaps in the calcareous layer before entering spaces in the shell membrane (Lillywhite and Ackerman, 1914; M.J. Packard *et al.*, 1982a, b, c; Solomon and Reid, 1983). The route followed by gases diffusing through the calcareous layer of the rigid-shelled eggs of geckos has not been established (M.J, Pakard *et al.*, 1982c).

The pores in testudinian and crocodilian eggshells are filled with liquid at oviposition, and therefore afford a substantial barrier to diffusion of respiratory gases (Thompson, 1985; also Feder *et al.*, 1982). However, water is removed from these pores during the first 2 weeks of incubation by an as yet unidentified mechanism, leading thereby to a progressive increase in conductance of eggshells to both oxygen and carbon dioxide (Thompson, 1985). The drying of the shell begins at the top of the egg and then spreads downward, forming a chalky-white band that is characteristic of eggs containing viable embryos (Ewert, 1985; Ferguson, 1985). The band does not form in infertile eggs. Thus, conductance of eggs to respiratory gases changes during the course of incubation owing to intervention of the embryo itself.

Lepidosaurian eggshells do not have pores, but hydration of shells may nonetheless affect transport coefficients for gases and, therefore, gas exchange (Black *et al.*, 1984). Eggs in wet environments have lower conductances for oxygen do eggs in dry environments, presumably because water occludes spaces between fibres of the shell membrane (Black *et al.*, 1984; also Feder *et al.*, 1982). There is no indication that conductance of lepidosaurian eggs is affected by developing embroys in the manner described for testudinian and crocodilian eggs.

B. Gas Tensions in Nests

The immediate source of oxygen for reptilian embroys developing in subterranean nests is air that was trapped inside the egg chamber (or between individual particles of soil) when the nest was closed. As oxygen is withdrawn from this pocket of air, however, it is replaced by oxygen diffusing into the cavity from air-filled pores in the adjacent soil (Marshall and Holmes, 1979; Prange and Ackerman, 1974; Taylor and Ashcroft, 1972), Diffusion proceeds at a rate proportional to the difference in partial pressure between air in pores in adjacent soil and that in the nest cavity; it depends also on the volume of air in these pores, the tortuosity of the pore channels, and the presence of constrictions in the channels that prevent free movement of oxygen molecules (Flegg, 1953; Marshall, 1959; Millington 1959; Reibel and Shair, 1982; van Bavel, 1952). Wind at the ground surface and infiltration of water following heavy rainfall may occasionally induce a mass flow of air through the upper reaches of the soil profile that partially ventilates some nests (Prange and Ackerman, 1974), but molecular diffusion is the only continuously operating mechanism for replacing oxygen) (Marshall and Holmes, 1979; Taylor and Asheroft, 1972).

Air trapped inside a nest at the outset of incubation probably has the same composition as air occupying pores in adjacent soil. Owing to aerobic respiration of soil microbes and plant roots, the partial pressure of oxygen in this air is below that of the atmosphere whereas that of carbon dioxide is higher than in atmospheric air (Marshall and Holmes, 1979; Taylor and Ashcroft, 1972). However, specific values vary among soils because of differences in oxygen demand of the soil flora and fauna and because of differences in the impedance of soils to diffusion of gases. For instance, tensions of oxygen and carbon dioxide in air held in pores of sandy loams depart only slightly from atmospheric levels, whereas those of gases in clay soils usually depart considerably from that standard (Taylor and Ashcroft, 1972).

Reptilian embroys consume relatively, little oxygen during

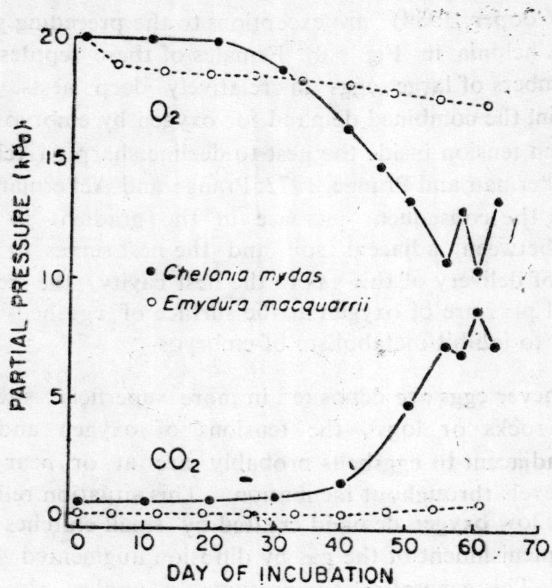

Fig. 5.8. Partial pressures of oxygen and carbon dioxide inside nests of two species of turtles. Note that tensions of these gases change only slightly over the course of incubation in the relatively shallow nests of Emydura macquarrii, but that major shifts occur in the deeper (and larger) nests of Chelonia mydas.

the first half of incubation also Ackerman, 1981a; Clark, 1953c; Dmi'el, 1970; Lynn and von Brand, 1945; Thampson, 1983, 1985), and oxygen in the nest cavity is replenished without undergoing an appreciable change in partial pressure (Ackerman, 1977; Lutz and Dunbar-Cooper, 1984; M.B. Thompson, 1981, 1983). Even in late stages of incubation, when oxygen requirements of individual embryos are high, the combined uptake of oxygen by embryos in a clutch usually does not cause a major decline in partial pressure of this gas in the nest cavity (Emydura in Figure 5.8; Lutz and Dunbar-Cooper, 1984; M.B. Thompson, 1981, 1983).

Nests of turtles of the families Cheloniidae and Dermochelyidae (and possibly the hole nests of some crocodilians: Lutz and

Dunbar-Cooper, 1984) are exceptions to the preceding generalization (Chelonia in Fig. 5.8). Females of these reptiles deposit large numbers of large eggs in relatively deep nests. Late in incubation, the combined demand for oxygen by embryos causes the oxygen tension inside the nest to decline sharply (Ackerman, 1977; Ackerman and Prange, 1972; Prange and Ackerman, 1974). Although the consequent increase in the gradient in oxygen tension between adjacent soil and the nest serves to increase the rate of delivery of this gas to the nest cavity, the reduction in partial pressure of oxygen at the surface of eggshells may be sufficient to inhibit metabolism of embryos

Whenever eggs are deposited in more superficial sites (*e.g.*, beneath rocks or logs), the tensions of oxygen and carbon dioxide adjacent to eggshells probably are at or near atmospheric levels throughout incubation. This situation reflects the generally low oxygen demand created by small clutches and the rapid replenishment of the gas by diffusion augmented by convection. This generalization presumably applies also to the mound nests of many crocodilians, because low relative humidity inside these nests indicates that they are partially ventilated.

C. Physiological Effects of Gases

Only one investigation has addressed the possible effects of altered tensions of oxygen and carbon dioxide on reptilian embryos. In that study, eggs of the sea turtles Caretta caretta and Chelonia mydas were incubated in sealed containers that were vented to the atmosphere through columns filled with dry sand (Ackerman, 1981b). The amount of sand in the columns was varied among containers to produce different resistances to diffusions of oxygen and carbon dioxide between the interior of containers and surrounding air. Embryos in containers with conductances to respiratory gases similar to those of nests in beach sand had higher rates of growth, shorter incubation periods, and higher hatching success that did embryos in containers with lower couductances (Ackerman, 1980, 1981b). Although data for partial pressures of gases were not reported, air in containers with low conductances probably had high tensions of carbon dioxide and low tensions of oxygen in the week or so before hatching (see Fig. 5.8). Thus, low tensions of oxygen and/or high tensions of carbon dioxide seemingly inhibit metabolism of reptilian embryos, probably by reducing the differences in partial pressure of these gases across eggshells. The increase in duration of incubation and decline in survival under conditions of hypoxis and hyperearbia are presumably manifestations of reduced metabolism. The relation between the metabolism of embryos and the gas tensions late in incubation needs to be quantified, however, so that the influence of varying gas tensions in natural nests can be evaluated more fully.

INTERDEPENDENCE OF ENVIRONMENTAL VARIABLES

A. Effect of One Variable on Another

We have discussed each of the major components of the physical environment individually, but it was somewhat artificial to do so. The several variables are highly interdependent, and a change in one usually is accompanied by a change in one or both of the others (Taylor and Ashcroft, 1972). Indeed, this interdependence of environmental variables markes it virtually impossible to design laboratory experiments in which one factor is varied while the others are held constant. This difficulty heightens the uncertainty associated with interpretation of results (G.C. and M.J. Packard, 1984). A few examples of this interdependence of variables will illustrate our point.

Water potential of soils is influenced by temperature, even if water content remains constant. Consequently, an increase in temperature will elicit a small increase in water potential, and a decrease in temperature will cause water potential to fall slightly (Taylor, 1958; Taylor and Stewart, 1960). The phenomenon seemingly is an outcome of thermally induced changes in surface tension of water trapped in capillaries in the soil (Cary and Hanks, 1972), and is more pronouced in dry soils than in wet ones (Taylor, 1958; Taylor and Stewart, 1960). A change in temperature could therefore influence exchange of liquid water between an egg and its nest environment by altering substrate water potential.

A change in temperature also affects vapor pressure of air in the nest, primarily by altering the saturation vapour pressure of that air (Taylor and Ashcroft, 1972). Also, the rates of molecular diffusion by water vapour, oxygen, and carbon dioxide are increased at higher temperatures and decreased at lower ones. Thus, a change in ambient temperature inside nests affects diffusion of all gases; the effect on diffusion of water vapour is likely to be the most pronounced.

Similarly, an increase in water content of a substrate has effects beyond the expected one on water potential. Wet substrates have higher thermal conductivities than do dry ones (Marshal and Holmes, 1979; Taylor and Ashcroft, 1972), so transport of heat between the ground surface and lower levels in the soil profile is fecilitated. Temperature of nests and eggs may thereby be affected. Furthermore, to the extent that water displaces air from pores in the soil and leads to a greater constriction at bottlenecks, diffusion of respiratory gases is impaired (Flegg, 1953; Millington, 1959; Reible and Shair, 1982). Metabolism of embryos could be inhibited if such impediments to gaseous diffusion were to cause the oxygen tension in a nest to decrease or the carbon dioxide tension to increase.

Changes in the tensions of carbon dioxide and oxygen in nests do not cause changes in temperature, and effects on the solute component of water potential probably are negligible. However, changes in gas tensions frequently signal a change in one or both of the other variables, as was indicated in prior discussion.

B. Interaction of Variables on Eggs

The effects of environmental variables on eggs and embryos may also be interactive. We illustrate this point by citing an experiment in which flexible-shelled eggs of snapping turtles (Chelydra

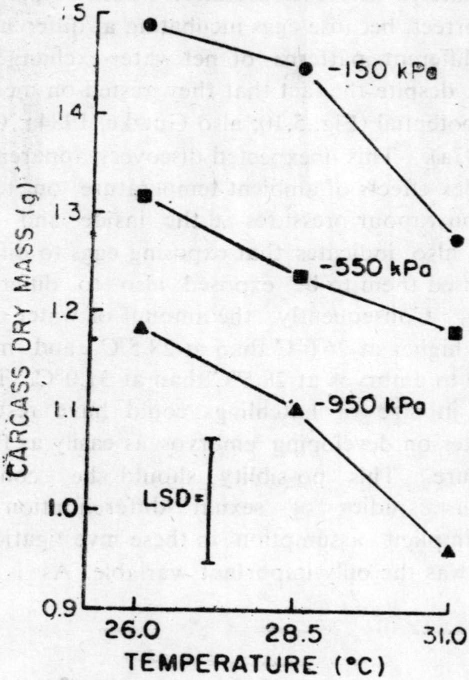

Fig. 5.9. Dry mass of carcasses of hatching snapping turtles (Chelydra serpentina) emerging from eggs incubated on vermiculite at different temperatures and water potentials. Means differing by the least significant difference (LSD) are significantly different at alpha=0.05.

serpentina) were incubated in the laboratory at constant temperatures 26.0°C, 28.5°C, and 31.0°C, and at water potentials of—150 kPa, —550 kPa, and—950 kPa (G.C. Packard *et al.*, 1987). At any given water note potential, turtles hatching at 31.0°C were smaller than those hatching at 28.5°C, and animals hatching at 28.5°C were smaller than those emergins at 26.0°C (Fig. 5.9). Because water potential of substrates was not a variable in these comparisons, a reasonable infrerpretation for the findings is that temperature intervened directly in physiology of developing embryos to elicit the observed variation in size of emergent young.

Unfortunately, this interpretation would be premature and possibly incorrect, because eggs incubating at different temperatures had different patterns of net water-exchange with their environment, despite the fact that they rested on media at the same water potential (Fig. 5.10; also Gutzke, 1984a; Gutzke and Packard, 1987a). This unexpected discovery apparently reflects on the complex effects of ambient temperature on temperature of eggs and on vapour pressures at the inside and outside of eggshells; it also indicates that exposing eggs to different temperatures caused them to be exposed also to different hydric environments. Consequently, the amount of water available to embryos was higher at 26.0°C than at 28.5°C, and more water was available to embryos at 28.5°C than at 31.0°C. The observed variation in size of hatchlings could have resulted from effects of water on developing embryos as easily as from effects of temperature. This possiblity should be considered in the numerous studies of sexual differentiation in turtles because the implicit assumption in these investigations is that temperature was the only important variable. As is illustrated

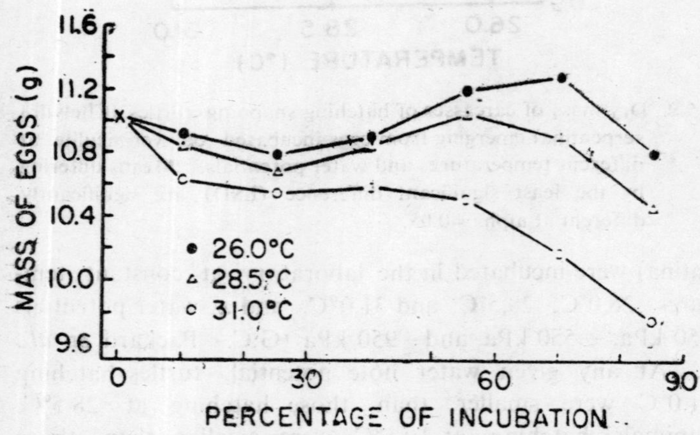

Fig. 5.10. Temporal changes in mass of eggs of snapping turtles (Chelydra serpentina) incubating on substrates of vermiculite at the same water potential (—150 kPa) but at different temperatures. Note tha eggs at 31.0°C lost more mass (largely water) than did eggs at 26°0.C or 28.5°C, and that eggs at 28.5°C lost more mass than did eggs at 26.0°C.

here, however, at least two important factors must have varied among experimental treatments.

SIGNIFICANCE OF PHYSICAL ENVIRONMENT

A. Introduction

Extremes of temperature, moisture, and gas tensions can lead to abnormal development and mortality of reptilian embrpos, as was discussed in preceding sections. There can be little doubt concerning the significance of serious deformity or death to embryos. Within the range of tolerance to each of the aforementioned environmental variables, however, a spectrum of subletbal effects may also provide raw material for the forces of natural selection. Reptilian embryos seemingly incubate longer, consume more of their yolk during development, and hatch at a larger size when they incubate in cool conditions and/or at high water potentials than they do in warm conditions and/or at low water potentials. Large reductions in partial pressure of oxygen from atmospheric levels probably do not occur commonly in reptilian nests, but may be a normal feature of incubation of sea turtles and some crocodilians. Low tensions of oxygen (or high tensions of carbon dioxide) may prolong incubation, inhibit metabolism, and cause hatchlings to be smaller than normal. The possible importance of these sublethal effects is explored briefly in this section.

B. Size of Hatchlings

One possible product of variation in physical conditions in nests is variation in size of hatchlings, elicited by differences in rates of growth of embryos, by differences in the duration of incubation, or by a combination of these factors. Attainment of large size before hatching may be advantageous to emergent young, by conferring them with a degree of protection against predation. According to this hypothesis, larger hatchlings are more difficult for predators to capture and/or to handle and swallow than are smaller conspecifics, so the former enjoy higher rates of survival than do the latter. Survival in the wild of hatchling lizards (Ferguson and Bohlen, 1978; Ferguson and Fox, 1984; Fergusion *et al.*, 1982; Fox, 1978; Snell, 1984) and turtles (Barrows and Schwarz, 1895; Swingland and Coe, 1979) is consistent with this proposition.

Several laboratory studies lend support to this hypothesis. The relatively large snapping turtles (Chelydra serpentina) hatching from eggs in wet environments are faster than the small turtles hatching from eggs in dry settings both when running on land and when swimming in water (Fig. 5.11). Superior performance by relatively large hatchlings could be the basis for superior survival by these animals in the wild. Also, larger hatchlings may be more aggressive than smaller hatchlings in pursuing and capturing food, and larger young may therefore sustain higher rates of growth than smaller animals in the cohort (Froese and Burghardt, 1974). High rates of growth may enable larger hatchlings to pass more rapidly through the range of body sizes in which predation is most intense, thereby contributing also to their superior survival.

On the other hand, larger offspring produced in environments favouring rapid growth and/or longer incubation have less residual yolk to sustain them after hatching than do smaller animals (Gutzke, 1984a; Morris *et al.*, 1983; G.C. Packard *et al.*, 1983, 1985a, 1987; M.J. and G.C. Packard, 1986). Because possession of a relatively large nutrient reserve may enhance survival mose than large size under some circumstances, small size at hatching conceivably is beneficial. For example, young green iguanas (Iguana iguana) depend to a large extent on residual yolk to sustain them for the first month of life outside the nest (Troyer, 1983), so survival may be enhanced in those individuals having a large yolk, irrespective of their body size. Furthermore, hatchlings of several freshwater turtles may spend their first winter in the nest cavity without feeding (Gibbons and Nelson, 1978; Wilbur and Morin, this volume), during which time they are supported largely by lipids in the retracted yolk (Congdon and Tinkle, 1982). Again, the size of the energy

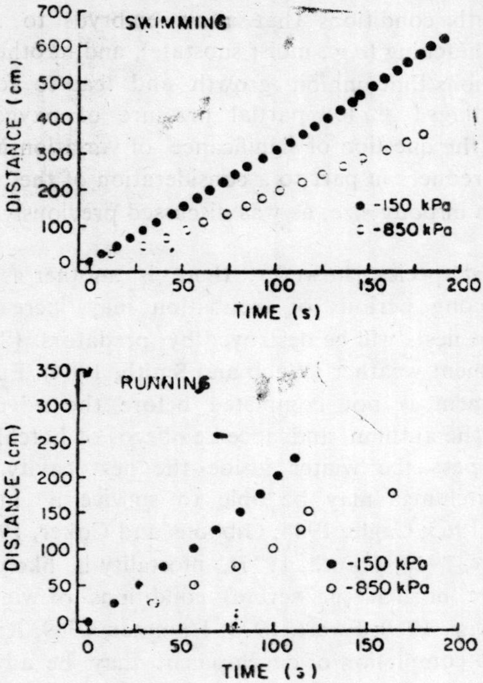

Fig. 5.11. Distance moved by hatchling snapping turtles (Chelydra sepentina) as a function of time swimming in water or running on land. Solid symbols represent turtles hatching from eggs on wet substrates (—150 kPa); open symbols represent animals emerging from eggs, on dry substrates (—850 kPa). The slopes of the lines represent speeds for animals in the samples. The large turtles emerging from eggs in wet environments were able to swim and to run faster than siblings from eggs in dry environments. (Redrawn from Miller *et al.*, 1987).

reserve in the retracted yolk may be a more important determinant of survival than body size at hatching. Such possibilities need to be investigated.

C. Length of Incubation

Variation in physical conditions in nests may also influence the duration of incubation. Longer incubation is associated in some instances with conditions that allow embryos to attain large size before hatching (*e.g.*, moist substate), and in other instances with conditions that inhibit growth and lead to formation of small hatchlings (*e.g.*, low partial pressure of oxygen). In all such cases, the question of significance of variation in length of incubation reduces in part to a consideration of the significance of variation in body size, as was discussed previously.

For most species, however, there is another aspect to the question. Long periods of incubation may increase the probability that nests will be destroyed by predators (Snow, 1982) or by inclement weather (Webb and Smith, 1984). Furthermore, if development is not completed before the advent of cool weather in the autumn, andvanced embryos or hatchling may be forced to pass the winter inside the nest cavity. Although embryos/hatchlings may be able to survive a mild winter (Bleakney, 1963; Cagle, 1944; Gibbons and Coker, 1977; Sexton, 1957: Toner, 1940; Trauth, 1977), mortality is likely to be high under more normal (or severe) conditions of winter weather (Breckenridge, 1960; Ewert, 1979; Hammer, 1969; Risely, 1933). Thus, rapid completion of development may be advantageous, even when conditions necessary for this outcome result in hatchlings of suboptimal size.

Duration of incubation may establish limits of distribution for oviparous reptiles in temperate or montane regions. Ground temperatures may be too low in parts of the northeastern United States to permit embryos of the terrestrial turtle Terrapene carobina to complete incubation during the limited growing season, thereby establishing the northern limits to distribution of this species (Allard, 1935). Similar arguments have been adduced for distribution of the lizard Dipsosaurus dorsalis in deserts of the American Southwest (Muth, 1980).

D. Sexual Differentiation

Environmental control of sexual differentiation in crocodilians, many turtles, and some lepidosaurians has important implications for the ecology and evolution of the species in which it occurs (Bull, 1980; Vogt and Bull, 1982a). For example, many species of turtles grow appreciably over the course of their reproductive lives. Shortly after attaining reproductive maturity, females tend to lay relatively small eggs (Congdon and Tinkle, 1982; Congdon *et al.*, 1982, 1982a) in relatively shallow nests, where generally high temperatures may promote differentiation of females (Withoft *et al.*, 1983). Later in life, the same females tend to lay relatively large eggs (Congdon and Tinkle, 1982; Congdon *et al.*, 1982, 1983a) in relatively deep nests, where generally cool conditions may promote differentiation of males (Wilhoft *et al.*, 1983). However, the small eggs produced early in the reproductive life of a female have a lower probability of hatching than do larger eggs produced later (Gutzke and Packard, 1985), and the small eggs give rise also to small hatchlings (Burger, 1977; Congdon *et al.*, 1983a; Morris *et al.*, 1983; G.C. Packard *et al.*, 1987) that may suffer disproportionately high mortality (Barrows and Schwartz, 1895; Swingland and Coe, 1979). Conseqnently, first efforts at reproduction by a given female may be devoted to the relatively unsuccessful production of female offspring, whereas later efforts may be devoted to the more successful production of males. Such possibilities, if realized, would have important implications for evolution of life histories (see Congdon *et al.*, 1983a). In particular, a high probability of producing female offspring early in reproductive life may explain why maturity is reached at relatively small body size in these species when reproduction is unlikely to be successful.

Turtles and crocodilians with protracted nesting seasons may produce offsprig primarily of one sex early in the season and of the other sex later in the year, owing to seasonal changes in temperature at the level of the nest cavity (Mrosovsky *et al.*, 1984a, b; Vogt and Bull, 1982a, 1984: Webb *et al.*, 1983a). If catastrophic mortality from heavy rainfall (Ragotski, 1959) and run-off (Plummer, 1976) or from high tides (Fowler, 1979) were to occur in the middle of the nesting season, offspring of one sex might suffer disproportionately. Such differential mortality would have consequences for sex ratio of the hatchling cohort and possibly for a subsequent adult population as well.

Environmental control of sexual differentiation has special importance for conservation of sea turtles. For a number of years, a common practice among conservationists has been to remove eggs from nests and to incubate them in styrofoam boxes held inside a shelter near the beach (Mrosovsky, 1983). The rationale for this practice is that nests would otherwise be destroyed by prodators, humans, or inclement weather. However, the conditions of artificial incubation have been relatively uniform, and relatively cool, and most turtles hatching from eggs handled in this manner probably have been males (Morreale *et al.*, 1982; Mrosovsky, 1982). Clearly, if use of artificial hatcheries is to continue, more attention must be given to the physical environment (hydric as well as thermal; Gutzke and Paukstis, 1983) to which eggs are exposed so as to assure the young of both sexes are produced.

Finally, sea-turtle rookeries frequently are confined to very restricted stretches of beach. Because thermal conditions vary among beaches, however, offspring of one rooker may be predominantly males whereas those of another may be largely females (Limpus *et al.*, 1983b). Although any single beach may attract large numbers of females for nesting, preservation of that beach alone may not be sufficient to assure survival of the population. Viability may depend instead on provision of enough different beaches as to assure that there is sufficient environmental heterogeneity to elicit differentiation of adequate numbers of both males and females for later recruitment into the breeding population.

6

Ecology of Digestive Tract

Alimentary canal is the main organ of digestive system. It is adapted variously in different animals. It is proportionately small in carnivorous animals and is quite long in herbivorous animals. On the basis of its development if can be divided into for gut (ectodem), mid gut (endodem) and huid guit (ectodem). But as the basis of functions of various parts the gut can be divided into four zones:

1. Ingressive zone (Mouth, Buccal Cavity and Pharrymx).
2. Progressive zone (Oesophagus and stomach)
3. Degressive zone (Small intestine.)
4. Egressive zone (Large intestine).

Oesophagus

With the exception of the Chelonioidea, the oesophageal relief patterns of turtles are extremely similar. The papillate pattern found in sea turtles (see below) is, however, unique among reptiles. Most turtles have fairly smooth, straight, unbranched longitudinal folds of different widths.

Anteriorily in Chelus fimbriatus, there are prominent, closely spaced longitudinal folds with deep clefts between them. The surface of the wall is smooth, but the surfaces of the folds are grooved. Folds have different diameters, but the diameter of an individual fold is constant throughout its length. Further posteriorly, some of the folds become slightly wavy. However, near the end of the esophagus, the folds become very broad and flat and their eages are very wavy. The flattened tops of the folds are grooved and ridged; they closely resemble lasagna noodles.

Chrysemys picta, Chelydra serpentina, and Emydoidea blandingii nave exactly the same pattern as the anterior region in Chelus as, judging from Bojanus' (1819-1821) figure does Emys arbicularis. The folds in Geochelone elegans are wavier than in Chelus, while those in Podocnemis expansa are wider, less well defined, and more often fused. The folds in Pseudemys scripta are narrower than usual, and occasionally branch and fuse. Those in Terrapene carolina are wide, often fused to one another, and branched. In addition the clefs between folds are unusually deep. In Kinosternon scorpioides, the clefts are deeper and tend to have narrower longitudinal folds running along their tops.

Lissemys punctata warrants special consideration. The folds are fine and straight as in the other trionychid studied, Trionyx spiniferus, but they are wipely and equally spaced. In Lissemys, but not Trionyx, the folds bear striations which give them a rough appearance. Between these major folds is either a single low fold or a series of small folds from which low, oblique side branches run to the major folds.

The esophagus of Chelonia mydas has long, conical, cornified papilae which project posteriorly. Anteriorly these papillae are so close together that their bases touch and are entirely confined. Posteriorly, the papillae are less numerous and incompletely cornified. Near the stomach, the papillae end, being replaced by a few wide, very irregular longitudinal folds. These folds anastomose, and have convolutions and ridges on them as well as deep clefts between them. They overlie a network of much finer folds and extend only a very short distance, ending in blunt, posteriorly extending papillae. At this point, the tract constricts and the rugae of the stomach begin. Eretmochelys, and presumably all other cheloniids, are similar (Pernkopf and Lehner, 1937).

Cuvier (1835) and Burne (1905) agree that the papiliae in Dermochelys coriacea are mueh longer than those in Chelonia. The large papilae in Dermochelys average 4.0 cm in length, and much smaller papillae lie both between and attached to the bases of the larger ones (Burne, 1905). This is in contrast to Chelonia in which all the papillae are about the same length. Finally, the papillae are all single anteriorly in the esophagus of Dermochelys. but posteriorly they become stouter and have two or even three points. These papillae are also mentioned and figured by Heldt (1933) and Villiers (1958).

Stomach

The pattern of gastric relief is consistent throughout almost all turtles, with large, broad, unbranched, parallel longitudinal folds. The folds are regular and often round or square in cross section. Their surface is usually smooth, and the wall between folds is flat except for small crevices or bumps. The folds are usually straight, but become wavy or even convoluted at the cardia in the transition from the oesophageal pattern or at the pyloras in the transition to the intestinal pattern. This pattern occurs in Chrysemys picta, Chelydra serpentina, Geochelone elegans, Kinosternon scorpioides, Pseudemys scripta, and Terrapene carolina. Emys orbicularis appears to have straight, unbranched folds (Bojanus, 1819-1821).

Anteriorily in the stomach of Chelodina longicollis, the folds are square in section and conform perfectly to the general description. Posteriorly, however, the folds branch and fuse extensively and, although they have flattened surfaces, are not necessarily square in section.

The longitudinal folds in Chelonia mydas have low lateral

branches that fade out quickly and disappear under the adjacent longitudinal fold. Cuvier (1835) notes that the stomach of Chelonia is divided into two regions; the second region is narrower than the first and lined by more numerous and pronounced longitudinal folds. In Dermochelys coriacca (Burne, 1905), the tubular stomach is divided into compartments by approximately thirteen low, irregular, transverse folds. Burne considers none of these large enough to be called "perforated diaphragms", a term used by some earlier authors.

Testudo sp, has thick, cushion-like folds starting above the cardia and extending posterinrly as wavy longitudinal folds (Botha, 1958). These folds are so tortuous at the gastro-oesophageal junction that they actually occlude the gut at the cardia, forming a sphincter. The folds eventually fade into a smooth-surfaced membrane.

In Chelus fimbriatus, the folds disappear in the pylorus. Among the folds are much finer longitudinal ridges; these are probably temporary structures since they are obliterated when the true folds are very close together.

Trionyx spiniferus also follows the general pattern except that, in the pylorus, the spaces between folds are covered by convoluted, papilla-like structures which mark the transition to the intestinal pattern.

Anteriorly the stomach of Malacochersus tornieri is very irregular. The usual longitudinal folds are covered by grooves and bumps of tissue which cause their diameters to vary. These folds are further a typical in that they are lower and less distinct than usual.

The eardiac region of Graptemys geographica is lined by folds the diameters of which vary considerably. In addition, the folds fade out or enlarge at random and are much closer together. Only a few distinct and widely separated folds extend into the pylorus. The portions of the wall between these folds may be raised slightly, and their surface is covered by a network of fine branching ridges.

The specimens of Lissemys punctata and Platemys platycephala lacked folds in the stomach. In Lissemys the thin-walled stomach was covered only by a fine pebbly pattern, while Platemys did not even have this. Possibly this simplified pattern merely indicates an especially full stomach and would be replaced by the more usual one in other individuals.

Small Intestine

The pattern found in the small intestine, and especially the duodenum, is a sharp, well-defined one unlike the generalized, simplified ones of the esophagus and stomach. In most turtles the type of pattern is fairly consistent, although some variation does occur.

The usual pattern in the duodenum is one of zigzag folds which are usually tall, very thin, and parallel-sided. Their widths are normally constaut; they run parallel to one another and are equally spaced. There are, of course, exceptions (see below). The zigzags range from very broad to very narrow, and the angles formed by the zigzags can be acute, right, or obtuse, depending on the species and the region of the gut. The zigzags usually change to undulgating waves posterior to the duodenum. Next the folds start to become straight and finally, as they near the colon, they become broader. Thus the pattern progresses from complexity in the duodenum to simplification near the colon; this is true not only for turtles, but for reptiles generally.

A variation of this pattern occurs in Geochelone elegans, Graptemys geographica, Kinosternon scorpioides, Pseudemys scripta, to a lesser extent Chelydra serpentina, and possibly Chelodina longicallis, Chelus fimbriatus, and Terrapene carolina. The duodenal pattern consists of more than one level of folds, or at least appears to do so. This fold-on-fold pattern disappears in the first part of the post-duodenal small intentine and is replaced by a single layer. From here to the colon the pattern described above is followed.

The duodenum of Chelydra serpentina contains broad,

straight longitudinal folds which are widely and equally spaced. Between the folds are delicate, parallel zigzag folds of the same height as the major folds, and underlying all of this is a fine network that is hard to discern. Closer examination shows that the broad longitudinal folds are made of compacted, slightly zigzagged folds. Behind the duodenum the small intestine contains wavy folds of various diameters; they are tall, close together, and parallel. In the posterior regions, where the folds are nearly straight, tall thick folds alternate with short thin ones. Here the folds are more widely separated, Jacobshagen's (1920) description of this species is similar.

In Kinosternon scorpioides the duodenum bears a series of wavy folds overlying one another. The pattern resembles a three-dimensional net. The superficial layer, although fused to the lower ones, does not branch, but the underlying folds form a true net. Further posteriorly the folds form a single, unbranched layer and proceed as very acutely angled zigzags.

Numerous emydines have been studied. Chrysemya picta has a distinct net pattern consisting of zigzagged longitudinal folds. These change to fairly straight longitudinal folds posteriorly. Emydoidea blandingii also follows the general pattern, but Graptemys geographica has a variation of the fold-on-fold pattern in which the zigzags give off side branches that run to or under adjacent folds; this gives the surface a woven appearance. Just posteriorly, the folds straighten and come closer together so that the side branches cannot be seen and the woven effect is lost. The folds then become narrower, so that the side branches are again visible. The posterior end of the small intestine looks like a straight tube surrounded by bulbous rings. Internally the pattern consists of broad transverse folds composed of irregular projections from the wall. Between these folds are more projections of the wall. The folds nearly occlude the lumen.

Pseudemys scripta has a duodenal pattern very similar to

that os Graptemys: obtusely angled zigzags with side branches that join adjacent folds. A series of fine folds underlies the major ones; it is difficult to tell whether these folds are branches of the major folds or a separate underlying layer to which the major folds are attached. Posterior to the duodenum the folds form only one layer; they are very low, but occassionally a short, very thick, tall fold arises and as quickly disappears. Posteriorly the folds are broader, flatter, and equally spaced in the usual fashion. Our observations agree with those of Jacobshagen (1920).

The duodenum of Terrapene carolina has the usual zigzag folds which form obtuse angles, but between these folds are much lower, wavier, frequently branching folds. As in Preudemys, we cannot tell whether these are branches of the major folds or a separate layer underlying the major folds. The rest of the small intestine shows the usual arrangement. The folds become irregularly wavy, and then straight near the colon; they are also broader and more widely separate posteriorly.

In Malaclemys terrapin the folds are irregular, branched, and wider at the base than the apex. As they progress posteriorly, the number of fine ribbonlike side branches increases, and small, variably shaped projections extend from the edges of the folds. Near the colon, the folds are broad and straight, although they still rise to points in the center. The edges of the folds are fluted, each fold interdigitating with the next.

Cuora amboinensis has longitudinal folds located between diagonal and transverse folds (Jacobshagen, 1920). The longitudinal folds are wavy anteriorly, but straight further posteriorly. They are always taller than the transverse or diagonal folds, although even the longitudinal folds decrease in height posteriorly, There are only logitudinal folds in the posterior half of the small intentestine. Jacobshagen reports a similar pattern in Emys orbicularis, except that here the longitudinal folds traverse a plain mesh pattern in the anterior part of the duodenum instead of diagonal and transverse folds; Bojanus

(1819-1821) shows a similar pattern of longitudinal ridges with smaller transverse folds between them. The meshwork quickly fades out, and only the longitudinal folds continue.

In Mauremys caspica, too, Jacobshagen (1920) found a net-like pattern in which the longitudinal folds are taller than the others. The longitudinal folds continue throughout the small intestine. For a short disance they are straight, then very convoluted, and finally straight again. The transverse and diagonal folds disappear more gradually in Mauremys than in the other two genera. In Clemmys insculpta, the transverse folds disappear quite abruptly about 10 cm behind the pylorus, and the longitudinal folds continue in a slightly coiled course (Jacobshagen, 1920).

In the testudinine Geochelone elegans, the zigzag folds in the duodenum are not of constant width and form angles varying from acute to obtuse. Intermittently one fold fuses with its neighbour to form a net-like pattern. Closer examination shows that these folds overlie other folds and in some cases merge with them, so the pattern is a complex one. In the posterior part of the duodenum, as the folds become straighter, the zigzags form only obtuse angles and tend to run parallel to each other. In the jejunum and ileum, the folds fuse and it is impossible to distinguish individual ones. In addition, individual folds are grooved and ridged. Jacobshagen (1920) studied G. denticulata and his results generally agree with ours for G. elegans. He speaks of a net of low folds instead of irregular zigzag patterns, but aside from this the descriptions are the same.

The lining of the duodenum in Malacochersus tornieri shows a honeycomb arrangement of fine ridges. In addition, there are three large longitudinal folds separated by grooves in the wall. These may be temporary. In Testudo graeca, there is a simple, net-like pattern, but posteriorly the longitudinal folds become larger and more prominent than the other folds although the latter never completely disappear (Jacobshagen, 1920).

In the sea turtle Chelonia mydas (Jacobshagen, 1920; our own work), the anterior duodenum contains a complex network of ridges. The meshes are polygonal anteriorly but soon become elongated. The pattern seems to be several layers deep, but the irregular longitudinal folds are the thickest and tallest. The upper, transverse to oblique folds appear to be branches of the longtudinal folds, but are very fine and considerably lower. Beneath the upper layer of folds, lie at least one perhaps two other layers of fine, delicate folds. Near the posterior end of the duodenum, the longitudinal folds are much more prominent; they are quite wavy and thick and have rounded tops, while the transverse folds are lower, finer and haphazardly arranged. The pattern no longer resembles a net. In the rest of the small intestine, only the longitudinal folds are seen. They are taller and narrower than previously, but there is much variety in the width of the folds, and thick ones alternate with thin ones. The folds are irregularly wavy, close together, and parallel for most of their length.

Eretmochelys imbricata is similar except that the longitudinal folds are more numerous, much taller, and narrower than in Chelonia mydas (Jacobshagen, 1920). The folds between the longitudinal ones, however, are proportionately lower. The pattern becomes thicker and lower further posteriorly. In another, and unidentified cheloniid, Jacobshagen describes the anterior part of the small intestine as being composed of narrow, parallel longitudinal folds having crinkled edges. These folds may be coiled but they rarely branch.

Caretta caretta too has a plain, tall, net-like pattern, but the height of the folds varies so that there is a false impression of a double pattern (Jacobshagen, 1920). Posteriorly, the longitudinal folds become straight and are more prominent than the others. Finally only much lower longitudinal folds remain. Next there is a region in which the wall becomes very thick; the longitudinal folds also increase in thickness, and bump-like projections are found on the free edges.

The Trionychoidea have the most unusual intestinal pattern

of all, and of these Lissemys punctata is the most deviant. The duodenum has no folds at all but is covered instead by fine striations. These striations tend to be transversely oriented, but they often form whorls just like those on finger-prints. A short distance posterior to the pylorus a single thick fold runs along the dorsal surface. Its path is irregular is one specimen, but quite straight in the second. There are also regular transverse ridges here which may be temporary since they are formed by transverse grooves in the wall. Posteriorly, the dorsal ridge becomes very convoluted and the grooves in the wall are no longer parallel, so the ridges that were transverse further anteriorly here resemble a single, much folded ridge that covers the entire wall. Thus the dorsal ridges and the pattern on the rest of the wall are quite similar. Next, the dorsal ridge becomes a broad, flat-topped, straight fold with fluted edges. The rest of the wall is covered by lower, narrower folds of the same description. Finally, all these folds become the same width as they get more distinct, taller, and develop rounded tops. They are also narrower and better described as wavy rather than straight with fluted edges.

The duodenum of Trionyx spiniferus is covered by narrow oblique folds. These folds are parallel to one another, low, and close together. Near the bases are small lateral projections which may extend to adjacent folds. A large, ventral longitudinal fold appears in the small intestine; it is straight and round in cross section. The oblique folds run more longitudinally than previously and are not so distinct, but appear more like villi than folds. More folds like the ventral longitudinal one form near the posterior end of the duodenum and cover most of the wall of the small intestine, but the smaller oblique folds are still found between these larger ones.

Jacobshagen (1915, 1920) reports a very complex pattern in Trionyx triunguis and T. ferox. In most ways is resembles that just described, but the major longitudinal fold is said to be spirally arranged. Further posteriorly in the small intestine, the folds form a very distinct network.

Chelodina longicollis has distinct, short, thick folds that are directed obliquely and intersect each other in X-shaped patterns. Fine, very short ridges project at right angles from the folds.

In Chelus fimbriatus, the zigzag folds in the duodenum are very low, little more than striations, and form acute angles. Next is a short section with branched, leaf-like, villiform projections that entirely cover the wall. Then the folds reappear, but they are taller here and form obtuse angles. They are very broad and close together with fine ridges joining adjacent folds to give a cross-hatched appearance. These fine ridges could be branches of the zigzag folds or a fine network underlying the zigzags.

Platemys platycephala also has very low and indistinct zigzag folds. They fuse more often than usual and occasionally disappear altogether. Near the colon, the folds become very broad and straight; they have fluted edges and at times give way to randomly directed folds which may mark the beginning of the colon. Jacobshagen's (1920) description of Platemys agrees substantially with ours.

Among the Pelomedusidae, we studied only Podocnemis expansa. Anteriorly the zigzags are quite broad and of a variable diameter, but they are otherwise normal. Small cross-branches between the folds and shapeless projections from the folds give a generally irregular appearance. As the folds straighten posteriorly, they became finer and branch more frequently. At the end of the small intestine the low, widely separated, flat-topped folds are all of the same width. The irregular projections fill the spaces between folds. Pelusios has a simple net-like pattern anterioriorly which soon gives way to longitudinal folds (Jacobshagen, 1920).

Colon

The pattern in the colon is very difficult to describe because it is rarely distinct; as the patterns are so general, we find it nearly impossible to correlate the findings for different turtles or other reptiles. Furthermore, the colon is distensible, and the pattern can completely disappear when the colon is full. Since the folds are temporary, many of them may well be artifacts of the fixation and handling of the animals; two specimens of the same species need not be very similar. Thus the following discussion must be used with much caution.

If the relief in the colon can be obliterated by stretching the gut, do the folds reform randomly or do they reform in the same pattern as they held previously when the pressure is taken off the gut? If the former is true, then a discussion of patterns in the relief is obviously pointless. Thus we did not concentrate heavily on the colon and make few comparisons between species.

Jacobshagen (1920) apprently had many of the same problems. He describes a net of folds with smooth edges. This net is narrow in Chelydra serpentina, wider in Clemmys insculpta, and much wider and irregular in Platemys. He also notes that Testudo graeca and Mauremys caspica lack a constant pattern of relief, but instead show large, temporary bumps.

In no case did we find a network of folds; longitudinal folds are usually present, often as continuations of those in the small intestine. In the colon these folds are fewer, broader, lower, and more rounded. In some cases, they are still distinguishable as folds, but some of them are most certainly artifacts. The folds can be obliterated if the wall is stretched. The surface of the folds and the wall between them are usually smooth.

In Chelodina longicollis, the pattern is as described above, but the folds are straight and covered by small irregular projections. Posteriorly, there are no folds, and the wall is smooth except for some larger bumps which may be temporary. The folds of the small intestine disappear in the anterior part of the colon of Chelus fimbriatus. The relief consists of irregular ridges which cannot be stretched out. The wall is again covered by small, probably temporary bumps. In Platemys platycephala, the folds are very narrow and frilly. They cover the entire surface of the colon, follow irregular paths, and have rough surfaces with numerous projections.

Chelydra serpentina has low, branching, smooth-surfaced, longitudinal folds.

The pattern in Emydoidea blandingii is also normal, except that the wide folds have fluted edges and adjacent folds interdigitate. Each fold is flat along its crest, and from the crest the folds slope to the wall of the intestine. The anterior end of the colon of Graptemys geographica closely resembles the posterior part of the small intestine. The folds cover the entire wall, branch profusely, and are of varying diameters. Further posteriorly, the folds become narrower, evently spaced, and unbranehed; all have the same diameter. Two specimens of Malaclemps terrapin were studied. The first shows a fairly smooth-walled colon with a few, probably temporary, folds; the second specimen has tall, narrow, branched longitudinal folds.

Pseudemys scripta has a varied pattern, either of low branching ridges which seem to be temporary, or of distinct folds, which can be tall, broad, and straight, or be lower and branched with clefts between them. Terrapene carolina has very prominent longitudinal folds. Anteriorly, these are tall and flat-topped, follow an irregular course, and branch frequently. The diameters vary, and the folds are often convoluted. Posteriorly, the folds are less numerous, straighter, and of more constant diameters. Only the bases os these folds are fluted, are the wall between them is covered by bumps and grooves.

In Geochelone elegans, the folds of the small intestine disappear quickly in the anterior end of the colon. Most of the latter is smooth-walled, although there are bumps and low ridges posteriorly. The colon in our specimen was quite distended, so that there was less internal relief than one would expect.

The colon of Lessemys punctata is virtually smooth-walled (*i.e.*, without folds), but there are some randomly distributed grooves which divide the wall into areas of various shapes. These may be transversely or longitudinally oriented; occassionally the prooves define small circles. The rectum has two longitudinal folds of unequal width, but is otherwise smooth-walled.

In Trionyx spiniferus, the folds are low, rounded, and branched. Their widths vary, and their courses are irregular. The surfaces of the folds are covered by tiny depressions as through made by a pin-head. Jacobshagen (1920) noted similar depressions in Trionyx triunguis.

RHYNCHOCEPHALIA

Sphenodon punctatus has been studied by Gunther (1867) and Jacobshagen (1920). According to the former, the stomach contains a few longitudinal ridges which end at the pylorus. The duodenum has folds forming a simple network with a small polygonal mesh (Jacobshagen, 1920). Both mention longitudinal folds along most of the length of the small intestine; according to Jacobshagen they are highest anteriorly where their wavy free edges bear short, broad, tongue-shaped processes. There are no transverse folds. Gunther does not mention the small processes and reports that the longitudinal folds do not reach the posterior end of the small intestine. The colon has a complex pattern, but this may well depend on the volume of its contents. Jacobshagen found an anterior region, including the caecum, in which the walls are smooth; another smooth-walled region occurs at the posterior end. Between these, he reports a simple network of folds. Both authors mention a circular ridge forming an incomplete value between the intestine and the cloaca.

SAURIA

Oesophagus

Lizards, like turtles, have a fairly consistent pattern of esophageal relief: large, well-spaced, smooth-surfaced, longitudinal folds which are quite straight and branch infrequently. The wall between the folds is covered by fine longitudinal striations or other similar modifications.

Exceptions do occur. The longitudinal folds may be slightly wavy as in Acontias percivalii, or they may have transverse or oblique sections as in Crotaphytus collaris. The folds may be more widely spaced as in Eumeces sp. or very close together as in Cyclura sp. or Cordylus giganteus. Chamaeleo jacksoni, Uromastyx sp., and Eumeces sp. (all vary similar) have reliefs that are rough surfaced, and the longitudinal folds vary in diameter. The wall between the folds may be pebbled (Uromastyx sp., Sceloporus occidentalis), smooth (Crotaphytus collaris), or striated. The striations may be parallel (Sauromalus obesus, Iguana iguana) or anastomosing and net-like (Ctenosaura sp., Phrynosoma cornutum, Trachydosaurus rugosus, Tupinambis, sp., Acontias percivalli).

In Cordylus giganteus, the folds are low and close together. Adjacent folds are frequently fused, and the folds are not always the same width. Eumeces sp. has fewer but thicker folds near the stomach; their diameters vary. In addition, the entire surface of the folds is pebbled, but this is expected at the transition from esophageal to grastic relief. In Sauromalus obesus, fine parallel striations frequenty occur on the surface of the folds as well as between them.

There is one exception to this general rule—an unidentified agamid. This has a pattern similar to that of the vipers (discussed below). The folds are very rough-surfaced and close together. They cover the entire wall, are of several different widths, and fuse and branch randomly.

Stomach

The stomachs of lizards have the simplest relief found in any reptilian gut. Their stomachs are mueh more prominent than those of snakes (described below). They have larger lumina than the oesophagus or intestine. The stomach also seems to be fairly thin-walled, and thus highly distensible, so that it is not surprising that the relief is very smooth.

Most of our specimens have no rugae at all in the stomach, while others have regions that are free of folds. When folds are present, they tend to be widely separated, longitudinally oriented, and few in number. Folds disappear when the stomach is stretched, and, since many stomachs were distended by their contents, many folds ordinarily present may be missing.

The wall between the folds usually has the pebbled appearance so common in stomachs, and this may extend to the folds as well. Occasionally the wall is striated or completely smooth.

Our specimen of Corythophanes cristatus had a badly mutilated stomach due to autodigestion. Apparently the whole of the stomach down to the pylorus was free of folds except for one prominent dorsal ridge, while the wall itself has a rough, pebbled appearance. The stomach narrows in the pyloric region and longitudinal folds become more apparent; they are low, branching, and irregular.

Cyclura has very prominent longitudinal folds which are nearly circular in cross section. They are smooth-surfaced, have constant diameters, and branch occasionally. The wall between folds resembles a sponge, being formed by a network of small crevices. The folds fuse posteriorly and only a few enter the narrow pyloric region.

The oesophageal folds of Parynosoma cornutum end shortly after entering the stomach. The stomach lacks rugae, and the wall is coverted by small, randomly arranged crevices that delimit cell-like regions.

The cardiac region of the stomach of Sauromalus obesus has thick, evently spaced, longitudinal folds, which are quite constant in diameter. The wall between these folds is raised into areas that resemble broader folds. These areas are covered by smaller grooves and ridges. A dorsal longitudinal fold runs the entire length of the stomach and is especially evident in the fundus which has few other folds (much as in

Corythophanes mentioned above). The wall of the fundus is covered by fine transverse ridges which, in some regions, are high enough to be considered folds.

In Crotaphytus collaris the stomach contains longitudinal ridges very similar to those in the esophagus. They are constant in diameter, unbranched, parallel, and straight. The wall between them has a spongy appearance caused by a network of small crevices.

The esophageal relief of Ctenosaura changes gradually to that of the stomach. The folds become less regular in size and direction as the anastomosing striations between the folds become for pronounced. The diameters of some of the folds become so small that they cannot be distinguished from the striations between folds, while others are much greater than any in the esophagus. As the stomach narrows to the pylorus, the longitudinal folds again become tall, well-defined structures with parallel sides. Occasionally, ridget run along the tops of these folds.

Sceloporus occidentalis has only two or three longitudinal folds, each having a few side branches that soon fade out. The wall itself is covered by fine, parallel, longitudinal grooves.

In Chamaeleo jacksoni distinct longitudinal folds sink slightly into the wall and have shallow clefts on either side. The folds vary in diameter and are not parallel. Individual folds are straight for part of their course and wavy for part. The wall between them has a fine pebbled surface.

The definite folds in Tupinambis sp. tend to be radially arranged around a central groove. As the folds are very thick and close together, it is often difficult to see the clefts between them. The entire wall is covered by papilla-like structures which make the surface exceedingly rough and irregular, but these may well be the result of post-mortem breakdown.

Two specimens of Acontias percivalli were studied: one

specimen had no rugae at all while the other had several prominent longitudinal folds, which were wavy and occasionally branched but constant in diameter. The surface of the wall had the rough, pebbled appearance.

The folds in the stomach of Eumeces sp. resemble those in the esophagus but follow a less regular course. They give off branches some of which continue and some of which disappear. Only two or three of the major folds enter the pyloric region where they are straight and unbranched. The walls of the stomach, as well as the folds themselves, are covered by shallow pits and projections.

Trachydosaurus rugosus has no folds in the anterior parts of the stomach. Longitudinal folds start just anterior to the pyloric region in which they become quite distinct. Their course is irregular anteriorly but becomes straighter posteriorly. The surface of the wall is pebbled in both the anterior and posterior regions.

The gastric wall of Cordylus giganteus is very thin and distensible. There are no folds, and the surface is not even pebbled. A few very faint longitudinal ridges exist where the stomach narrows towards the pylorus.

Small Intestine

Two patterns are common in the small intestines of lizards. One consists of tall, closely packed, well-defined longitudinal folds which are very thin and ribbon-like. This pattern is best typified by Crotaphytus collaris, Phrynosoma cornutum, and Sceloporus occidentalis. The second pattern is a more generalized and more common one, at least in the animals we studied. Here the folds are lower, thicker, and either wavy or straight with fluted edges. The folds are poorly organized and are often no more than short ridges. They tend to bear many small projections which makes the relief more complex and disorganized. This pattern is best typified by Eumeces sp. The other species tend to fall somewhere between the two extremes.

Some, like Corythophanes cristatus and Iguana iguana, are more similar to the first pattern while others, like Uromastyx sp., Trachydosaurus rugosus, and Cyclura, are closer the second pattern. No sharp distinction between the two types of relief is possible since some species, or at least regions of some species, are half-way between the two.

Nearly all lizards have longitudinal intestinal folds. Transverse folds are present in certain regions of some animals, but are the predominating type of fold only in Tupinambis sp.; they are rarer than in turtles, and net-like patterns are also rare. Jacobshagen (1920), however, reported a network of folds to be a common pattern in lizards, for example Phrynosoma cornutum, Chamaeleo chameleo, Cordylus cordylus, and Lacerta. The longitudinal folds are much reduced in Agama sp., Calotes sp., Gekko gecko, Hemidactylus sp., Ptychozoon kuhli, Chlamydosaurus kingi, Egernia depressa, Tiliqua nigrolutea, and Lacerta viridis. Finally transverse folds are said to be absent in Metopocerus, cornutus, Agama marmorata, Agama stellio, Uroplatus fimbriatus, Pygopus lepidypodus, Eumeces sp., Scincus officinalis, and Lacerta agilis.

We agree that the longitudinal folds are rarely straight. None of the species was studied have straight folds, and Jacobshagen (1920) found them only in Hemidactylus sp.

Crotaphytus collaris has the typical type one pattern. The folds are tall, ribbon-like, and of remarkably constant diameter. They are tightly waved, but posteriorly the waves become more undulating as the folds begin to straighten out. Between the tall folds are much lower, straighter, longitudinal ridges, but there are no net-like or transverse ridges.

Corythophanes cristatus also has the first basic pattern. The folds are tall and thin, but the free border is very irregular and the folds often appear to run randomly. In these cases, the pattern resembles that described below for Chamaeleo jacksoni. Small cross-folds can be seen joining the major folds.

Further posteriorly, the folds become straighter, lower, and more regular and the cross-folds disappear.

Jacobshagen's (1920) and our specimens of Iguana iguana are quite similar. First, there are tall, longitudinally oriented processes in the duodenum. These processes are joined to one another by much lower longitudinal ridges. There are also both longitudinal and transverse ridges, but they are lower than the processes. Posteriorly, the folds become higher and cannot be distinguished from the processes. Jacobshagen says the transverse folds become the more obvious, but in our specimen the longitudinal ones are more prominent. Finally, only the longitudinal folds remain and form a type-one pattern; they are tall, prominent, of constant diameter, and have a tightly crimped, wave pattern. Taller, thicker folds alternate with lower, thinner ones. These folds straighten slightly near the end of the small intestine.

Phrynosoma cornutum also has a type-one pattern. The duodenal folds are tall, thin, and all of the same diameter. They are also slightly wavy and usually run parallel to each other. Posteriorly they become lower and straighter but are still very distinct. In Phrynosoma Jacobshagen (1920) found transverse folds in the most anterior intestinal regions, and these disappeared posteriorly; we found no such folds, Sceloporus occidentalis is identical to Phrynosoma cornutum except that the folds are much lower and less distinct in the posterior regions of the small intestine.

The duodenum of Chamaeleo jacksoni has an irregular pattern of longitudinal folds. These folds are tall and often so thin as to appear membranous. Their free edges bear projections and the borders are crenulated. The paths of the folds are not straight, and they often run together. Convolutions are common; when they occur the usually distinct folds cannot be distinguished from one another. The spaces between folds are very rough, and occasionally fine longitudinal ridges can be seen between the folds. Posterior to the duodenum, the folds are thicker and their free borders even more irregular. Fine, low transverse folds join the longitudinal ones; the former are probably modifications of the rough surface in the duodenum. This agrees well with Cuvier's (1835) description for chameleons in general and with Meckel's description for Chamaeleo pumilis, but not with Jacobshagen's Meckel's for Chamaeleo chamaeleon (all cited in Jacobshagen, 1920).

The unidentified agamid we studied has numerous, very fine folds in the duodenum which are usually longitudinally oriented. These folds are so close together and so covered by fine projections that it is almost impossible to tell one from the next. The folds are more clearly discerned posteriorly where they are thin, wavy structures with crenulated edges. They completely cover the wall of the gut. This description agrees only with that of Meckel (cited by Jacobshagen, 1920) for Agama stellio; Agama sp. and Calotes marmoratus have much more prominent longitudinal folds in the anterior regions (Jacobshagen, 1920).

Our specimen of Uromastyx sp. was poor, appearing at first to have villi and no folds, but closer examination shows that it has a very disorganized version of the Chamaeleo jacksoni pattern. The processes are so long that they nearly obscure the folds. The latter are longitudinal but close together and frequently branched. Adjacent folds are joined by cross-folds of the same height and almost the same diameter as the major folds themselves. The longitudinal folds become more pronounced posteriorly in our specimen, while Jacobshagen (1920) found transverse and oblique folds to predominate in his Uromastyx aegyptius.

Cordylus giganteus has low, narrow, longitudinal folds. Their path is irregular, and they branch and anastomose frequently but are numerous enough to cover the entire surface of the wall. The surface of the folds is rough.

Eumeces sp. has a pattern similar to that in Chamaeleo jacksoni, but no transverse folds are visible anteriorly. The longitudinal folds are quite thick and rough-surfaced, but they proceed in a regular wavy, almost zigzag fashion, and adjacent folds are more often parallel than in most lizards. The fact that the folds are so densly packed and covered by processes makes the pattern look disorganized, but it is really quite regular. In the middle of the intestine, the gut seems distended and there are fewer longitudinal folds; Jacobshagen (1920) also found this, but attributed it to gut contents. In this region, too, there are a few transverse folds similar in appearance and height to the longitudinal ones. Finally, at the posterior end of the small intestine there are transverse folds only. They resemble the more anterior folds.

Trachydosaurus rugosus has a good type-twopattern for most of its length. Anteriorly the folds are low, thick, densely packed, and often discontinuous. They tend to run longitudinally, but in some regions turn and run obliquely. Posterior to the duodenum, the folds become straighter and more distinct. These folds are wide at the base and pointed at the apex. There is a distended portion in the intestine in which folds disappear; however, they are re-established near posterior end of the intestine. There they are straighter and more widely spaced, although of a height comparable to that in the anterior regions.

Three species that we studied have irregular patterns. The duodenum of Ctenosaura has tall, densely packed, tongue-shaped villi which are oriented transversely. Occasionally these villi are arranged in rows and resemble transverse folds except that they are not usually joined to one another. At the posterior end of the duodenum, the villi become lower and are arranged in transverse rows. Next, the villi begin to turn longitudinally and adjacent ones fuse partially to form irregular longitudinal folds. As the progress posteriorly these folds become lower and more widely spaced. They also adopt a regular zigzag pattern. Near the end of the small intestine, the wall becomes thin and distended and the folds have extensive and irregular transverse segments. In some areas there are so many of these segments that it is difficult to tell that the folds are still progressing longitudinally. Next to the colon, the gut constricts again and the relief consists of low, crinkled, widely separated longitudinal folds.

Cyclura sp. resembles Ctenosaura in having papilla-like projections that merge into folds. The papillae are tall, sharp, and form transverse folds in some regians of the duodenum especially next to the pylorus. Posterior to this is transitional region in which the papillae form oblique folds. Finally they form definite and continuous, but fairly low, longitudinal folds. To this point, all the folds are densely packed, rough-surfaced, and parallel and form tight waves. Posteriorly, however, the longitudinal folds become gently undulating, more widely separated, and smooth-surfaced. Most of the folds disappear in a distended area near the posterior end of the small intestine.

Jacobshagen (1920) also discusses several specimens that have irregular villus-like processes with either transverse or longitudinal axes; these can merge into folds. Ameiva ameiva, Ophisaurus ventralis, and Anguis fragilis are in this list. Darco volans is also unusual having a net formed by a longitudinally stretched meshwork of ridges to which short processes are attached at the anterior end of the intestine (Jacobshagen, 1920). Posteriorly there are only longitudinal ridges.

Tupinambis sp. was the one animal in which we found only transverse folds. Jacobshagen (1920) says that longitudinal folds are the most prominent ones anteriorly in T. tequixin, but that transverse ones soon overshadow them. In our specimen these transverse folds are straight, parallel, and narrow; both the folds and the spaces between them are covered by irregularly shaped processes which may obscure the folds. The pattern tends to disappear towards the posterior end of the intestine as the folds get finer, lower, and closer together. In the transition between small and large intestine, the folds disappear altogether and are replaced by a few randomly arranged folds of an otherwise smooth wall. Next to the colon, three of these large foods are oriented transversely and may act as valves.

Colon

Many lizards have a structure resembling a spiral valve in the colon. A suggestion of this pattern can be seen in Chamaeleo jacksoni, which has huge, thick, transverse folds separated by deep defts. These folds are so large that they resemble overlapping plates of tissue rather than folds. Smaller longitudinal folds run along the surfaces of the plates. These folds are few and haphazardly arranged. The surface of the plates is very rough, being covered by grooves and small villus-like projections.

The colon of Cyclura has five tall, thick, transverse folds on one side of the colon only. These folds get smaller and lower posteriorly and probably serve as valves partially closing the colon. Otherwise the wall of the colon is smooth except for a few shallow, randomly arranged grooves. The colon of Ctenosaura is very similar to that of Cyclura except that there are only four instead of five folds. One would except them to serve the same purpose.

Iguana iguana has an unique colonic pattern. An elongated valve leads from the small intestine into an expanded, smooth, thin-walled antechamber. A part of this chamber is partitioned off to form a caecum. A conspicuous valve leads to the next section of the colon. In this region, five large transverse folds divide the colon into five consecutive pockets. The folds start immediately behind the valve and decrease in size posteriorly. The rest of the colon is also expanded thin-walled, and free of permanent folds. The few irregular folds that are present can be obliterated when the wall is distended.

Guibe (1970) notes that there is a greatly enlarged region around the anterior end of the colon in Iguana iguana and Conolophus cristatus. He also reports an incomplete transverse valve, followed by four semilunar folds; this agrees well with our findings in Iguana.

Anteriorly the colon of Sauromalus obesus is dilated and

smooth-walled. More posteriorly, the few fairly wide longitudinal folds are straight and constant in diameter.

The other specimens studied have relatively indistinct patterns in the colon. The unidentified agamid has a wall free of folds, but covered by a fine pebbled patten. The colon of Uromastyx sp. is smooth except for a constricted region with a few low, evenly spaced, longitudinal folds of constant diameter. The entire surface has a pebbled appearance.

Cordylus giganteus has broad smooth-surfaced folds running at oblique angles to one another. These folds all have the same diameters and many fade out as quickly as they begin. Much of the wall is smooth-surfaced, but some regions are covered by rough, irregularly shaped projections.

The anterior part of the colon in Eumeces sp. has a few folds which tend to run longitudinally, but branch occasionally and have variable diameters. Deep clefts divide the wall into fold-like regions, while the wall itself has a rough, almost papillose appearance. Posteriorly the colon shows an interesting pattern with folds like those described above radiating from a central mass of tissue. These branch and follow apparently random courses.

There is nothing to be gained from a discussion of the animals dealt with by Jacobshagen (1920). He concentrates on the small intestine, as do we, and little about the colon. We generally seem in agreement. All lizards, except Ptychozoon kuhli, are reported to have a low, narrow-meshed, smooth-edged net pattern (P. kuhli has low longitudinal folds). He feels there are temporary bulges under this net, a circular fold before the cloaca, and perhaps two or three other circular folds at various places. In many cases he could distinguish no pattern of relief, and many of the so-called folds or ridges are no doubt temporary bulges. The fine pebbled or striated appearance we frequently found is probably what Jacobshagen referred to as the net-like pattern.

AMPHISBAENIA

Oesophagus

In all the amphisbaenians we studied, the oesophagus bears longitudinal ridges or folds of various sizes. They are relatively large in amphisbaenids, and may be closely spaced as in Amphisbaena alba or more widely separated as in A. darmini. In both there may be smaller ridges between them; these may anastomose. The main ridges are essentially hemispherical in section and of approximately equal diameter. The posterior part of the esophagus is dilated to form a saccular portion which lacks the longitudinal folds and has very thin walls.

Anops kingi has five large folds like those in Amphisbaena. These diminish in size and disappear near the stomach. There is no dilated portion.

The ridges in the wall of the oesophagus of Trogonophis wiegmanni are fine, densely packed, and of variable diameters. These ridges are more widely separated but still present in the dilated portion near the entrace to the stomach.

Stomach

The stomach of amphisbaenians is greatly distensible and details of its pattern vary with the amount of food it contains. Typically the relief is low in most of the stomach. Two patterns are seen. In Amphisaena alba and Anops kingi there are numerous small, low, anastomosing, more or less longitudinal ridges that form a longitudinally stretched network; in Amphisbaena darwini and Trogonophis wiegmanni ridges are lacking and the wall bears numerous small, irregular, villiform projections. With extreme stretching these patterns may disappear and the wall be smooth.

In Trogonophis, but not the amphisbaenids, the longitudinal folds of the esophagus continue into the anterior part of the stomach. There they become larger and more uniform in diameter.

The pylorus is very muscular and well marked in all four forms, and the pyloric portion is the stomach is usually less distended and thicker-walled than the other parts of that organ. The pyloric region bears rather large transverse ridges in Amphisbaena alba and Trogonophis wiegmanni, but not in Amphisbaena darwini or Anops kingi.

Small Intestine

The small intestine of amphisbaenians typically has markedly zigzagged longitudinal folds which are closely packed and cover almost the entire wall. In the anterior part of the duodenum of Amphisbaena alba and Anops kingi the folds may not be continuous, but even when they are quite villiform, the zigzag pattern is obvious. Anteriorly the ridges tend to be tallest and the zigzag most marked, with very acute angles. Father posteriorly the ridges are lower and straighter. In Amphisbaena alba and A. darwini taller ridres alternate with lower ones. Smaller transvers ridges joining the longitudinal ones occur in Amphisbaena alba, but were not obvious in the other specimens we studied; the wall between the ridges was smooth in our Tregonophis wiegmanni.

Jacobshagen (1920) reported a network of folds, like that he found in many lizards, in Amphisbaena alba, Blanus cinereus, Leposternon microcephalum, and Trognophis wiegmanni. However, in Bipes canaliculatus he found irregular, villus-like processes which partially merged to form folds; this pattern is what we observed in the duodenum of Amphisbaena alba and Anops kingi. As usual Jacobshagen stressed the net-like aspects rather than the larger longitudinal ridges.

Colon

The pattern seen in the colon appears to vary, but some of this could well be due to the degree of distension.

In Amphisbaena alba the colon is thin walled with very low, broad, indistinct, zigzag longitudinal folds anteriorly and low transverse folds posteriorly. There is an abrupt change to this pattern at the ileocolic valve. In Amphisbaena darwini, on the other hand, the longitudinal folds from the small intestine continue into the colon; they are low and nearly straight.

Anops kingi has a different pattern with an abrupt change at the ileocolic valve. The colon bears discontinuous longitudinal ridges which may anastomose and are connected by a series of much lower, zigzag, transverse ridges. This pattern is also present in the small caecum.

Trogonophis wiegmanni has a thicker colonic wall, which is completely covered with short, longitudinal fold-like masses of variable size. Like those in Anops these may branch and anastomose.

OPHIDIA

Oesophagus

All snakes have longitudinal folds in the esophageal wall, but we have found three common patterns. Some forms have a thin-walled oesophagus in which the folds are quite fine, straight, widely spaced, unbranched, and of the same diameter. The wall between the folds can be smooth, but is more often pebbled or with fine longitudinal striations. In the second type, the folds are close together, tall, very rough-surfaced, and branched; they have deep but narrow clefts between them. The widths of the folds vary. Type three is intermediate between types one and two. Here the folds are quite tall, widely spaced, unbranched, and rough-surfaced. The spaces between folds are filled by smaller longitudinal ridges comparable to either the fine striations of type one or the smaller folds of type two.

The distribution of these types corresponds roughly to taxonomic groups. Nearly all members of the family Viperidae (Agkistrodan rhodostoma, Crotalus atrox, Crotalus mitchelli, Bitis arietans, and Echis carinatus) have type-two relief. Variation among them is slight: the major difference is that the average width of folds is greater for some than for others. Sistrurus catenatus is the only exception amongst the viperids we saw. In it the major folds are narrow, low, well spaced, straight, and unbranched. The esophageal wall seems thin and fairly smooth. Between the major folds in Sistrurus are fine longitudinal ridges that are little more than striations. This is really a type-one pattern.

Members of the Elapidae typically have type three, but considerable variation occurs. The major longitudinal folds are exceptionally straight. In addition, they are of constant width, do not branch, are equally if closely spaced, are smooth-surfaced, and have straight, parallel sides. Bungarus caeruleus and Hydrophic brooki are very similar except that the latter has fewer and more widely spaced longitudinal folds with larger, more distinctive striations between them. In Hydrophis ornatus the folds are even more equal in height, width, and spacing than in the other two species. In addition, the spaces between folds are greater and the folds lower. Naja naja does not fit into this pattern. The oesophagus seems to be fairly smooth-walled, and there are no large folds anteriorly. The wall is covered by fine longitudinal ridges like those between the folds in Hydrophis brooki. These ridges are very narrow, very densely packed, and frequently branched. More posteriorly, the wall is nearly smooth, and ridges are present in only a few regions. Near the stomach, the ridges are covered by projections from the wall, and a few, much taller, irregular folds are also present.

The pattern found in Thamnophis sirtalis resembles that of the viperids in that the folds are very close together, are much branched, and have different widths; however, unlike those in vipers, these folds are fine and smooth-surfaced, and individual folds are constant in diameter. Posteriorly, there are fewer, wider folds which branch less frequently and which have deeper, wider clefts between them. Masticophis flagellum, Elaphe sp., and Heterodon platyrhinos all have similar thin-walled esophagi with a simple pattern of fine, dense, branching longitudinal ridges like that seen in Naja naja.

In Xenodon merremii, Ferri (1971) found variably sized

longitudinal folds. His photo shows high but thin folds with irregular but wide spacing. These folds are covered by very small, wavy, longitudinal ridges, which may fuse to produce a pattern of small cell-like depressions. Thus this has a slightly modified type three.

We studied no representatives of the Scolecophidia. According to Robb (1960), the esophageal relief in Typhlops schneideri consists of a series of longitudinal folds stretching the length of the organ. The wall is very thin and transparent between the folds, but it becomes thicker near the stomach.

Among the Henophidia we studied three bodies and Acrochordus javanicus. The folds in Acrochordus are wide, but all the same width, straight, rounded, and close together, with deep, narrow clefts between them. The surface of the folds tends to have a pebbled or striated appearance, and occasionally a fold even has fluted edges.

Boa constrictor has a pattern very similar to that of the viperids, but has larger, more irregular, and rougher-surfaced folds. Folds branch and fuse so extensively that it is difficult to follow an individual fold. In Corallus enhydris, the major folds are widely if unvenly separated. They are straighter and more numeraus anteriorly and are of many different widths. Posteriorly there are fewer, lower, quite wavy folds. The folds themselves seem to be quite smooth, but the wall between them bears low striations that almost give the appearance of a fine net.

Python sp. has an extensive series of wavy longitudinal folds that become wavier posteriorly. The folds are fairly close together, unbranched, and usually parallel. Their tops are generally flat, and the entire surface of the folds and the walls between them is covered by fine striations like those in Corallus enhydris.

Stomach

All the species we studied have rugae, broad longitudinal folds which usually run throughout the length of the stomach. These rugae are broader, more distinct, and more widely spaced than the longitudinal folds of the oesophagus. A second common feature of ophidian stomachs is the rough, spongy appearance of the wall between the folds. Many variations on these characters occur, but all can be derived from these basic traits. For example, in Agklstrodon rhodostoma, Naja naja, and Python sp., the folds are so close together that little space can be seen between them. In Bitis arietans, Corallus enhydris, Crotalus mitchelli, and Sistrurus catenatus the folds are straight and unbranched. In Acrochordus javanicus, Agkistrodon rhodostoma, Echis carinatus, Heterodon platyrhinos, Hydrophis brooki, and Python sp. the folds are wavy but usually unbranched. Finally, in Bungarus caeruleus, Crotalus atrox, and Thamnophis sirtalis the folds are irregular and much branched.

Most snakes have thick-walled, muscular stomachs, in which the lumina are not much greater than those of either the oesophagus or small intestine. It is not surprising, therefore, that the relief consists of rather thick longitudinal folds, as compared to lizards which have thin-walled stomachs and either fine, narrow longitudinal folds or no folds at all.

In Typhlops schneideri the gastric folds are very deep, irregular, and branched. They remain deep in the pyloric region and appear somewhat spiralled (Robb, 1960).

At its junction with the oesophagus, the stomach of Python sp. is smooth and thin-walled. Further posteriorly, deep, narrow, longitudinal clefts separate very broad, folk-like masses of tissue. These masses are basically flat on top, but their surfaces are covered by shallow grooves and ridges. Smaller, rounded folds of considerable width extend along the top of the broad folds. Occasionally these smaller folds give off transverse side branches which curve down the sides of the broads folds. Posteriorly, these major folds become even broader and also develop wavy edges. The smaller folds follow the contours of the broad ones, especially along the wavy edges, but occasionally they give off side branches that enter the clefts. In the pyloric region, both the broad and smaller folds become lower and straighter.

The first region of the stomach of Corallus enhydris has no rugae, but five quite tall, fairly straight, nearly parallel rugae occur in the next section. These are discontinuous irregular folds, only one of which continues into the next region of the stomach where it becomes wavy. The last region has no rugae. The spaces between folds bear fine, nearly parallel striations which rarely anastomose. These are especially prominent in the posterior regions of the stomach.

In Acrochordus javanicus, the esophageal folds continue a short distance into the stomach and become more widely separated and wavy. The folds in the lateral surface of the stomach are thick and thrown into waves which become hard to distinguish further medially. Here bumps and ridges look like remnants of the folds found laterally. All of the gastric wall has a very rough, sponge-like texture.

Anteriorly short, irregular, widely spaced folds line the stomach of Agkistrodon rhodostoma. Their surface is bumpy and covered by large tongue-like processes. Between the folds is a low, fairly regular mesh-work. Posteriorly the folds consolidate into broad, rounded, closely packed, longitudinal folds which are wavy but parallel.

The folds in the anterior part of the stomach of Crotalus atrox do not differ greatly from those in the oesophagus. They are closely packed, of variable width, and often branch or fuse together; the clefts between them are very narrow and deep. The surface of the folds is covered by very small convoluted ridges which make it look like the surface of a brain. Posteriorly, the folds are more distinct and their surfaces are covered by fine striations which follow the contours of the folds C. mitchelli is similar to C. atrox, except that there are small irregular projections on the wall and the surface of the folds is pebbled.

Sistrurus catenatus has large longitudinal folds which are fairly straight and constant in diameter, but often discontinuous. The wall between the folds is covered by fine wavy ridges, which branch and anastomose extensively to form a fine mesh-work pattern.

In one specimen of Bitis arietans the esophageal folds end at the stomach and, for a distance, the gastric lining is flat. The surface has the spongy appearance caused by the pebbled surface and the fine network of small grooves in the wall. In the middle part of the stomach, very broad, tall, parallel folds lie close; they are often discontinuous. The folds are usually smooth, but the narrow spaces between them have the grooved, spongy relief described above. The stomach narrows as the folds become narrower, straighter, and more regular; they are very sharply demarcated with perpendicular sides and flattened tops. The wall between them consists of a network formed by the anastomosis of fine striations.

The folds in Echis carinatus resemble those found in the oesophagus, but are larger and more widely spaced. The folds are wavy and sometimes fuse together, but their diameters tend to stay constant. The surface between the folds is usually smooth, but occasionally it or the folds bear projections.

The gastric relief varies considerably from region to region in Hydrophis ornatus. Anteriorly, a few, slightly wavy, longitudinal folds anastomose intermittently. The wall between these folds bears fine transverse ridges which increase slightly in height as they fuse with the logitudinal folds. The entire middle region is lined by convoluted, mainly longitudinal folds. Some are narrow, wavy, and unbranched; others, broad, indistinct, and branched. The pyloric region bears three poorly defined, lumpy, longitudinal folds.

The only fold of any size in the stomach of Hydrophis brooki runs along the dorsal wall. It is thick and wavy for part of its length and straight for the rest. The remaining surface of the wall is covered by smaller, closely applied ridges. These

TABLE 6.1

Histological and Histochemical Peculiarities of the Cloacal Glands in Female Squamates

Order Famlly Genus	*Anatomical characteristics*	*Cell type*	*Erythrophilia*	*Cyanophilia*	*PAS-reaction*	*Alcian blue (pH=2.6)*	*Danielli's coupled tetrazolium reaction*
1	*2*	*3*	*4*	*5*	*6*	*7*	*8*
Sphenodon	Few lateral tubular glands; a single cell type	M	0	+	+	+	0
Gekkonidae	Two ventral and two latero-dorsal groups of glands; a single cell type	M	0	+	+	+	0
Xantusiidae	Few unbranched tubules; a single cell type	M	0	+	+	+	0
Pygopodiidae	Two coalscent ventral glands and two elongate laterodorsal glands; a single cell type	M	0	+	+	+	0
Iguanidae	A conspicuous flat ventral glandular mass and two small lateral glands; a single cell type	M	0	+	+	+	0

Agamidae	Unbranched glandular tubes disposed about the proctodaeum; a single cell type	M	0	+	+	+	0
Chama-eleonidae	Unbranched grandular tubes; a single cell type						
Scincidae	Unbranched glandular tubes; a single cell type	M	0	+	+	+	0
Feyliniidae	Two lateral ovoid glands; a single cell type	MS	0	+	+	+	+
Lacertidae	Largely branched glandular tubes; two cell types	S	+	0	0	0	+
Teiidae	A flat ventral grandular mass; a single cell type	SM	+	0	+	0	+
Anguidae	Conspicuous tubular or sacci-form glands; two cell types	S	+	0	0	0	+
Anniellidae	8 small ventral glands and 2 dorsolateral groups; two cell types	M	0	+	+	+	0
Amphisbaenia	Two dorsal glands and, in some species, a ventral gland; a single cell type	M	0	+	+	+	0

(*Contd.*)

TABLE 6.1 (*Contd.*)

1	*2*	*3*	*4*	*5*	*6*	*7*	*8*
Typhlopidae	A conspicuous ovoid gland; a single cell type	SM	+	0	+	0	+
Boidae	Glands located near the dorsal depression of the proctodaeum; two cell types	M	0	+	+	+	0
		SM	+	0	+	0	+
Colubroidea	Two dorsolateral groups of glands; a single cell type	MS	0	+	+	+	+

M, Mucocytes;
MS, muco-serous cells;
S, serous cells;
SM, sero-mucous cells;
0, negative results;
+, weakly positive results;
+strongly positive rosults.

ridges branch and fuse extensively, and, although there are fewer major folds than usual, these smaller ridges are quite large and more extensive than usual.

The folds in Bungarus cacruleus resemble the creases in a piece of erumpled material. They are very sharply defined and run more or less longitudinally, but their height varies. The folds often run obliquely while some side branches run transversely. The spaces between the folds are exceptionally smooth.

The gastric relief in Naja naja is very irregular, but apparently consists of large longitudinal folds with clefts between them. These folds are covered by smaller convoluted folds which proceed in all directions.

Anteriorly the stomach of Heterodon platyrhinos has a thin wall; the folds, which are lows narrow, and widely separated, follow a wavy, irregular, and often convoluted path. These folds are very irregular, being fused or densely packed in some regions and lacking in others. The wall between folds bears fine parallel situations anteriorly. The striations give way to the pebbled appearance further posteriorly in the stomach. The pattern in the pyloric region is entirely different: there are a few longitudinal folds like those described above, but most of the wall is covered by short, jagged ridges which run obliquely. These ridges are sculptured with pits and clefts lying between them.

The stomach of Xenodon merremii appears to show the usual pattern (Ferri, 1971). There are numerous longitudinal folds which are low and anastomosing.

The cardiac region of the stomach of Thamnophis sirtalis has irregular longitudinal folds, which are quite widely spaced and branch occasionally but are of constant diameters. Between the folds are delicate, slightly wavy ridges which run parallel to one another. In the fundus, the folds are large and prominent, but very irregularly oriented. They are often convoluted and their diameters change at random. In the pyloric region, the folds are more longitudinal, but still convoluted or wavy. On occasion, these irregular folds seem to be overlying a smaller, more regular network of folds.

Masticophis flagellum has a few major longitudinal folds which are discontinuous, variable in diameter, and rough-surfaced. Between the folds, the wall is rough and covered by small bumps which give it a pebbled appearance.

Small Intestine

The small intestines of the snakes we studied, with the possible exception of the viperids, have little in common. Generally longitudinally folds predominate, but they show all possible variations. As in the other reptiles, the folds tend to become straighter and lower as they course posteriorly. Like us, Jacobshagen (1920) finds few outstanding characteristics in the small intestines of snakes. He states that all ophidians except the Typhlopidae have variations of a simple network of folds. The greatest variation in this pattern is shown in the Henophidia as discussed below.

Jacobshagen (1920) calls any pattern that has both transverse and longitudinal folds a net. Usually the longitudinal folds are more pronounced than the transverse ones; in many cases, the transverse folds may be so small or obscure that the effect is one of simple longitudinal folds rather than of a net. We use net only for those patterns which actually look like a net. As a result, Jacobshagen considers most snakes to have a net-like pattern, while we consider them to have longitudinal folds. For example, many viperids have similar looking patterns even though some have transverse folds while others do not; when present, the transverse folds are quite obscure and do not make the pattern look like a net. We think is less confusing, therefore, to refer to modifications of a pattern of longitudinal folds than to a net-like pattern and a pattern of longitudinal folds.

Typhlopids hove an entirely different pattern of intestinal

relief than other snakes. Villiform structures predominate, and folds and ridges, when present at all, are low and poorly defined. Typhlops schneideri has very dense, flat, triangular villi (Robb, 1960). These villi thin out near the colon, and a space in the duodenum, where the bile and pancreatic ducts enter, is free of folds. According to Jacobshagen (1920), Typhlops lineatus has tall, tongue-shaped villi in the duodenum which may be linked by low transverse or longitudinal folds. Further posteriorly the height of the folds increases, and the pattern consists of zigzag bands with only the tips of the villi visible as projections. All of these bands tend to be longitudinal and, as they progress posteriorly, become broader, flatter, straighter, and closer together.

Anteriorly the duodenum of Python sp. bears tall, leaf-like, villiform projections. The are very thin, densely packed, and their dimension is usually transverse. At first they do not seem to be arranged in any pattern, but, as Jacobshagen (1920) points out for Python sebae, the projections are processes arising from iow transverse folds and are part of a net with very small meshes. A little further posteriorly, there are fewer villi, but they still form a dense eover over all regions of the wall. Here more of the villi are longitudinally oriented, and it is impossible to visualize their organized arrangement into a network. Following this, a transitional zone has fairly broad, slightly wavy, longitudinal folds with deep, narrow clefts separating them. These folds are of different lengths and merge extensively. Their sufaces very rough, apparently due to the presence of very low villi. The logitudinal folds become broader further posteriorly, and the villi disappear almost entirely, leaving only intestine, the folds with smooth surfaces. Towards the posterior end of the folds develop an irregular honeycombed pattern on the surface. This pattern is formed by low, narrow ridges, both transverse and longitudinal, which delimit poloygonal areas Jacobshagan (1920) notes that Python sebae, Boa constrictor, and Eunectes notaeus all have the basic pattern of net-like folds found in other snakes, but that the peojections of these folds predominate over the folds themselves, often obscurring the net-like pattern.

Acrochordus javanicus has a most unusual duodenal pattern, with the major folds running obliquely along the intestine for a certain distance, then turning at right angles, and continuing obliquely. All along the folds, delicate cross-ridges connect the major folds. The pattern is somewhat modified posterior to the duodenum. The major oblique folds are still present, but they run in one direction only and are often discontinuous; cross-ridges are still numerous. Near the end of the small intestine, the pattern becomes a cross-hatched one. The major folds are still present, but they are finer, lower, and continuous. Conversely the cross-folds are taller and more widely spaced, so that they are about the same size as the major folds. In fact, the only important difference between them is that the cross-folds are not continuous and extend only from one oblique fold to the adjacent one. Near the colon, this fine net-like pattern is entirely lost in a constricted region with four very broad but low longitudinal folds of equal length.

All viperids have patterns of intestinal relief that are relatively similar to each other. The longitudinal folds are usually tall and fairly narrow, but they follow an irregularly wavy course and have quite variable diameters. The folds all have an irregular appearance, tending to be wider at the base than at the apex or to have extensive side branches.

In Agkistrodon rhodostoma the folds are tall, ribbon-like, closely applied to one another, and often irregular. Their path is irregular, but side branches are rare although the folds occasionally fuse.

The duodenum of Crotalus atrox bears very wavy, often convoluted longitudinal folds; they are very tall, ribbon-like, and of constant diameter. There are no obvious branches, but adjacent folds often fuse, especially in the convoluted areas. The folds become a little striaghter posterior to the duodenum, and the wall between the folds is smooth. Near the colon, the folds are quite low, widely separated, and unbranched, and the wall between the folds is quite smooth. The pattern in C. mitchelli does not differ greatly from that in C. atrox. Jacobshagen (1920) reported that the longitudinal folds in Crotalus sp. are low and straight and have short, wide processes attached to their free edges; there is also an inconspicuous series of transverse and oblique folds. These observations do not agree with ours. Even though he may well have studied a different species of Crotalus, the differences seem extreme and the pattern reported is unlike that in a variety of viperids.

In the duodenum of Bitis arietans, the folds are wavy and, although they come to a point along the free edges, they give off extensive, thick lateral branches which are lower than the major longitudinal folds. This produces a complex, net-like pattern. Posteriorly, the folds become much thicker and the fine, pointed crest is lost. The extensive lateral branches are retained, and they, with the varying diameters of the folds and their irregular course, present a complicated and irrigular pattern. The thick longitudinal folds become even more prominent in the posterior part of the small intestine, while the cross-branches become lower, finer, and less distinct. The the appearance of a net is lost. Another specimen showed the duodenal folds less pointed, thicker, and more constant in diameter, with the cross-branches lower and narrower. The structure, therefore, does not appear like a net. Posteriorly, the folds are widely separated, but still wavy and irregular. Transverse folds of about the same height and thickness connect them at fairly regular intervals. The spaces between folds are covered by very fine papillose structures which make the surface appear very rough and porous.

The pattern in Echis carinatus is remarkably similar to that in Bitis arietans. The only difference is that the folds in Echis lack lateral branches. The small intestine of Sistrurus catenatus also resembles that of Bitis arietans, but here the folds are more widely separated, and, although the folds are quite wavy, they proceed in a fairlp direct manner. There are very few transverse folds, and the folds are generally wider at their base than along the free edge.

In a very general way, the pattern seen in the viperids is followed by two elapids—Bungarus caeruleus and Hydrophistall brooki. Bungarus has longitudinal duonenal folds. These are wavy, ribbon-like, parallel, and all of the same diameter. Taller, slightly thicker folds alternate with shorter, thinner ones. Posterior to the duodenum, the folds resemble those of Bitis more closely. They are still tall and ribbon-like, but they are irregularly wavy and follow very wandering courses. Although still close together, they are no longer parallel. Unlike those in viperids, the folds are smooth and the spaces between them free of projections. In Hydrophis brooki, the duodenal folds are wavy, parallel, low, and slightly rough-surfaced, but otherwise follow the description for Bungarus. Surprisingly, the folds near the colon become tightly zigzagged rather than straighter; however, they are still low and close together.

The duodenum of Hydrophis ornatus is unusual in being extremely smooth. No folds are present, and the only distinctive features are transversely directed projections which resemble dripping paint, a pattern which is a continuation of the gastric relief. Further posteriorly, the small intestine has longitudinal folds which are low, only slightly wavy, and widely but equally spaced (a pattern very similar to that in the small intestine of Crotalus atrox). The wall between the folds varies from smooth to pebbled. Next to the colon, the small intestine is smooth, distended, and free of folds.

Heterodon platyrhinos has a saccular intestine which is divided into compartments by transverse folds. These folds apparently act as diaphragms with central holes through which the food can pass. The wall of the intestine is covered by many other projections and ridges which appear randomly oriented. Towards the posterior end of the small intestine, the folds are flattened. There are thirteen to fourteen of these folds in a length of eighteen centimeters.

In Xenodon merremii, Ferri (1971) found a complex and unsual structure. Anteriorly there are longitudinal, transverse, and oblique ridges which form a honeycombed pattern. Further posteriorly there are only the longitudinal ridges which are irregular, wavy, and widely spaced. There are low bumps between them. Ferri notes that these ridges do not disappear when the intestine is distended.

Thamnophis sirtalis also follows the general viperid pattern. The relief is quite constant throughout the small intestine. The tall, wavy duodenal folds are quite narrow and their diameter does not vary much; they run parallel and are occasionally joined to one another by oblique or transverse branches. Posteriorly in the duodenum, the folds assume a zigzag pattern but otherwise remain unchanged. The folds become lower and straighter posterior to the duodenum, but sharp-angled lateral bulges still occur. The folds are here very close together. This pattern continues to the end of the small intestine; the only changes that occur are that the folds get progressively lower and straighter. Near the colon, the folds are nearly straight, broad, and low. The surface of the folds and of the wall between them is smooth.

Masticophis flagellum has a pattern more similar to that of the typhlopids than other snakes. The small intestine is covered by vllliform processes. These usually appear randomly arranged but, on occasion, are loosely strung together to form irregular transverse ridges.

Colon

Like other reptiles, snakes have no distinctive pattern of colonic relief. Patterns of folds, when present, are simple. The thin walls have few, if any, permanent folds. They are distensible, and most folds are obliterated when the wall is stretched. This has surely prevented resolution of the true pattern in many cases. Transverse ridges and folds are common and may be found anywhere along the wall. Anteriorly they tend to be fine and quite numerous, while posteriorly they are broader and seem superficial, almost like bands of tissue laid on the wall.

Jacobshagen (1920), too, found few distinctive patterns. He reports that Meckel found longitudinal folds in Typhlops oxyrhynchus, and he himself found low networks of folds in Cylindrophis melanotus, C. rufus, Boa constrictor, Eunectes notaeus, Bitis cornuta, Vipera berus, Natrix natrix, Philothamnus natalensis, and Coronella austriaca.

Anteriorly the colon of Boa constrictor is distended and smooth-walled. The nex region has fine transverse ridges which are relatively close together, smooth, and sharply defined. In this region, too, there are rather tall, broad, longitudinal folds spaced at regular intervals. Further posteriorly, the transverse ridges become lower and less distinct. They may disappear together and be replaced by bumps and knobs of tissue which form patches on the wall. In the rectum, the wall becomes quite constricted, and there are large, broad, and closely packed transverse folds.

The more anterior regions of the colon of Python sp. are thin-walled and smooth-surfaced. The most prominent feature is a series of transverse bands of tissue; these resemble strapping on a package in that they are narrow, evenly spaced, and continue across the longitudinal grooves in the wall. In this region the few longitudinal folds are low, narrow, and discontinuous; they occasionally merge with the transverse bands. These transverse bands are also present posteriorly in the colon, but are less prominent. Fairly straight and tall longitudinal folds are the most obvious features here. These are widely spaced and rarely branch, but tend to be discontinuous. Very fine, transverse ridges cross the wall and are especially prominent on either side of the longitudinal folds. The latter are rough-surfaced with the ridges continuing right over them.

The colon of Acrochordus javanicus is saccular and distended, without any folds other than the wrinkles expected in a thin-walled structure. As the colon constricts slightly near its posterior end, there are two longitudinal ridges, but these could be artifacts of the fixation.

The colon of Bitis arietans is thin and covered by broad, flattened folds. They branch and fuse frequently, fold back on themselves, and follow apparently random paths. They look like strips af tissue laid on top of the intestinal wall. The colon of another specimen has three regions. The first of these is thin, distended, and without real folds, and is covered by very fine transverse ridges. The edges of the ridges and the surface between them bear tiny papillate projections. In the second region, the transverse ridges persist, but there are also taller longitudinal folds at regular intervals. These folds have a lumpy appearance as if the transverse ridges continue over their tops. The third region is also very thin-walled, and its relief may possibly be a modification of the transverse ridges: it has the wrinkled appearance of well-worn leather, and the small islands of tissue between the fine cracks protrude slightly from the wall.

The large intestine of Echis carinatus is also very thin-walled nearly transparent at times. At the anterior end of the colon, there are five or six narrow but wavy longitudinal folds. Their diameters are the same and they are equally spaced. These wavy folds disappear tn the mid-colon and are replaced by three or four very broad, tall, straight longitudinal folds. Towards the posterior end of the colon, the few ribbon-like folds tend to be longitudinal but are not very straight. In addition, there are many transverse ridges which are quite short, narrow, and randomly distributed.

Anteriorly the colon of Crotalus atrox has a very complicated system of transverse folds. These are very thin quite tall, convoluted, and often folded back on themselves. The folds frequently branch and fuse, and are much more closely packed in some regions than in others. Posteriorly the folds are predominantly longitudinal. They vary in size, but all are thin and relatively tall. They branch frequently and diverge from a longitudinal path. Between these widely spaced folds, the wall is covered by fine, closely packed transverse ridges which continue up and over the longitudinal folds.

Sistrurus catenatus has a fairly extensive network of relatively tall, rounded longitudinal folds in the colon. They branch frequently and lie closer together than the longitudinal colonic folds of most snakes. A branching network of fine transverse ridges covers the wall and continues over the folds.

In Trimeresurus sp. the anterior part of the colon is entirely covered by papilla-like structures that occasionally form short, randomly directed ridges. Posteriorly, the wall is smoother; there are a few wrinkles and scattered patches of the papillae described above, but the most distinctive feature is a single large transverse fold. This is tall, broad, and uneven in diameter. It is covered by spirally arranged ridges that give it the appearance of a braided rope.

The colon of Naja naja looks like a continuation of the small intestine. Fine crevices divide the wall into irregular areas which could be considered longitudinal folds. These folds, like those in the small intestine, are covered by bumps and ridges which give them a very rough appearance.

The wall of the colon in Heterodon platyrhinos is thin and smooth textured and is thrown into a series of folds, grooves, and ridges. These tend to be longitudinally arranged, but are not necessarily so; there are a few transverse ridges as in the colons of most snakes. Next to the cloaca, the pattern is the same except that two transverse folds project above the rest, as if lying on top of the other irregularities in the wall. They are even more complex than the rest of the wall, and broad and covered by grooves and folds. They may function as valves.

Anteriorly the colon of Thamnophis sirtalis is covered by a series of short, low ridges that are thick relative to their length and appear randomly placed. They branch and curl back on themselves extensively, covering the wall so that no spaces can be seen between them. More posteriorly there are longer, more distinctly longitudinal folds. These are also wider, straighter, and taller than most of the other folds, although they still may be discontinuous. The dense network of randomly arranged ridges is still present among these longitudinal ones. The middle region of the colon is smoother than the anterior one. It bears a series of tall irregular longitudinal folds of many different widths; they are often covered with grooves and branch extensively. The rest of the wall is covered by fine, rough longitudinal striations. Here some of the longitudinal folds turn and run transversely, and very rarely there are individual transverse folds. Near the posterior end of the colon, there are no folds, but the wall is covered by fine longitudinal striations.

The relief in Elaphe sp. consists of broad, very irregular longitudinal folds, which are grooved and ridged transversely and have fluted edges to give the appearance of lasagna noodles. The folds are sometimes discontinuous, and occasionally one merges with its neighbour. Clefts between the folds are frequently, but not always, fairly deep. Often there are pits, not clefts, separating the folds.

CROCODILIA

As is usual, most of the specimens we studied were small. Our Alligator had an approximate total length of 60 cm, the Caiman crocodilus one of 100 cm, the two C. c. crocodilus and two C. c. fuscus ones of 40 cm, the C. latirostis one of 45 cm, the Crocodylus one of 160 cm, and the Tomistoma one of 75 cm.

Oesophagus

The common pattern of oesophageal relief is long, straight, smooth-surfaced longitudinal folds which are very prominent and unbranched. The patterns of individuals within a subfamily appear similar, but our sample is extremely small. In all crocodilians studied, the relief is a simple generalized one of longitudinal folds. The sole exception is Caiman crocodilus, crocodilus which has define transverse ridges, a rare pattern among reptiles. Other animals have, in the basic longitudinal folds, irregularities that proceed in a transverse direction, but these are rare and inconsistent even in one individual.

In Alligator mississippiensis and Caiman latirostri, the folds are very straight with perpendicular sides and deep, narrow clefts between them; the wall between the folds is covered by fine, parallel, longitudinal striations. In small Caiman crocodilus, the folds are not quite as close together as in the other two species nor are they as straight. The wall between folds is quite smooth anteriorly, but next to the stomach there are low, though well defined, transverse ridges between the longitudinal ones. These are close together and very regular. In the large specimen of Caiman crocodilus these small ridges are replaced by small irregular bumps. In all of the Alligatorinae studied, the longitudinal folds are of the same diameter and remain so for their entire length.

Crocodylus porosus and Tomistoma schlegelii (Crocodylinae) have similar patterns of relief. The folds are generally broader, rougher, and more irregular than in the Alligatorinae. They branch and fuse more frequently and have variable diameters. The clefts between the folds are narrow and deep; the surface of the folds is rough in Tomistoma schlegelii, and they are subdivided by shallow longitudinal ridges in Crocodylus porosus.

Stomach

The stomachs of crocodilians, like those of lizards, are large, sac-like structures. They are divided into cardiac, fundic, and pyloric regions, and the relief varies among these areas. Generally the texture of the gastric wall is smooth except for the pebbled appearance so common in stomachs. The cardiac and pyloric regions have few folds, but the fundus has an extensive network of thick rugae separated by deep, short, random crevices in the wall.

The gastric wall of Alligator mississippiensis is thick and muscullar, and the pattern is more pronounced in the fundus than in the cardiac or pyloric regions. The rugae are thick, branching structures, but can be obliterated if the stomach is distended. There are no folds in the cardiac or pyloric regions, but their surface is rough.

There are prominent rugae anteriorily in the stomach of Caiman crocodilus. These thick, short, branching structures disappear near the posterior end of the stomach where the wall has a pebbled appearance. The pyloric region is thick-walled, narrower than the rest of the stomach, and has a few very thick folds. In our largest specimen of this species, the rugae are very large with lower ridges between them as well as transverse folds joining them. In pyloric region, the walls lack good ridges but have small villiform projections into a sac-like chamber.

Posteriorly the stomach of Crocodylus porosus has an interesting pebbled pattern. The deep, variously shaped pits give the folds a net-like appearance.

Small Intestine

All crocodilians have a distinctive intestinal relief pattern. The folds are well defined and zigzag patterns predominate, although some of these patterns are complicated by transverse folds, protruberances, and other irregularities. Jacobshagen (1920) says that crocodiles retain a primitive pattern of net-like folds as manifested by the zigzags; it seems more appropriate to note that the pattern is zigzagged.

The duodenum of Alligator mississippiensis has an obtusely angled zigzag pattern. The folds are moderately tall, and each follows the contours of its neighbours. Transverse ridges between adjacent folds occur frequently but inconsistently. These ridges are lower than the longitudinal folds. Occasionally, this clearly defined pattern breaks down into a disorganized net-like pattern. Further posteriorly, the folds become taller, thinner, and straighter, so the zigzag pattern is replaced by one of wavy, ribbon-like folds. There are no transverse ridges between them, and the pattern has fewer irregularities than anteriorly. This pattern continues for most of the small intestine; however, the folds become more irregular and tend to fuse with one another posteriorly. The pattern changes again at the end of the small intestine where the folds are rounded, thicker, much lower, and thrown into compact waves. Between the longitudinal folds are scattered remnants of other, lower longitudinal folds.

Jacobshagen (1920) described an animal he studied and repeated a description of Meckel. Our results differ slightly from both. Jacobshagen found that the numbers of transverse and oblique folds increase in the region where the longitudinal folds become straighter. In addition, he found that a net-like relief is retained throughout the small intestine. In contrast, Meckel found that the net-like pattern disappears in the middle of the small intestine and does not reappear posteriorly at all. He does agree that there are remnants of longitudinal folds between the major longitudinal folds in the postetior part of the small intestine. It seems that there is much individual variation in Alligator mississippiensis. The pattern is primitive and fairly complex, and all the folds are probably essential. However, all authors found a net-like pattern with fairly intensive transverse folds in the anterior part of the small intestine.

The duodenum of Caiman crocodilus has low zigzag folds, but the zigzags are very broad and form acute angles. In the duodenum, especially its posterior part in small specimens, prominent transverse ridges connect the zigzags, and the latter form more acute angles and are less regular. The just posterior to the duodenum, the folds are fewer, and their zigzag pattern is much more regular. Small ridges fuse with the zigzags at right angles. There are other, less regular longitudinal folds which tend to be straighter and often lower than the zigzag folds. Further posteriorly, the intestine is thin-walled and the ridges are fine, low, and wavy, quite regular, in their pattern and widely separated. There is some variation in the degree to which the net-like pattern persists posterior to the duodenum. Near the colon, the folds become taller, and their diameter is less constant. Both the folds and the wall between them become rough-surfaced. Jacobshagen's results (1920) are very similar except that the zigzag folds are very tall at the start of the intestine and the longitudinal folds at the end of the small intestine were apparently no taller than in the preceding region.

The pattern in Caiman latirostris is essentially as in C. crocodilus, but the whole intestine has a rougher appearance. In addition, the posterior parts have small, lumpy, almost villus-like protruberances from the wall and lack any clear pattern.

Jacobshagen reports Rathke's finding a "double" net pattern of folds in Crocodylus acutus. The folds of this net come closer together posteriorly and are fused closely to one another by short low transverse folds. These transverse folds are found in the angles of the longitudinal ones. Further posteriorly, the longitudinal folds become more separate and distinct as the transverse ones become lower and finally disappear.

Crocodylus niloticus seems to be even more variable than Alligator mississippiensis. We did not study Crocodylus niloticus, but Jacobshagen (1920) concludes finally that Meckel's description is the most accurate. Apparently, tall folds delimit four-sided cell-like areas at the beginning of the intestine. The folds become lower posteriorly before becoming undulating longitudinal folds, and finally broad, straight strips. Cuvier (1835), in contrast, says that there are no cross-folds in the first part of the intestine but that the second half has zigzag longitudinal folds joined by cross-folds. Gegenbaur and Haler evidently support these observations (all references and descriptions from Jacobshagen, 1920).

The pattern in Crocodylus porosus is complex. The duodenum has a suggestion of a fine zigzag or net-like pattern overlying a complex pattern of randomly arranged ridges and deep pits of various diameters. This pattern continues posteriorly, but in the middle of the jejunum wide zigzag folds start to predominate. This pattern is regular and the folds are parallel; deeper, irregular transverse ridges lie between the zigzags, and give a complex disorganized appearance. Finally, near the posterior end of the small intestine, the longitudinal folds become straight. Here the folds are tall and quite narrow, and the underlying relief has faded to a rough pebbling of the wall between the folds.

Osteolaemus tetraspis also has a net-like pattern of folds (Jacobshagen, 1920) composed of tall longitudinal and transverse folds. The longitudinal folds have rough edges and short tongue-shaped processes attached to them, but both of these become scarcer posteriorly as the longitudinal folds become smoother, lower and straighter. The transverse folds also decrease in size posteriorly.

Throughout much of the small intestine, Tomistoma schlegelii has the broad, low, zigzag folds described for the duodenum of Caiman crocodilus. As is usually the case, the zigzag pattern is lost near the posterior end of the intestine; the folds become taller, straighter and more irregular.

Colon

The wall of the crocodilian colon is generally smooth-surfaced and if any pattern exists at all, it is one of broad, rounded, low, widely separated longitudinal folds. All the folds in the colon appear in the same position whenever the gut is not under tension. As might be expected, Jacobshagen found a pattern of net-like folds or a modification thereof in all, the animals that he studied. The net-like pattern for all crocodiles is low, smooth, and has fine meshes. Longitudinal wrinkling of the wall also occurs frequently.

The colon of Alligator mississippiensis has a much larger lumen than the small intestine, but it has no distinctive pattern of relief. There is one large, broad transverse band near the posterior end, but except for small bumps and mounds of tissue, the colon is smooth-walled. A sphincter leads to the cloaca.

The short longitudinal folds in the collon of Caiman crocodilus are slightly wavy, broad, low, evenly separated, and with constant diameters. Between them, the wall is covered by knobs of tissue which occasionally become quite large. These knobs do not appear in the largest specimen we studied.

The colon of Tomistoma schlegelii has several low, broad, widely separated folds. They branch frequently, but their widths remain constant; the wall between the folds is covered by shallow depressions as though made by pin heads. This latter could conceivably be considered as a net fold pattern.

7

Ecology of Digestive Tract (Contd.)

The macroscopic morphology of the digestive tract of reptiles has been widely by a number of workers. The microscopic morphology of this structure has also been dealt in a variety of older individual studies and in more recent works of Luppa (1961, and unpublished investigations), Gabe and Saint Girons (1964), Ferri (1971), and Dilmuhamedov (1975). See Diefenbach (1974) for a summary table of diverse studies on the reptilan gut. A chapter on the digestive process will appear in the next volume on physiology. The results of some of the older investigations may appear dubious because of the use of outmoded techniques. The careful investigation of the digestive apparatus of Lacerta agilis by Krause (1922) forms the fundamental basis of comparison for histological studies. Development of morphological techniques has recently accelerated, but such approaches have scarcely been utilized in the study of the alimentary canal. Histochemical and enzymological investigations on the reptilian gut were carried out by Arvy and Gabe (1952), Gabe (1971a, b) and Gabe and Saint Girons (1972a, b), Luppa (1961) applied both histochemical and histoenzymatic methods to the gastric epithelium of various adult and embryonic reptiles.

The present study summarizes past investigation and surveys the histo-chemistry and fine structure of the digestive apparatus in different reptiles. Emphasis is on the study of the stomach, since this shows the most marked histological specialization among the vertebrates. The functional meaning of different types of cells within the stomach receives primary emphasis.

ORAL CAVITY AND PHARYNX

A bony palate roofs the oral cavity of reptiles. This is distinct from the base of the braincase and separates the nasal and oral cavities (Krause, 1922). The differentiation of this oral roof varies widely (Fuchs, 1908; Parsons, 1970). Crocodilians have a secondary palate consisting of two parallel producesses that originate from the upper jaws and fuse along the midline to create a secondary nasal cavity, longer than the primary one; the oral cavity is secondarily reduced (Barge, 1937).

Many reptiles have palatal glands. Lizards have azygous, median, and paired lateral glands (Barge, 1937). These are asbent in snakes but are well developed in turtles. According to Gaupp (1888) crocodilians lack palatal glands, but Rose (1893) reported their existence in Crocodylus porosus. Woerdemann (1921) and Reese (1925) agree on the existence of palatine glands grouped into a median and a lateral series. These, and other oral glands (Fahrenholz, 1937), are considered in more detail by Kochva elsewhere in this series (Physiology B). See Edmund (1969) for reptilian dentitions.

The oral cavity is lined by a muscous membrane consisting of a compound, squamous, non-keratinized epithelium and a connective tissue lamina propria, but lacks a muscularis mucosa. The epithelium regenerates by division in the basal cell layers. In Lacerta the cells of the squamous epithelium are separated by wide intercellular spaces and always possess a strongly staining ectoplasmic layer (Krause, 1922).

The epithelial lining of the oral cavity shows considerable variation, both regional and specific. A single species may have

TABLE 7.1

Histochemical Methods Used for the Detection of Compounds in the Digestive Tract of Reptiles

Substance or group	*Croco-dilia*	*Rhynchoce-phalia*	*Gekkonidae*	*Lacertidae*	*Agamidae*	*Ophidia*	*Testudines*
			Family or Order				
1	*2*	*3*	*4*	*5*	*6*	*7*	*8*
A. Carbohydrates PAS-reaction[1] (McManus, 1945)	stomach (Gabe, 1971b)	stomach, small intestine (Gabe and Saint Girons, 1964) cloacal glands (Gabe and Saint Girons, 1965)	small intestine (Kanwar 1961); cloacal glands (Gabe and Saint Girons, 1965)	main gastric glands, pyloric glands small intestine, large intestine[2] cloacal glands (Gabe and Saint Girons, 1965)	main gastric glands, pyloric glands, large intestine;[3] cloacal glands (Gabe and Saint Girons, 1965)	main gastric glands pyloric glands, small intestine large intestine[2]; cloacal glands (Gabe and Saint Girons 1965) oesophagus,	lingual glands, main gastric glands, pyloric glands, small intestine, large intestine[2]

					stomach, small intestine, large intestine (Ferri, 1971)	
Astra blue (Pioch, 1957)			Pyloric glands[3]	pyloric glands[3]	pyloric glands[2]	lingual glands, pyloric glands[2]
Astra blue at pH 0-2 (Bloom and Kelley, 1960)			oesophagus, main gastric glands, small intestine, large intestine[3]	oesophagus[3]	oesophagus main gastrtc glands small intestine, large intestine[3]	lingual glands, oesophagus, small intestine, large intestine[3]
Alcian blue at pH 2.6[1]	cloacal glands (Gabe and Saint Girons 1965)	cloacal glands (Gabe and Saint Girons 1965)	cloacal glands (Gabe and Saint Girons 1965)	cloacal glanps (Gabe and Saint Girons 1965)	cloacal glands (Gabe and Saint Girons, 1965)	

(*Contd.*)

TABLE 7.1 (*Contd.*)

1	*2*	*3*	*4*	*5*	*6*	*7*	*8*
PAS and alcian blue combined.				main gastric glands[2]	main gastric glands[2]	main gastric glands, large intestine[2]	main gastric glands[2]
Toluidine blue at 0.1% (Kramer and Windrum, 1955)				main gastric glands[2]	main gastric glands[2]	main gastric glands[2]	main gastric glands[2]
B. *Proteins and Amino Aaids* Alloxan-Schiff/ Ninhydrin-Schiff (Pearse, 1968)				main gastric glands, small intestine[2]		main gastric glands, small intestine[2]; oesophagus, stomach, small intestine, large intestine (Ferri, 1971)	small intestine[2]

Million reaction of Pollister and Ris (1947) and of Bensley and Gersh (1933)	main gastric glands, small intestine[2]	main gastric lands[2]	main gastric glands, small intestine;[2] oesophagus, stomach, small intestine, large intestine (Ferri, 1971)	main gastric glands, small intestine[2]
Tyrosine (Morel-Sisley Lillie 1957)	main gastric glands, small intestine[2]		main gastric glands, small intestine;[2] stomach (Manfredi Romanini, 1959)	small intestine[2]
Tryptophane (Adams, 1957)	main gestric glands,	main gastric glands[2]	main gastric glands,	main gastric glands,

(*Contd.*)

TABLE 7.1 (*Contd.*)

1	2	3	4	5	6	7	8
				small intestine[2]		small intestine[2]	small intestine[2]
Arginine Sakaguchi, modificated by Deitch and Terner, 1965)						oesophagus, stomach, small intestine, large intestine (Ferri, 1971)	
SS and SH groups (Barrnett and Seligman, 1952 and 1954)				oesophagus, main gastric glands, pyloric glands, small intestine[2]	oesophagus pyloric glands[2]	oesophagus main gastric glands, pyloric glands, small intestine[2]	oesophagus, stomach pyloric glands, small intestine[2]
SS groups and SS+SH groups (Adams, 1956)[1]		cloacal glands (Gabe and Saint Girons, 1965)	cloacal glands (Gabe and Saint Girons, 1965)	cloacal glands (Garbe and Saint Girons, 1965)	cloacal glands (Gabe and Saint Girons, 1955)	cloacal glands (Gabe and Saint Girons, 1965)	

C. *Nucleic Acids* Fuelgen-reaction for DNA RNA and DNA, method with chrome alum and gallocyanin (Einarson, 1951; Sandritter, 1952)	pyloric glands oesophagus, main gastric glands, pyloric glands, small intestine[2]	pyloric glands[3] oesophagus, pyloric glands[2]	pyloric glands[1] oesophagus, main gastric glands, pyloric glands, small intestine[2]	pyloric glands[3] oesophagus, pyloric glands, small instestine[1]
D. Lipids Sudan black B of McManus (1946) Phospholipids with acid hematein of Baker (1946)	oesophagus. main gastric glands, pyloric glands, small intensine[3]	oesophagus main gastric glands, pyloric glands[2]	oesophagus, main gastric glands, pyloric glands small intestine[3]	oesophagus, main gastric glands, pyloric glands small intestine[3]

(*Contd.*)

TABLE 7.1 (*Contd.*)

1	2	3	4	5	6	7	8
				small intestine[2]		small intestine[2]	small intestine[2]
Danielli's coupled tetrazonium reaction[1]		cloacal glands (Gabe and Saint Girons, 1965)	cloacal glands (Gabe and Saint Girons, 1965)	cloacal glands (Gabe and Saint Girons, 1965)	cloacal glands (Gabe and Saint Girons 1965)	cloacal glands (Gabe and Saint Girons, 1965)	

1. PAS-reaction, alcian, blue (pH 2.6), SS—groups and SS+SH-groups and Danielli's coupled tetrazonium reaction were also used on the cloacal glands of following squamates: Xantusiidae, Pygopodidae, Iguanidae, Chamaelcodidae, Scincidae Feyliniidae', Tediidae, Anguidae, Annielliadae, Varanidae, Helodermatidae, Amphisbaenidae Trogonphidae.
2. Luppa 1961, and unpublished.

compound squamous epithelium, ciliated epithelium, goblet cells, and simple non-ciliated columnar epithelium. In Lacerta the compound epithelium of the oral cavity gradually changes posteriorly into a compound, ciliated, columnar epithelium. The anterior edge of the M. pterygoideous forms the approximate margin between the two epithelial types (Krause, 1922).

Taste buds are scattered throughout the oral epithelium. In Lacerta they are most numerous laterally and on the palatal folds. Reptilian taste buds show no significant difference in their cellular composition from those of mammals. Sensory neuro-epithelial cells (=Schmeckzellen of Krause, 1922) and supportive cells occur in both.

The oral cavity of reptiles is partly lined by ciliated epithelium with goblet cells. In contrast to the condition in fishes and amphibians this epithelium is variably displaced by a compound squamous epithelium. This displacement reaches its maximum in birds and mammals and allows stronger mechanical trituration of food in the oral cavity (Fahrenholz, 1937). It is not known whether these goblet cells have disappeared completely in any reptile.

The reptilian tongue varies extensively in shape (it will be dealt with in a separate volume). In most reptiles the dorsal surface of the tongue bears variably formed papillae; its sides and ventral surface are smooth (Gnanamuth, 1937). Except for the tips of the papillae all of these surfaces are pigmented in Lacerta.

Reptilian tongues generally have a large number of glands (Gabe and Saint Girons, 1969). The glands are fully developed in species with a voluminous and fleshly tongue (Gabe, 1971b). Indeed, turtles, geckos, and Pygopodidae, as well as species in the groups Inguania and Anguoidea have a tongue that contains numerous glands and even the tongue of crocodilians is highly glandular. The sublingual glands of Lacerta form a doubly layered, cylindrical, strongly projecting ridge on each side of

TABLE 7.2

Histochemical Methods for the Localization of Enzymes in the Digestive Tract of Reptiles

Enzyme	*Crocodilia*	*Family or order*			
		Lacertidae	*Agamidae*	*Ophidia*	*Testudines*
1	*2*	*3*	*4*	*5*	*6*
A. *Hydrolases*					
1. Non-specific alkaline phosphatase (a) method of Gomori (Pearse, 1960) (b) method of Gossner (1958)	esophagus, stomach, pyloric, glands (Arvy and Bonichon, 1958a, b)	esophagus, pyloric glands, small intestine, large intestine[2]	esophagus, pyloric glands, small intestine large instestine[1]	esophagus, pyloric glands, small instestine large intestine[3] esophagus, stomach, small intestine, large intestine (Ferri, 1971)	esophagus, main gastric glands, pyloric glands, small intestine, large intestine[2]

2. Non-specific acid phosphatase (a) Burstone, 1961 (b) Pearse, 1960 (c) Eranko (cit. after Arvy *et al.*, 1962b)		esophagus pyloric glands, small intestine, large intestine[3]	esophagus pyloric glands, large intestine[3]	esophagus, pyloric glands, small iniestine, large intestine[2]; esophagus, stomach, small intestine, large Intestine (Ferri, 1971)	esophagus, main gastric, glands, pyloric glands, small intestine, large intestine[3]
3. Non-specific esterases (a) Shnitka and Seligman, 1961 (b) Holt, 1958 (c) Gossner, 1958	stomach (Arvy and Bonichon, 1958b)	esophagus, pyloric glands, small intestine, large intestine[3]	esophagus, pyloric glands, stomach large intestine[3]	esophagus, pyloric glands, small intestine[3]; esophagus, stomach, small intestine, large	esophagus, pyloric glands, small intestine[3]

(*Contd.*)

TABLE 7.2 (*Contd.*)

1	2	3	4	5	6
				intestine (Ferri, 1971)	
4. Adenosinctri-phosphatase (Wachstein and Meisel, 1957)				esophagus, stomach, small intestine, large intestine (Ferri, 1971)	
5. 5-nucleotidase (Wachstein and Meisel, 1957)				esophagus, stomach, small intestine, large intestine (Ferri, 1971)	
6. Leucine amino-peptidase Nachlas *et al.*, 1957)		small intestine[2]	main gastric glands[2]	main gastric glands, small intestine[2]	main gastric glands, small intestine[2]

B. *Oxidoreductares*

7. Cytoehrome oxidase (Burstone, 1959)	small intestine[2]	main gastric glands, small intestine, large intestine[3]	main gastric glands, small intestine[2]	esophagus, main gastric glands, small intestine[2]
8. Succinate dehydrogenase[1]	esophagus, small intestine[2]	esophagus, small intestine[2]	esophagus, main gastric glands, small intestine[3]	esophagus, main gastric glands, small intestine[3]
9. NADH and NADPH diaphorases	small intestine[2]	main gastric glands, small intestine[3]	main gastric glands, small intestine[2]	main gastric glands, small intestine[3]
10. Isocitrate dehydrogenase	esophagus, pyloric glands[3]	esophagus, pyloric glands, large intestine[3]	esophagus, pyloric glands,	esophagus, pyloric glands,[2]

(*Contd.*)

TABLE 7.2 (*Contd.*)

1	2	3	4	5	6
11. Glucose-6-phosphate dehydrogenase		esophagus, small intestine, large intestine[2]	esophagus, main gastric glands, large intestine	esophagus, main gastric glands, small intestine, large intestine[2]	esophagus, main gastric glands, small intestine, large intestine
12. Glutamate dehydrogenase			main gastric glands small intestine[3]	main gastric glands, small intestine[2]	main gastric glands, small intestine
13. Malate dehydrogenase			main gastric glands, small intestine[3]	main gastric glands, small intestine[2]	main gastric glands, small intestine[2]
14. glycerpphosphate dehydrogenase		large intestine, small intestine	main gastric glands, small	main gastric glands, small	main gastric glands, small

		intestine, large intestine	intestine, large intestine	intestine, large intestine
15. hydroxy-butyrate dehydrogenase	esophagus, pyloric glands, small intestine	esophagus, main gastric glands, pyloric glands,	esophagus, main gastric glands, pyloric glands, small intestine	esophagus, main gastric glands, pyloric glands, small intestine
16. Alcohol dehydrogenase	small intestine, large intestine	main gastric glands, small intestine	main gastric glands, small intestine, large intestine	main gastric glands, small intestine large intestine
17. Lactate dehydrogenase	esophagus, pyloric glands, small intestine, large intestine	esophagus, pyloric glands, small intestine, large intestine	esophagus, pyloric glands, small intestine, large intestine	esophagus, pyloric glands, small intestine, large intestine

(*Contd.*)

TABLE 7.2 (*Contd.*)

1	*2*	*3*	*4*	*5*	*6*
18. Monoamine oxidase (Glenner *et al.*. 1957)		esophagus, small intestine large intestine	esophagus, main gastric glands, small intestine, large intestine	esophagus. main gastric glands, small intestine, large iatestine	esophagus, main gastrlc glands, small intestine, large intestine
C. *Lyases*					
19. Carbonic anhydrase (Hausler, 1958)		small intestine	main gastric glands small intestine	main gastric glands, small intestine	main gastric glands, small intestine

1. Barka and Anderson, 1963.
2. Luppa, 1961 and unpublished.

the lingual base, and are composed of two topographically and structurally distinct parts (Krause, 1922).

The tongue is largely covered by compound, partially keratinized squamous epithelium. In turtles this epithelium is pierced by the ducts of the very common mucous glands. The cylindrical tongue of the Varanidae is keratinized (Gabe, 1971b) as is that of Lacerta. In the posterior portion of the tongue of Lacerta, the compound epithelium in the groove between the papillae is transformed into a two-layered columnar epithelium with numerous goblet cells. Ventrally on the tongue there is an equivalent progressive transition from keratinized, compound, squamous epithelium to non-keratinized epithelium and finally to a columnar epithelium with goblet cells. The weakly developed lamina propria is gradually transformed into the intermuscular connective tissues.

The compound tubulo-alveolar glands, which in Testudo hermanni extend down into the musculature, are lined by cuboidal epithelium which becomes columnar in the region of their duct. The cytoplasm of the prismatic cells shows strongly alveolar structure when fixed. Connective tissue septa passing between the glands lead to their segmental structure.

The sublingual gland of Lacerta consists of separate, branching tubular mucous glands. The secretory tubules of the median portion are relatively thick and formed of high columnar glandular cells (Krause, 1922). The paired ventral portion of the sublingual glands is also formed from mucous glands. The glandular cells composing it strain more darkly than those of the dorsal portion. Krause (1922) ascribes this to a lower mucous content.

The lingual glands of Testudo hermanni show a relatively strong reaction to PAS. They also react with alcian blue (Pioch, 1957); some of the secretory cells stain with alcian blue at pH 0-2 (method of Bloom and Kelly, 1960). The total analysis for polysaccharide suggests that two kinds of mucoid substances are produced in the lingual glands of Testudo hermanni: most of the secretory cells contain neutral mucopolysaccharides and acid mucopolysaccharides with weakly acid carboxyl groups, but others contain secretions with neutral mucopolysaccharides and acid mucopolysaccharides containing phosphate groupings. In most cases, protein-like substances do occur in the secretory product, either with neutral mucopolysaccharides (seromucous type) or with mucins which are more or less acid (muco-serous type).

THE FOREGUT

General Morphology

The postbranchial alimentary canal is delimited and subdivided like that of the amphibians. The oral plus pharyngeal cavity narrows posteriorly and is not sharply separated from the alimentary canal. The reptilian alimentary canal proper begins posterior to the opening of the trachea with the narrowing of the throat to a tube of constant diameter (the esophagus). In crocodilians the esophagus is separated from the oral cavity and the entrance to the trachea by a transverse fold which projects from the floor of the mouth and joins a shorter fold, descending from the cranial floor (Gans, 1976). As in the Amphibia, the foregut is generally divided into an esophagus, which serves as a supply tube, and a clearly differentiated stomach, which serves for food storage (Pernkopf and Lehner, 1937).

Oesophagus

Morphology

The reptilian esophagus may be subdivided into two sections of different lengths; this division is particularly marked in the forms with a long, extended, serpentiform body. The more extensive anterior section is sharply differentiated from the shorter, posterior esophageal section which enters the visceral cavity dorsal to the pericardio-peritoneal septum. In many snakes the border between the esophagus and stomach is indicated only by the nature of the mucous membrane. In turtles and crocodilians, on the other hand, the border is clearly marked by the significantly greater diameter of the stomach.

Except for the regular appearance of longitudinal folds, the internal structure of the esophagus varies markedly. Most notable are the papillae with keratinized tips in the posterior half of the esophagus of marine turtles. The posterior esophagus of Python shows a network-like arrangement of folds of the mucous membrane. These structures are considered in more detail by Persons and Cameron (this volume).

Histology

The esophageal epithelium is mainly two cells thick and contains variably distributed ciliated cells and goblet cells. In snakes (Gianelli and Giacomini, 1896). Typhlops vermicularis (Heyder, 1974), and the alligator (Beguin, 1904a) the posterior section of the esophagus appears to have mainly goblet cells. The esophagus of the Cheloniidae, with its keratinized papillae, is lined by stratified squamous epithelium, the top layer of which may be keratinized. In the Testudininae (Testudo graeca, Testudo horsfieleii; Jeksajewa and Koloss, 1964; Dilmuhamepov, 1975) the epithelium is also stratified. The more basal layers of cells are characterized by longitudinally oval, strongly staining nuclei and by a more markedly basophilic cytoplasm. The surface of the epithelium is formed by prismatic cells. The middle layer of cells are strongly flattened with rounded nuclei.

The subepithelial tunica propria consists of collagenous and elastic fibres and, in turtles and Uromastyx acanthinurus, contains mucous glands that vary in number, size and distribution (Beguin, 1904a). In Testudines this layer also forms papillae that project into the epithelium. The ducts and terminal portions of the tubulo-alveolar glands are lined by prismatic cells, mainly with flattened basal nuclei. Esophageal glands are particularly frequent in turtles (Gabe, 1971b). Oppel (1900), who draws up a long list of species the esophagus of which is provided with glands, emphasizes that there is no link between the systematic position of the animal and the occurrence of glands. Thus, the glands occur in the esophagus of Alligator, and are absent in that of Crocodylus. In any case, the general morphological character of these glands is very different from that of the esophageal glands of the Amphibia (Gabe, 1971b). In most reptiles lymph nodes penetrate, either individually or in aggregates, from the tunica propria into the epithelium (Prenant' 1896; Gianelli and Giacomini, 1896). There are also capillaries and larger blood vessels. The muscularis mucosae is ordinarily found only in the posterior section of the reptilian esophagus; it is completely lacking in turtles (Gianeli and Giacomini, 1896: Gianelli, 1900). In the gecko Tarentola annulata. the muscularis mucosae forms embryonically as mesenchymal condensation but, during the later development, is completely lost in the region of the esophagus. The tunica muscularis consists of two layers of smooth muscle fibres (Gegenbaur, 1901); the thin layer is circular, and the external longitudinal one may be completely missing in the anterior portion of the esophagus. The circular layer is always wider than the longitudinal layer.

Histochemistry

(a) *Ciliated cells.* The supranuclear regions and the cilia contain alpha amino acids, proteins with reactive SH and SS groups and ribonucleic acids The reactions are weaker in the infranuclear regions. PAS-positive mucoproteins oceur in smaller quantities, while sudanophilic lipids are more abundant in the supranuclear regions of the ciliated cells.

The supranuclear cytoplasm contains a high level of non-specific acid phosphate and esterase bound to corpuscles. In contrast, the non-specific alkaline phosphatase is found only in the subepithelial vessels of Lacerta, Natrix, Agama and Chryemys, while it occurs in the subepithelial tissues of Testudo graeca and T. hermanni. The same enzyme has been found in the esophageal epithelium of Crocodylus (Arvy, 1962a; Arvy and Bonichon, 1985b).

The stratified epithelium of Testudo graeca and T. hermanni contains non-specific acid phosphatases and non-specific esterases, particularly in the superficial layer of cells, while they are more diffusely distributed in the remainder of the epithelium. The activity of both enzymes is restricted mainly to the base of the cells of the tubulo-alveolar glands.

Oxidoreductase activity is most obvious in the supranuclear zone of mitochondria within the ciliated cells of Lacerta, Agama, and Natrix. A second region with marked enzymatic activity lies at the base of the epithelium. The entire ciliated cell shows a strong positive reaction to the test for monoamino oxidase.

The stratified, as well as the glandular, epithelium of Testudo shows relatively strong oxidoreductase activity. Particularly intense reactions are given for the basal layer of cells, while the superficial layers provide a diffuse reaction. The glucose-6-phosphate dehydrogenase is more or less diffusely distributed throughout the epithellal region.

(b) *Goblet cells.* The goblet cells react positively with PAS (Manfredi Romanini, 1953; Luppa, 1961; Gabe and Saint Girons, 1964) as well as with tests for acid mucopolysaccharide (Luppa, 1961). The positive reaction of the astra blue stain (method of Bloom and Kelly, 1960) suggests the presence of sulphated acid mucopolysaccharide. In contrast to the ciliated cells, there is a relatively low enzymatic activity within the goblet cells. Only in the basal region, for instance, is there a positive reaction with tests for succinate dehydrogenase and β-hydroxybutyrate dehydrogenase. Tests for non-specific acid phosphatase and non-specific esterase give negative results.

Fine Structure

(a) *Ciliated cells.* Each cilium consists of eleven microfibrils; in cross section nine of them are peripherally and two centrally located. A basal plate lies near the cell's surface and is separated from the rod-like basal body by a relatively thin layer of crytoplasm. This body is characterized by a noticeably osmophilic external layer. Root-like fibrils descend from the basal bodies into the mitochondrial cytoplasmic region. Variably sized extensions of the cytoplasm project between the cilia. The basal bodies lie immediately above the mitochondrial region, which extends basally towards the nuclei. The mitochondria are of cristate type and occur along the longitudinal height of the cell. Also found there are lysosomes of different shapes, canals formed by membranes covered with ribosomes, and individual lipid inclusions. The mitochondria are often very long and characterized by noticeable differences in form. The rarely noticed Golgi apparatus lies immediately distal to the nucleus. The nuclei are egg-shaped to longitudinally oval, occasionally kidney-shaped; they generally contain a nucleolus. The continuous plasma membrane of the ciliated cells shows well marked desmosomes in the apical region.

(b) *Goblet cells.* The goblet cells lack cilia, though their free surface may, depending upon their functional state, bear microvilli. The goblet-like, enlarged, supranuclear cell region is normally filled with mucous granules of variable structure. The individual granules are limited by a thin membrane. The mucous secretion is surrounded by dark supranuclear cytoplasm, which contains ER, free ribosomes and cristate mitochondria. The Golgi apparatus lies between the goblet portion and the ovoid to cup-shaped basal nucleus. The structural differences between the individual goblet cells, noted in both light and electron microscopy, are to be traced to different secretory states.

Stomach

Morphology

In spite of the notable differences in form, the reptilian stomach always contains two clearly delimited sections: the corpus and the pars pylorica. The corpus is also called fundic region or short fundus. In most limbless forms (*e.g.*, Anguis, Ophisaurus, Naja, Enhydris, Typhlops) the stomach is elongated, corresponding to the body shape. A pyloric construction does not sharply separate the stomach and midgut. Marked bending of the stomach is found, first of all, in turtles in adaptaion to body form; they have a larger left convex and smaller right concave curvature. The corpus and the generally longer pars pylorica then lie immediately adjacent to each other. The stomach of Dermochelys consists of an oval sac and a more dorsally positioned tubular section with an S-shaped curve.

The stomach of crocodilians is clearly subdivided into a corpus and a pars pylorica. The thin-walled, highly granular corpus and the following curved portion form a slightly dorsoventrally flattened sac into which the narrow esophagus inserts anteriorly and to which the short pyloric section attaches posteriorly.

The difference between the epithelium of the corpus and that of the esophagus can be seen macroscopically, appearing as a serration in snakes and turtles. Contraction causes the gastric wall to fold into longitudinal pleats, that are particularly marked in the region of the pars pylorica. The transition zone between the gastric and intestinal epithelium ordinarily shows a ring-fold or funnel-shaped pyloric value that is lacking in Caretta.

Corpus of the Stomach

(a) ***Histology and cytology af the main glands of the stomach.*** The corpus and parts pylorica of the stomach contain main glands and pyloric glands respectively. In reptiles the glandular portion of the corpus is formed by thickly packed, straight, tubular and ramified glands (fundic glands), each of which may be divided into a pit region, a glandular neck, and a glandular body.

The tubular glands of the turtle Caretta are combined into groups separated by connective tissue (Oppel, 1896). Clear (mucoid) neck cells are especially common in those tubular glands lying in the centre of the group. Oppel interpreted this as an analogue of the groupings of glands, which reaches its apex in the glandular stomach (proventriculus) of birds. This grouping ultimately leads to the intercalation of a preliminary chamber lined with mucoid cells and separating the glandular tube from the lumed of the stomach. Motta and Renaud 1878) also found some groupings of these glands in Emys orbicularis.

The fundic glands may be divided into three anatomical types (Gabe and Saint Girons, 1972a). The first type (Sphenodon punctatus and saurians, except for most Anguimorpha) lacks specialized neck cells and has two cell types in the glandular epithelium, the erythrophilic, PAS-negative (dark, serous) and cyanophilic, PAS-positive ("clear"; "light"; mucous) cells. The second type (most of the Anguimorpha, Booidea, Colubroidea), has specialized neck cells, but only the erythrophilic (dark, serous) cells occur in the glandular epithelium. The fundic glands of the third type (Leptotyphlopidae, Typhlopidae) also lack mucous ("clear") cells and specialized neck cells. According to the literature and to some unpublished data (Gabe), the first type occurs also in Testudines, the third in crocodilians (Staley, 1925; Eisler, 1889). In the snake, Xenodon merremii, paraffin sections show body, argentaffin and argyrophil cells (Ferri *et al.*, 1974a). Clear cells were recognized only after araldite embedding (Ferri *et al.*, 1974b).

The multiply-branched tubular glands end individually or in pairs in a common pit.. The cylindrical to funnel-shaped gastric pits are lined by cuboid cells with basal nuclei. Towards the lumen, the cells gradually change into a simple, high columnar epithelium in which the longitudinally oval nuclei lie close to the base of the cells. This epithelium is fundamentally the same in the different regions of the stomach. The acidity of the secreted mucins shows some interspecific variability (Gabe and Saint Girons, 1972a). The superficial epitheliul cells have specialized, cap-like, apical expansions. Empirical staining methods permit the recognition of two zones of cells: an apical one corresponding to the secretion-collecting zone in the superficial epithelium of the mammalian stomach, and a basal one containing the nucleus and reaching from the Golgi zone to the

base of the cell. The secretary material of the apical epithelial zone stains faintly with acid dyes, while the basal region primarily shows a basophilic reaction. The neck cells preceding the body of the gland show two variations: barrel-shaped cells with flattened nuclei pressed against the base of the cell and cuboid cells with rounded nuclei.

The mucin-containing, barrel-shaped "clear" cells of the body of the fundic glands usually have flat nuclei located near the base of the cell and a faintly reticulate cytoplasm, The cells always appear unstained after exposure to acid dyes. The cuboidal to round "dark" cells always contain scattered round to avoid nuclei with large basophilic nucleoli. The cytoplasm of the "dark" cells shows a double staining reaction with combined basic and acidic dyes according to more recent investigations of the entire cell with the basophilic fraction slightly higher than the acidophilic. When the "dark" cells are stained using the method of Hamperl, the apical region shows blue-violet granules. The supranuclear cytoplasmic region of the "clear" cells reacts with none of the above stains (Table 7.3). The occurrence and distribution of these cells types differ. Although the body of the gland in the Sauria and Testudines consists of "dark" and irregularly distributed "clear" cells, snakes and agamids show only "dark" cells. In the latter group the glandular neck, lined with mucous cells, is very noticeable. The evidence for "parietal" cells in the fundic glands of Testudo is without doubt the result of deficient techniques.

(b) *Histochemistry*. (i) Superficial and pit cells. In the Sauria, Testudines and Sphenodon punctatus the secretion of the apical zones of the epithelial cells contains PAS-positive, neutral mucopolysaccharides and proteins with reactive SH and SS groups. According to Gabe the same is true for Crocodylus niloticus. The secretary products of the surface epithelium of these animals is thus represented by neutral mucosubstances. In snakes the secretions consist mainly of acid mucopolysaccharides, with neutral mucopolysaccharides occurring in several spots. The basal region of the superficial columnar epithelial cells and those of the pits always react

TABLE 7.3

Staining Properties of the "Clear" and "Dark" Cells of the Main Glands of the Stomach

Stain	*"Clear" Cells*	*"Dark" Cells*
Eosin	—	+
Congo red	—	+
Aurantia	—	+
Hemalum	—	+
Heidenhain's hematoxylin	—	+
Methyl violet	—	+
Hamperl (1926)	—	+
Toluidine blue 0.1%	+	+
Mucicarmin	+	—

+ Positive; — negative.

negatively with specific stains for polysaccharides, but stain with basic dyes because they contain ribonucleic acid. Staining with Sudan black B demonstrates the presence of lipids in the same region of the cells. Enzymes of the oxidoreductase group react mostly positive. The demonstration of glucose-6-phosphate dehydrogenase is strongly positive, whereas the reaction for non-specific acid phosphate and non-specific esterase is either weekly or moderately positive in the superficial and pit epithelial cells. In Agama caucasica the superficial epithelium and that of the crypts give a strong, non-specific esterase activity with the indoxylacetate method of Holt.

(ii) Neck cells and "clear" cells. The topochemical analysis of the secretory products in the neck cells shows differences among turtles, lizards and snakes. Thus their secretion in Testudines and Sauria consists of neutral mucopolysaccharides and proteins with poorly reacting SS and SH groups, while that of the Ophidia consists mainly of acid mucopolysaccharides to which other mucopolysaccharides may be locally added.

The results of combinations of stains suggest that the secretion of "clear" cells consists mainly of neutral mucopolysaccharides with added acid mucopolysaccharides. In the Agamidae there is only a minor and unstable acid mucopolysaccharide fraction (Luppa, 1961). Enzymatic investigations show the content of enzymes to be very minor in both the neck cells and the "clear" cells.

(iii) "Derk" cells (main gland cells). The "dark" cells (Gabe and Saint Girons, 1964) are characterized by granules which become very conspicuous after staining with Heidenhain's iron hematoxylin. The cells are also called granular cells (Andrew, 1959), body cells, cells of the chief glands (Friedman, 1937), oxynticopeptic cells, oxyntic cells, acidophilic cells, chief cells, fundic cells. The granules of the dark cells show a strong specific blue stain with Adams (1957) reaction for demonstrating tryptophane. The reaction is characteristic for the zymogen granules of the main cells of the mammalian stomach. Proteins with reactive SH and SS groups could also be demonstrated in the zymogen granules of Lacerta agilis, Natrix natrix, and Testudo graeca (Luppa, 1961, 1963). Manfredi Romanini (1959) has, furthermore, demonstrated tyrosin by means of the Morel-Sisley reaction (method of Lillie, 1957) in the granules of Chrysemys scripta, Coronella austriaca, and Vipera aspis. The Millon and Xanthydrol reactions were also positive. Sudanophilic lipid granules are peripheral to the nuclei of the "dark" cells, Besides these there is supposed to be a second, alcohol-soluble type of granule (Manfredi Romanini, 1959). The alcohol resistant lipid granules are presumably mitochondria, which are very common in the "dark" cells. The cells are devoid of histochemically detectable polysaccharides. A slight PAS reactivity is to be noted, in the "dark" cells of Sphenodon punctatus, after fixation with Regaud's fluid and postchromatization (Gabe and Saint Girons, 1964).

The basal region normally lacks zymogen granules, but

reacts strongly with stains that demonstrate ribonucleic acids. This strong basal reaction is responsible for the "dark" quality of these cells.

In contrast to the mixed and "clear" cells, the "dark" cells show very strong enzymatic activity, especially of hydrolases, oxidoreductases, lyases, and synthetases. Among the hydrolases demonstrated, the non-specific acid phosphatase and the non-specific esterase show a clear, granular localized reaction in the "dark" cells. The reactions of both enzymes appear to occur in the same regions. The mitochondrial enzymes are characterized by strong activity, while the intermitochondrial, glucose-6-phosphate dehydrogenase reacts relatively weakly. In this connection the "dark" cells resemble the basal cells of the peptic cells of the mammalian stomach. The presence of cytochrome oxidase and lactate dehydrogenase indicates that both the aerobic and anaerobic paths occur in the "dark" cells of the reptilian stomach.

The enzyme carbonic anhydrase, so important for the characterization of gastric cells, yields a differential reaction in the "dark" cells. All cells of Natrix natrix react strongly and the reaction-product is diffusely distributed over the entire cell. The "dark" cells of Testudo hermanni and Agama caucasica, on the other hand, show only a slight darkening of the cytoplasm, with the black reaction-product in the lumen of of the glands. Similar relationships occur in Rana (Vollrath, 1959).

Arvy and Bonichon (1958a, b) demonstrated strong activity of non-specific alkaline phosphatase and non-specific esterase in the glandular epithelium of the gastric mucosa of Crocodylus niloticus. Ferri *et al.*, (1975) described acid phosphatase AMP-ase and esterase in surface epithelial cells, neck cells, body cells and pyloric cells of the glandular gastric epithelium of Xenodon merremii.

(iv) Argyrophil cells. In the stomach argyrophil cells were

demonstrated by Gabe (1971a) in the following 16 species of Lepidosaurians: (Gabe, 1971a; Table 7.4).

The spindle-shaped argyrophil cells are mainly localized in the fundic region and in the pyloric glands. On the other hand, the typical enterochromaffin cells occur throughout the proximal part of the duodenum and are argentaffin The argyrophil cells of the fundic region differ in chemical characteristics from the cells of the pyloric area. Both cell types correspond, according to their histochemical characteristics, to the enterochromaffin-like cell of Hakanson (1970a as well as Hakanson *et al.*, 1967, 1969, 1973).

TABLE 7.4

Histochemical Characteristics of the Argyrophil Cell Types of the Stomach of Reptiles (After Gabe, 1971a)

Reaction	*Cell Types* *Enterochromaffin-like cells located in main glands (glandular neck-region) of the corpus*	*Enterochromaffin-like cells, "Gastrin cells" located in the pyloric glands of the pars pylorica*
Argentaffin	—	—
Argyrophil	+	+
Diazoreaction	—	—
Xanthydrolreaction	—	—
Toluidine blue after HCl hydrolyse	—	metachromasia

\+ = positive; — = negative.

The enterochromaffin-like cells of the pyloric glands may be identical with the gastrin cells (G-cells according to the terminology of Solcia *et al.*, 1969a, b, 1970; Capella *et al.*, 1971). Hakanson *et al.*, (1973) suggest that similar cells of the cat pyloric gland area are also identical with the cells believed to store gastrin.

The enterochromaffin-like cells contain histamin but lack 5-hydroxytryptamine, and catecholamines, under physiological conditions, and are not detectable by the fluorescence microscopic procedure of Falck and Hillarp (Flack, 1962; Falck *et al.*, 1962). These cells can be called histaminergic cells together with the mast cells, which also contain histamine (Gabe, 1971a).

The enterochromaffin-like cells are able to store arylethylamin and to decarboxylate amine precursors. Therefore, they belong to the APUD-System (Pearse, 1969, 1971). It is still unclear whether the mast cells can also coordinate to this system.

It is suggested that the enterochromaffin-like cells may be endocrine cells and have a variety of functions. Further electron microscopic studies are necessary to define the endocrine cell types that occur in the gastrointestinal mucosa of reptiles.

(c) *Fine structure of the "dark" cells.* The "dark" cells, which are particularly interesting from a functional viewpoint, show ultrastructural characteristics suggesting that they have a double function. The surface of these cell along the lumen bears many microvilli, number and form of which are obviously dependent upon the functional state of the cell. Ferri *et al.*, (1974) in Xenodon merremii) observed a few short microvilli with a surface coat of glycocalyx on the apical membrane. The apical surfaces of adjacent cells form interdigitating folds. Besides these, desmoses occur at zozes of attachment between the surfaces of cells. The basis of the cells are often folded. Internally the cells show numerous granules.

These granules are, as in mammals, surrounded by membrances, and their interior contains a finely granular material of a characteristic density and arrangement. They are often

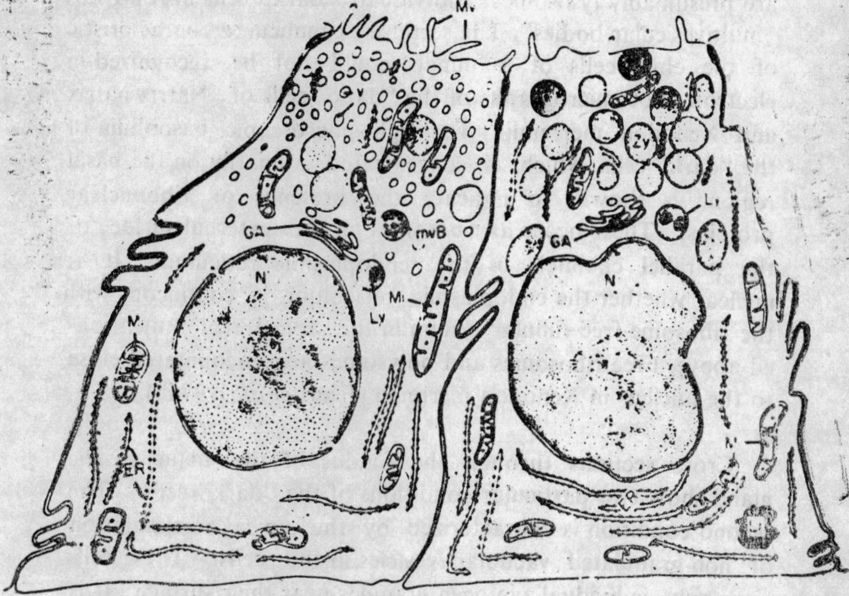

Fig. 7.1. Diagram of two "dark" cells of the main gestric glands) in different functional states as revealed by electron microscopy. ER, ergatoplasm; GA, Golgi apparatus; Li, lipid; Ly, lysosomes; Mi, mitochondria; My, microvilli; my B, multivesicular bodies; N, nucleus; V, vacuoles; Zy, zymogen granules. Note the variations in the concentration of zymogen granules in the supranuclear regions of the cells.

associated with the rough endoplasmic reticulum (so called RER). During extrusion the graules pass into the lumen of the gland, free of their surrounding membrane. A further fine-structural characteristic of the "dark" cells is a system of non-granular, vacuolar or tubular membranes that are mainly concentrated in the apical region of the cell. A vertical section through the cell shows mainly profiles of vacuoles.

Golgi complexes are found not only in the vicinity of the nuclei, but also in the lateral or basal regions. The very common, but morphologically very variable, mitochondria are all of cristate-type. Particles are included in almost all of the cellular organelles. Individually oucurring osmophilic bodies are presumably lysosomes. Individual "dark" cells also possess "multivesicular bodies". The secretory canaliculae characteristic of the chief cells of mammals could not be recognized in electron photomicrographs of the "dark" cells of Natrix natrix and Xenodon merremii. The light-microscopic basophilia of the "dark" cells, which is concentrated primarily on the basal region, is shown to indicate the presence of ribonuclear proteins. These occur as ribosomes on the external surface of the parallel channels of the endoplasmic reticulum. It is nuclear whether the endoplasmic reticulum, is contincous with the ribosome-free tubular system in the apical region mentioned above. Free ribosomes and polysomes are encountered close to the nucleus in Xenodon merremii (Feeri *et al.*, 1974b).

Cross sections through the fundus of the main gastric gland show two particular conditions of the "dark" cells. The second condition is characterized by the apical accumulation of non-granulated vacuolar vesicles in the cells. These cells have only individual zymogen granules near their surface. It is assumed that these ultrastructural differences in the "dark" cells represent functional stages of a single type of cell.

(d) *Function of the main gastric glands.* The granulated, so-called "dark" cells are of particular importance in a discussion of the function of the main grastric glanes of reptiles, since they have traditionally been assumed to be the digestive element of the gastric mucosa (Partsch, 1877; Edinger, 1879; Kahle, 1913). On the other hand, views regarding the localization of production of hydrochloric acid and pepsin in the glands of the reptilian stomach are diverse, as documented in past literature. Production of hydrochloric acid has been thought to be localized in the "dark" cells (=main gland cells) by Oppel (1896), Kranenburg (1901), Hirano (1926a, b, 1927), Krause (1922), and Staley (1925), but Jeksajewa (1958b) believed it to occur in the esophageal epithelial cells. Similarly, the production of pepsin has been postulated to occur in the neck cells of the main gastric glands (Motta and Renaud, 1878; Kranenburg, 1901; Hirano, 1926a, b 1927) or in the tubular glands of the fundic portion of the stomach (Jeksajewa, 1958a, b). Finaliy, some workers have suggested that the function of acid and pepsin production is combined in the main gland cells of the fundic portion of the stomach (Sacchi, 1886a, b; Plenk, 1932; Wright *et at.*, 1957; Luppa, 1961; Gabe and Saint Girons, 1964; Smit, 1968; Skoczylas, 1970; Ferri, 1971). The function of producing pepsin and hydrochloric acid is thus usually attributed to different cells of the main gastric glandular region.

The investigations of Partsch (1877), Edinger (1879), and Kahle (1913) indicate that the gastric secretion is acidic and contains free hydrochloric acid and a variable quantity of pepsin. According to Oppel (1896), the quantity of pepsin in the mucosa is related to the concentration of the granules in the cells and decreases during digestion, as does the number of granules in the secretory cells (Wright *et al.*, 1957).

Only Jeksajewa (1958b) assumes that acid is produced in the esophagus. Thus theory is, however, based only on material stained using Zimmermann's (1925) technique (hematoxylin, mucicarmine and aurantia) for the characterization of three types of cells in main gastric glandular region of mammals and on cells of the esophagus of Lacerta agilis. The apical portions of the cells in question stain yellow; their bases are stained light violet. The yellow stain was utilized to characterize these cell as acid producing chief cells. It is questionable whether a type of cell can be functionally charcterized simply by an empirical staining method. The production of both pepsin and hydrochloric acid by the "dark" cells (main gland cells) of the fundic portion of the stomach may be demonstrated by the following results of light and electron microscopical investigations.

1. Zymogen granules are demonstrated by the specific histochemical tryptophane reaction and in the electron microscopical picture. The clear basophilia in the infranuclear cell region is caused by the presence of

membrane-bound ribosomes and of a well-developed Golgi zone. These characterstics are clear in dicators of a pepsin-producing chief cell (Helander, 1962).

2. There are numerous large, dense mitochondria showing the strong activity of enzymes of the oxidoreductase type, which is typical for parietal cells. (Ito and Winchester, 1963). There are also many vacuoles and tubules with smooth surface in the apical cytoplasmic region. Both these characteristics may, according to present knowledge, indicate an acid-producing parietal cell of the mammalian stomach.

The presence of histochemically demonstrable carbonic anhydrase also suggests the function of producing hydrochloric acid for the "dark" cells of the gastric glands of the reptilian stomach. This enzyme can be regularly demonstrated in the parietal cells of the mammalian stomach (Kurata, 1953; Hausler, 1958; Schiebler and Vollrath, 1959; Vollrath, 1959; Korhonen *et al.*, 1966). Physiologists believe that it is associated with the formation of hydrochloric acid in the parietal cells (Davenport, 1939; Davies and Roughton, 1948; Berliner and Orloff, 1956; Leuthardt, 1959).

The predominance of either cross sections of vacuoles and tubules or of accumuluted zymogen in the apical region of the cells may suggest physiological conditions connected with the double function of the cell.

The results of this histological, histochemical, and ultrastructural investigation make it appear fairly certain that the functions of producing pepsin and of producing hydrochloric acid are combined in the "dark" cells of the main gastric glands of the reptilian stomach. It is, therefore, appropriate to substitute the term main gastric glandular cell (as suggested by Plenk, 1932) for the name "dark" cell. A similar double function, based upon ultarstructural and also partly on topochemical studies, is assumed for the acidophilic cells of the anuran stomach (Sedar, 1961; Vital and Orrego, 1960) and of the glandular stomach in the chicken (Luppa, 1962; Fantin, 1962; Selander, 1963).

Pars Pylorica

(a) *Histology of the pyloric glands.* The tubular glands of the pars pylorica are generally shorter and less branched than are those of the corpus of the stomach. Pyloric glands have been reported in all geougs of reptiles. The reports of Gargano (1907) and off Reese (1913) on alligator that the main glands and pyloric glands differ only in their lengths are apparently based upon the investigation of insufficiently fixed material (Plenk, 1932). The epithelium of the pyloric glands contains only mucous cells. The histological and histochemical peculiarities of these cells are closely smilar to those of the neck cells and the "clear" cells of the fundic glands (Arcangeli, 1907, 1908; Beguin, 1902; Kahle, 1913; Gabe and Saint Cirons, (1972a). Argentaffin and argyrophil cells are located been the pyloric cells in Xenodon merremii (Ferri *et al.*, 1974a). The pyloric cells of Xenodon merremii have also the same ultrastructural characteristics as the neck cells of the fundic glands (Ferri *et al.*, 1974b). The characteristics of the cells within the pyloric glands permit the differentiation of the gastric pit and the body of the gland. There is no morphologically differentiated glandular neck. The body of the gland consists of morphologically similar, clear, columnar cells, with basally positiond, rounded, loosely formed nuclei in the cytoplasm. Electron microscopy shows not only the mucoid glandular cells but characteristic basally granular (argentaffin) cells. The region of the pyloric glands is the richest region in argentaffin and argyrophil cells of the digestive tract of Xenodon merremii (Ferri *et al.*, 1974a). According to Kolossow and Subussow (1930), the cells of the pyloric glands in Emys orbicularis have more strongly flattened nuclei and are thus differentiated from the neck cells of the main glands. The superficial and epithelial cells and those of the pit appear structurally like the corresponding cells of the main gastric glands.

(b) *Histochemistry of the pyloric glands.* In the Sauria (including Agamidae), Ophidia, and Testudines, the PAS-reaction is equally positive from the surface of the gastric epithelium to the terminal portion of the glandular body. Tests for the sulfhydryl (SH) and disulfide (SS) groups, or for disulfide (SS) groups alone, also show intensive reaction in the same regions with the exception of the glandular body, in which the reaction is weakly positive to negative. Negative results are attained with processes demonstrating proteinbound sulfhydryl groups. Acid mucopolysaccharides are found only in low concentrations within the pyloric glands of the Ophidia based on alcian and astra blue staining as well as the Hale reaction.

Some ribonucleic acid is localized in the infranuclear region of the cells of the glandular body. Its concentration is slightly higher in the cells of the epithelial surface and of the gastric pits, where it also occurs in the supranuclear region. Desoxyribonucleic acid can be demonstrated only in the nuclei of the glands of the gastric mucosa. Unlike the glandular cells of the gastric corpus, those of the pyloric portion lack lipid granules that react positively with Sudan black. The lipid reactions in general are almost negative in the mucoid glands of the pyloric region. Only the superficial and pit cell are weakly positive.

All of the epithelium of the pyloric glands of Testudo graeca, T. hermanni, Lacerta agilis, Agama caucasica, and Natrix natrix shows n negative reaction for non-specific alkaline phosphatase. Arvy and Bonichon (1958b), however, report a relatively strong activity of this enzyme in the cells of the body of the gland in Crocodylus niloticus. Non-specific acid phosphatase is located in specific granular particles in the zone between the nucleus and the mucous secretion and more diffusely, in the remainder of the cells of the superficial and pit epithelium. The glandular cells show the greatest concentration of the reaction product near the base of the cell, and also sometimes along the margins of cell.

The indoxylacetate-reaction for non-specific esterase indicates specificity and localization of activity in the superficial and pit epithelium which agrees closely with the zone for the non-specific acid phosphatase. In the glandular cells this activity is more diffusely distributed. The superficial epithelium shows very strong glucose-6-phosphate dehydrogenase activity, while activity of β-hydroxybutyrate dehydrogenase is moderate. The glandular cells give a noticeably weaker reaction in both cases. The lactate dehydrogenase shows a definite reaction in the apical and basal sections of the superficial and pit epithelial cells, while the reaction is almost negative in the glandular cells.

The results of this histological and histochemical investigation do not show whether the pyloric glands only secrete mucus or have other, more specific, functions.

Almost all investigators (Oppel, 1896; Gianelli and Giacomini, 1896; Greschik, 1917; Krause, 1922; Staley, 1925; Kolossow and Sabussow, 1930; Plenk, 1932) have described a transitional zone between the region of the main gastric glands and the pars pylorica. Here the neck cells occupy a wider region and the granulated "main gland cells" are more and more restricted in the area they occupy.

Cardiac Region

Some reports note a differentiation of the glands of the anterior end of the reptilian stomach. These markedly shortened glands are lined either by main glandular cells or else by mucoid cells. In Testudo hermanni, the cardiac glands are lined only by mucoid cells; other reports speak of completely glandless pits.

The cardiac region generally occupies a restricted area. Thus Kolossow and Sabussow (1930) report a 2 mm wide region in Emys orbicularis, while Beguin (1904a) refers to a few glands lined only with clear neck cells at the beginning of the stomach of Lacerta. Greschik (1917) reports two or three alveolar cardiac glands, again lined only by mucoid cells, at the anterior end of the stomach of Ablepharus. Finally, Staley (1925) reports that the cardiac glands of the alligator are lined only by neck cells.

Connective Tissues and Musculature

(a) *Histology.* The tunica propria lies below the gastric epithelium and consists mainly of reticular connective tissues. Its intercellular spaces are filled with free cellular elements such as lymphocytes, plasma cells, eosinophilic granulocytes, other cells with eosinophilic granules, and mast cells with strongly metachromatic granules. The ovoid mast cells were demonstrated most frequently in the mucosa, the submucosa, the muscle coat, in the adventitia of major vessels and between the tubes of glands. They were never observed inside the gland epithelium (Gabe, 1971a). The cells are smaller than the same cells in amphibians. The tissues are also panetrated by blood vessels and lymphatic spaces.

The well-developed lamina muscularis mucosae may, as in mammals, be subdivided into internal, primarily circular layers, and an external longitudinal one. This arrangement, as well as the thickness of the mucous musculature, shows some variation in the different regions of the stomach. The intestine of crocodiles occasionally lacks a circular layer (Toro, 1930). The narrow tunica submucosa consists of loose connective tissue, penetrated by lymphatic ducts and larger blood vessels. There are alse lymphoid insertions, generally in the form of nodes, that penetrate into the mucosa (Pernkopf and Lehner, 1937). Finally this layer contains the ganglion cells of the plexus submucosus.

The tunica muscularis is always of smooth muscle and consists of two layers, the inner one circular and the outer longitudinal. The longitudinal layer, which represents a direct continuation of the longitudinal layer of the esophagus, is usually more weakly developed than the circular layer. In many reptiles the muscular layer decreases in thickness in the direction of the pylorus. The terminal portion of the pars pylorica ordinarily appears thin in these animals. There is generally no sphincter-like swelling of the musculature in the vicinity of the pyloric constriction.

The tunica muscularis is particularly strongly developed in the zone preceding the pyloric portion of the crocodilian stomach. As in birds, the very muscular stomach may be referred to as a gizzard (Pernkopf, 1929). This section of the stomach is also characterized by dorsally and ventrally positioned, radial sheets of connective tissue the central portions of which lie immediately on the submucosal connective tissue.

The serous covering of the stomach, formed by the peritoneum, surrounds its external surface. In most snakes and crocodilians, this serosa is incomplete because of a more or less complete funsion of the covering of the stomach and the other organs of the visceral cavity, especially the liver. The thin, simple, squamous epithelium of the serosa extends over a layer of connective tissue which is rich in elastic fibres.

(b) *Histochemistry*. Tests for non-specific acid phosphatases and non specific esterases lead to diffuse, corpuscularly localized reactions in the tunica muscularis. Negative or weakly positive reactions are obtained for the oxidoreductases such as glucose-6-phosphate dehydrogenase, cytochrome oxidase, malate dehydrogenase, and alcohol dehydrogenase. The reactions for isocitrate dehydrogenase, succinate dehydrogenase, and glutamate dehydrogenase are weak, while the β-hydroxybutyrate dehydrogenase and NADH and NADPH diaphorases yield slightly stronger reactions. The strongest reaction is shown by lactate dehydrogenase. The results of the enzymatic tests of the tunica muscularis of the stomach thus agree with those obtained for the tunica muscularis of the esophagus.

The walls of the blood vessels in the tunica propria mucosae and tunica submucosa give very strong reaction for succinate dehydrogenase.

The cytoplasm of the mast cells contains great amounts of

histamine and granules showing strongly metachromatic staining with basic dyes of the thiazin group (Arvy, 1962c; Gabe, 1971a). The granules also stain with copper phthalocyanin dyes, as for instance alcian blue. The diazo reaction and xanthydrol reaction which are positive in the mast cells of some mammals show a negative reaction in those reptiles thus far investigated.

The muscular bands of the tunica muscularis and the muscularis mucosae are penetrated by cholinergic nerve trunks (Arvy and Bonichon, 1958b). Fine chalinergic fibrils pass from the submucosa in parallel to the tubules of the glands. The perikaryon of the nerve cells in Auerbach's and Meissner's plexuses do not give a positive reaction for acetylcholineste (Arvy and Bonichon, 1958a, b).

Development of Gastric Epithelium

(a) *Histogenesis*. The development of the stomach of sauropsids was described by Sjogren (1945), and the torsion of the reptilian in estine by Grunwald (1931). The histogenesis of the gastric epithelium passes in cycles through several stages. Normally a three-layerd, high columnar epithelium forms first. In Lacerta the glands start to form at Stage 33 and become more noticecble near Stage 36 (Peter, 1904). At Stage 43 the glands are fully formed and between this time and hatching, they differentiate into the pit region, neck and body. The formation of glands involves two processes:

1. purely mitotic processes leading to the formation of a lumen (Lacerta agilis, L. trilineata, L. vivipara) as well as intercllular cavities and
2. the consequent regrouping of the compound epithelium into a simple one (Eremian velox, E. strauchii).

The processes leading to glandular formation occur earlier in that region of the stomach adjoining the duobenum than in the corpus. The glands of the main glandular portion of the gastric corpus differentiate earlier than the pyloric glands. The secretory products within the cells of the gastric glands of the main glandular region of the stomach differentiate during the terminal phases of embryonic development. The glands form in the embryonic stomach of snakes by the formation of pits and by the invasion of the subepithelial connective tissues by buds from the epithelium. Bauchin's valve or diverticulum generally forms an anatomical pars pylorica of the stomach of lizards after Stage 29.

Comparison of the development of reptilian gastic glands allows one to make certain fundamental generalities. Deviations occur in the genus Eremias, which the transition from a compound to a simple epithelium occurs by means of an intensive formation of spaces and the discharge of cells into the lumen of the stomach.

In two older works the development of the digestive canal of the American alligator was studied by Reese (1910), the histogenesis of the esophagus and stomach of the gecko was investigated by Schlesinger (1924).

(b) *Histochemistry*. Development of the mucopolyasccharide secretion in the "clear", or neck, and in the true glandular cells of the pyloric glands only begins when the glands are nearly differentiated morphologically in Sauria and Ophidia. Weakly acid and neutral mucopolysaccharides, which are not affected by digestion with diastase, may be demonstrated in the superficial and the pit epithelial cells of the corpus and the groups as well as associated ribonucleic acids. The apically positioned precursors of secretory granules contain also weekly acid and neutral mucopolysaccharides. The cells basis contains granules with reactive SH and SS groups at the boundary between the small and large intestines. When it is absent, the position of boundary may be determined only by the difference in intestinal diameter.

Protein and SH and SS groups, the concentrations of which generally increase to the hatching stage, occupy a topographic position similar to that of the mucopolyaccharides. The increase in the concentration of cystine is accompanied by a parallel increase and stabilization of metachromatically reacting substances.

While the diastase-resistant PAS-positive substances and proteins with reactive sulfydryl (SH) and disulfide (SS) groups gradually increase in the intensity of their reactions as the gland develops, there is a decrease in the number of glycogen granules which occur mainly in the apical region of the differentiating epithelium. The granules simultaneously decrease in diameter. Only a trace of the glycogen can be demonstrated once the gastric glands are morphologically differentiated.

THE HINDGUT

General Morphology

The reptilian intestine is short relative to body length. It is longest in turtles, shortest in snakes (Fig. 7.2) and intermediate in lizards (Fig. 7.3). According to Jacobshagen (1937), the largest relative length of the intestine in turtles is caused primarily by the widening of their trunk and only secondarily through the influence of their food habits.

The differences in the intestinal lengths are to be compared with differences in the position of the intestine. Jacobshagen (1937) has reported a straight intestine in Coronella and amphisbaenians). In most snakes the transversely undulating folds are joined by connective tissue and encompassed by the peritoneum. The curvatures of the intestine are arranged in a three-dimensional scheme in crocodilians and turtles (Jacobshagen, 1937).

The large intestine may be subdivided into three distinct sections: The colon (=coprodaeum, sensu Jacobshagen, 1937; coprodaeum is also used for a part of the cloaca), the cloaca proper (=urodaeum), and the posterior cloacal canal (=proctodaeum) (Figs. 7.2 and 7.3). The coprodaeum, which often has an asymmetrical caecum dorsale, is ordinarily the longest of these three sections. Various reptiles may lack a caecum.

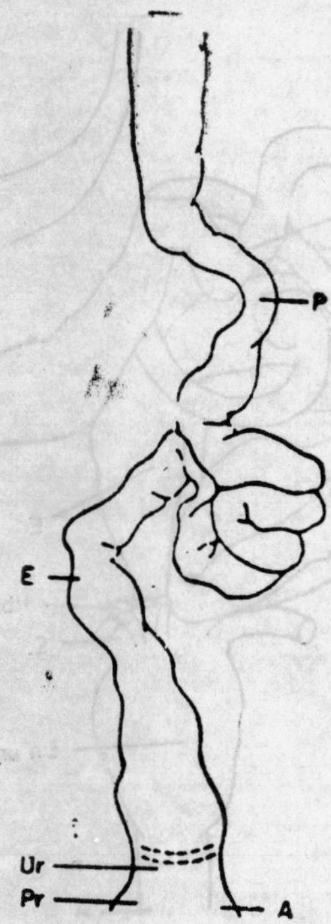

Fig. 7.2. Intestine of Cordylus giganteus. A, Cloacal aperture; E, ileocolic junction; P, pylorus; Pr, proctodaeum; Ur, urodaeum.

The proctodaeum of all reptiles, except the Rhynchocephalia, contains intromittent organs. In lizards, snakes and Sphenodon the opening of this section of the gut is followed by a transverse fold while the fold is longitudinal in the Testudines and Crocodilia (Jacobshagen, 1937).

The mucous relief of the small and large intestine varies notably (Jacobshagen, 1920). The small intestine of Sphenodon contains a simple network, while longitudinal zigzag folds are found in turtles. The Trionychidae are particularly specialized, having a round, sausage-like longitudinal projection on the ventral intestiual wall located opposite the attachment of the

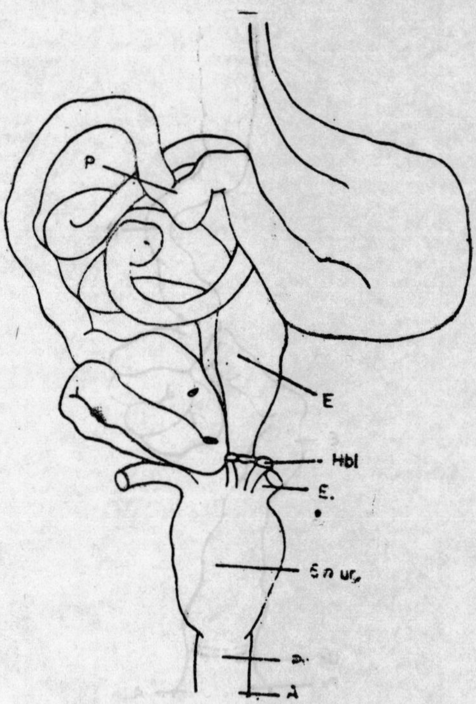

Fig. 7.3. Intestine and associated structures of Emys orbicularis (after Jacobshagen, 1937). A, Clocal aperture; E, ileocolic junction; El, oviduct; Hbl, urinary bladder; Ma, stomach; P, pylorus; Pr, proctodaeum; Sin. ur., urinary sinus.

dorsal fold. These patterns are described in more detail and illustrated by Parsons and Cameron (this volume).

Small Intestine

Histology

(a) *General.* That cylindrical epithelium covering the folds on the inside of the intestine is either simple (Natrix, Lacerta, Agama; or apparently arranged in a stratified pattern (turtles, crocodilians). In some reptiles the epithelium at the base of the folds forms crypt-like, occasionally branched pits which are similar to the glandular crypts of Lieberkuhn in birds and mammals, but apparently relatively undeveloped. The epithelium of these crypts is equivalent to that of the surface. Previous investigations indicate that intestinal crypts are found in crocodilians (Toro, 1930), turtles (*e.g.*, Trionyx ferox) and lizards (including agamids). They are supposedly lacking in snakes (Vialli, 1929b; Reis and Lyons, 1943. Renewal of the intestinal epithelium occurs in cells located along the basal membrane in the fresh-water turtle Chrysemys picta (Wurth and Musacchia, 1964). The time required for the replacement of the intestinal epithelium is approximmately eight weeks for turtles held at 20—24°C in winter. Loss of cells apparently occurs by a movement of the cells which pass randomly from the basal membrane toward the lumen. The cellular renewal of intestinal epitbelium in the western fence lizard, Sceloporus occidentalis, is described by Johnson *et al.*, (1967).

The columnar epithelium lies on a basal membrane and contains marginal cells, goblet cells, Paneth cells, and entero-chromaffin cells (=argentaffine, basally granular, or yellow). The cytology of the enterochromaffin cells as well as Paneth's cells was first described by Vialli (1929b).

(b) *Marginal cells.* Under the light microscope marginal cells consist of a striated border, a granular zone (Ludwig and Richterich, 1954), a "clear" zone (Baker, 1942, 1951; Ludwig and Richterich, 1954), and a zone of mitochondria. Deep to the vertically orientated oval nucleus is a second region containing many mitochondria. The marginal cell of the reptilian small intestine thus agrees completely in its light microscopic appearance with the marginal cells of the mammalian intestine, (Patzelt, 1930; Lunwig and Richterich, 1954).

(c) *Goblet cells.* The morphological characteristics of the goblet cells, which are irregularly placed between the marginal cells, have been characterized by Vialli (1929b). Their morpho-logy and histochemistry are described in a more recent investigation by Gabe and Saint Girons (1972b). The supranuclear cytoplasm of goblet cells is filled with mucus or its precursors. The relatively narrow base of the cells contains an ovoid nucleus. A well-developed Golgi apparatus lies between the zone in which the secretion is collected and the nucleus. The goblet cells of the small intestine of crocodilians are probably never completely filled with mucous secretion (Toro, 1930). The cytoplasm, surruunding the nucleus, stains more darkly than the remaining cytoplasm. There is no information available regarding the numerical distribution of goblet cells in the small intestine, but Baecker (1934) noted that the goblet cells are so numerous in certain sections of the intestine of Anguis fragilis that there are almost no marginal cells. Mucocytes are more abundant in the large than in the small intestine (Gabe, 1971b).

(d) *Cells of Paneth.* Mammalian Paneth cells are secretory active cells, which are differentiated in baso-apical direction (Otto, 1973). They occur primarily on the bases or sides of the glands of Lieberkuhn (Baecker, 1934; Hintzsche and Anderegg, 1938). It has been cliamed that in the reptiles they are almost always located in the superficial epithelium of the small intestine (Jacobshagen, 1937; Toro, 1930; Vialli, 1929b; Baecker, 1934), but this not true of all species. Thus Baecker (1934) found that in Lacerta agilis and Agama stellio the Paneth cells were limited to the depths of the crypts, while in Lacerta ptyusensis they were distributed almost uniformly across the entire mucus surface. Only in some crocodilians (Crocodylus niloticus, Crocodylus porosus, Alligator sinensis) do the Paneth cells occur primarily in the crypts of all the regions of the intestine (Taguchi, 1920). They increase in number posteriorly, and individual ones occur in the crypts of the coprodaeum. In reptiles the Paneth cells generally extend into the rectum.

The claim of Vialli (1929b) that the distribution of these cells is related to phylogeny cannot be confirmed; Paneth cells are missing in Agama inermis, but Baecker (1934) has definitely demonstrated them in Agema stellio. Similar instances are known for urodeles and birds (Baecker, 1934).

Those reptiles in which the occurrence of Paneth cells

has been investigated. This table shows that the intestines of turtles, Anguisfragilis, and of some snakes lack Paneth cells. The conflicting reports about their occurrence in crocodilians (compare Taguchi, 1920 and Toro, 1930) may have been caused by the use of acetic acid in fixatives by Toro as these fluids may dissolve the granulae of the Paneth cells (Baecker, 1934).

Paneth cells always occur singly and are generally quite slender. Nicolas (1891) reports that these cells may also have a striated border in lizards. The dense granules show a notable, but functionally variable, affinity to acid dyes. They are easily soluble in mixtures containing acetic acid and mercuric chloride. Secretion seems to be preceded by a dissolution of the cellular inclusions. Transitional patterns similar to those reported in goblet cells (Clara, 1926; Toro, 1930; Rogick, 1931; Luppa, 1958) have not thus far been observed. The Paneth cells of the reptilian intestine are fundamentally similary to those of other vertebrates. Electron microscopical, histochemical and experimental investigations, such as carried out on mammals (Patzelt, 1936; Hintzsche and Anderegg, 1938: Selzman and Liebelt, 1961, 1962a, b; Behnk and Moe, 1964: Linss and Geyer, 1965; Rieken and Pearse, 1966; Geyer, 1966; Spicer *et al.*, 1967; Meyer *et al.*, 1970; Otto, 1972), are necessary to increase our knowledge about the functional importance of reptilian Paneth cells.

(e) *Endocrine cells.* The endocrine cells of the duodenum of reptiles (classified as enterochromaffin cells, argentaffin cells or basally granulated cells) differ from the Paneth cells and apparently occur in all groups of reptiles (Nicolas, 1891; Kull, 1912, 1925; Dias-Amado, 1925a, b; Cordier, 1926; Vialli, 1929b; Kolossow and Sabussow, 1930; Toro, 1930, 1931; Clara, 1932; Citterio, 1935; Gabe, 1973). Although there are no confirmed quanitative statements about the distribution of these cells, lizards apparently always contain more cells than snakes. The cytoplasm is clear and only slightly stainable (Vialli, 1966). In dimension and shape, the round, chromatin-poar nucleus is similar to the nuclei of adjacent cells. The nuclei generally lack nucleoli. Other characteristics of the cells are the low number of mitochondria, and the absence of lipids and glycogen (Patzelt, 1936). The basal section of the cell vary considerably (Vialli, 1929b). Lacerta viridis has cells deviating markedly from normal (Clara, 1932).

Gabe (1972) detected two types of endocrine cells in the duodenal mucosa of 11 species of reptiles. One of these cell types corresponds to the known enterochromaffin cell and possesses its usual morphological and histochemical peculiarities. These cells are pear-shaped. Their secretory product is argyrophil and argentaffin, stored principally in the basal part of the cell body, and reacts positively to the histochemical methods for detecting indoles. Following hydrochloric acid hydrolysis, it displays a strong orthochromatic basophilia (Table 7.5).

The cells corresponding to the second type are spindle-shaped. In contrast to that of the first type their secretory product is argyrophil, but not argentaffin, and devoid of indolic corspounds, as well as metachromatically basophil after hydrochloric acid hydrolysis (Table 7.5).

Connective Tissue and Musculature

The tunica propria consists of lymphoreticular tissue and contains lymphatic follicles. These follicles show a definite concentration in the pyloric region of Natrix natrix, and Vialli (1929a) reported such a pattern in other forms. The lymphatic content of the reticulum apparently varies considerably, being relatively low in Trionyx but very high in Emys. Baecker (1934) observed extremely numerous eosinophilic leucocytes in the propria as well as between the cells of the intestinal epithelium of Emys orbicularis. He, furthermore, noted granulated wandering cells between the epithelial cells. Infoldings of the mucous membranes contain nodular assemblages of eosinophilic leucocytes.

TABLE 7.5

Main Histological Characteristics of the Two Endocrine Cell Types in the Duodenum of Reptiles (After Gabe, 1973)

	E.c.	*E.e.c.*
Localization	especially crypts	top of the villi
Shape	conic	Spindle-shaped
Hematoxylin phosphotungstic	+++	+++
Lead-hematoxylin	+++	+++
Method of Davenport	±	+
Silver proteinate (Grimelius)	+++	+++
Silver nitrate (Grimelius)	+++	+++
HCl-toluidine blue	+++a	+++b
Azoreaction	+++	0
Chromaffinreaction	++	0
Argentaffinreaction	++	0
Xanthydrolreaction	++	0

E.c.=Enterochromaffin cells; E.e.c.=Endocrine non enterochromaffin cells; a=orthochromatic stain; b=metachromatic stain; 0=negative; ±=result with few contrast; + to +++=results more or less strongly positive.

The muscularis mucosae varies in its structure and is not always clearly double-layered. Crocodilians, Varanus, and especially Testudo graeca are characterized by particularly numerous smooth muscle cells. Toro (1930) reported muscular bundles entering the intestinal folds of Crocodylus. Reis and Lyons (1943) found no muscularis mucosae in the duodenum of the common greater snake, Thamnophis sirtalis.

The tela submucosa consists of a loose network of bundles of thin collagenous fibrils; it lacks elastic fibres. Individual lymphocytes lie adjacent to the blood vessels within the network of bundles. Prominent lymph nodules are present at the bases of the folds of the mucosa in Thamnophis sirialis (Reis and Lyons, 1943). The muscularis consists of an internal circular layer and an external longitudinal one. The longitudinal layer is very thin in Emys. Toro (1930) found bundles connecting the circular and longitudinal muscular layers in Crocodylus and claimed that they were involved in the intestinal movement. Ganglionic cells of the plexus mysentericus lie in groups between the circular and longitudinal muscles. Similar groups of cells often occur between the longitudinal bundles in crocodilians.

The longitudinal layer of muscles is covered by a thin, loosely fibrillous serosa of simple squamous epithelium (Jacobshagen, 1937).

Histochemistry of the Epithelim of the Small Intestine

(a) *Marignal cells.* Summarizes the results of the histochemical investigations. The PAS-reactive neutral mucopolysaccharides, tyrosin, vnd those proteins with reactive SH and SS groups are largely localized in the striated border of Lacerta agilis, Natrix natrix, and Mauremys caspica. Acid mucopolysaccharides are found only in Natrix natrix. The "clear" zone (terminal network or terminal web) reacts only with the tests for proteins. Ribonucleic acids and lipids are conncentrated in the apical and basal sections of the cells. The granules react positively with the acid hematein test; this corresponds with the location of the mitochondria which are particularly numerous in the apical zone of the cell. The striated border and "clear" zone remain unstained with acid hematein.

Non-specific alkaline phosphatase is localized in the striated border. The same topographical distribution was observed in mammals (Jervis, 1963). In lizards, Natrix, and turtles non-specific acid phosphatase and non-specific esterase are more strongly concentrated in the apical mitochondrial zone as in mammals (Friedman *et al.*, 1966), but occur more diffusely below the nuclei where they are located in granulate structures. The reaction of non-specific acid phosphatse is totally inhibited by 0.02 M sodium fluoride. The esterase reaction in Lacerta agilis is notably inhibited by 10^{-3} M E600 (Paraoxon or Mintacol).

Leucine aminopepidase produces a strong reaction in the supra-nuclear region. Except for alcohal and β-hydroxybutyrate dehydrogenases, the oxidoreductases show a strong reaction in the supranuclear mitochondrial zone, while striated border and "clear" zones always react negatively Glucose-6-phosphate dehydrogenase and monoamine oxidase produce approximately equally strong reactions. The reaction is of equal strength regardless of the location of the marginal cells, from the tips of the folds to the bases of the crypts between them.

(b) *Goblet cells.* The histochemical nature of the goblet cells corresponds to that known for other vertebrates (Gabe, 1971b). Throughout the small intestine, the goblet cells of Mauremys, Testudo, Lacerta and Natrix react positively for PAS-positive neutral mucopolysaccharides and for acid mucopolysaccharides. The goblet cells of Sphenodon punctatus (Gabe and Saint Girons, 1964) and of Hemidactylus sp. (Kanwar, 1961) also give a strong PAS-reaction, take up alcian blue, and are intensely metachromatic. The positive reaction with alcian blue at pH 0.2 indicates that sulphate groups are present (Bloom and Kelly, 1960). All tests for enzymatic reactions give negative results in the region of the mucous secretion.

(c) *Cells of Paneth.* The Paneth cells of reptiles have not yet been investigated cytochemically to determine the chemical composition of the secretory granules. Only mammalian cells have been investigated in this fashion.

(d) *Endocrine cells.* The results of histochemical investigations are listed in Table 7.5. Furthermore, the granules of the endocrine cells (basally granulated cells after Clara) yield a positive reaction with nitroamine red and nitrazol CF (Clara, 1932). This reaction, however, is considerably weaker than in the granules in the basally granulated cells of birds and mammals. In Lacerta viridis no clear cut reation could be obtained with the acetic acid ammonium molybdate method of Quastel (1932). The positive result of the diazo-coupling reaction, the oxidation reaction with potassium dichromate and ammoniacal silver nitrate solution, and the negative acetic acid ammonium molybdate reaction jointly suggests that the granular substance in reptiles includes a dioxylbenzol derivative (Clara, 1932).

(e) *Connective tissue and musculature.* Granules in the connective tissue and musculature of the midgut yield a positive reaction for non-specific acid phosphatase. In contrast the indoxyl acetate method for demonstrating non-specific esterases produces a negative reaction. If, however, naphthol-AS-acetate is used as a substrate, the connective tissue and musculature form a corpuscular, widely distributed reaction product. Individual free cells of the tunica propria, particularly the mast cells, and the cells of the plexus myentericus, show a strongly positive reaction. Among the oxidoreductases, glucose-6-phosphate dehydrogenase, malate dehydrogenase, and alcohol dehydrogenase give negative to weakly positive reactions in the musculature of the small intestine, while lactate dehydrogenase, succinate dehydrogenase, and β-hydroxybutyrate dehydrogenase give clearly positive ones. Strong monoamine oxidase activity is present in the nerve cells of the plexus myentericus. The distribution of the acetylcholinesterase activity in the intestine is the same as in the stomach (Arvy and Bonichon, 1958a). Cholinergic lines pass into the folds of the mucosa.

Fine Structure of the Marginal Cells

The apical surfaces of the marginal cells are limited by thickly packed, fingerlike microvilli. The finer internal structure of the microvilli continues into a zone in which the filaments always appear in cross-section; this is the so-called terminal network or the "clear" zone of light microscopy. In reptiles, as in mammals and other vertebrates, this region contains no cellular organelles (Ludwig and Richterich, 1954; Zetterquist, 1956; Strauss, 1963; Sjostrand, 1963; Porter and Bonneville, 1965; Luppa, 1966). The zone lying below the terminal network contains cristate mitochondria, ribosome-lined canals of the endoplasmic reticulum, free ribosomes, differentially structured lysomoes, multivesicular bodies, and the supranuclear Golgi apparatus. Moreover, both the apical epithelial region and the basal region are filled with lipoid droplets. The ovata to longitudinally oval nuclei are delimited by a markedly porous nuclear membrane and usually possess a nucleolus. Mitochondria are concentrated basal to the nucleus, but less so than in the apical region. The surfaces of adjacent cells are laterally connected by finger-like processes. Desmosomes are clearly visible with light microscopy as further attachment sites in the region of the terminal network.

Large Intestine

Colon

(a) *Histology of the mucosa.* Taguchi (1920) indicates that in Crocodylus the colon has a thicker wall than the more anterior portions of the intestine, but that the mucosa is similar. Although the glandular crypts are reduced, some of them still retain Paneth cells. Crypts also occur in the dorsal wall of the colon of sphenodon (Osawa, 1897). Greschick (1917), however, found cryptlike depressions in only a few forms such as Anguis fragilis. Glands of Lieberkuhn have also been reported in Emys (Machate, 1879), Ophisaurus apodus (Oppel, 1897), and Chamaeleo dilepis (Esch, 1936). Mitoses are observed in the cells of these glands. The epithelium of the colon is the same as that of the small intestine (Jacobshagen, 1937). It consists of high columnar cells which lie on a basal membrane and show a partly formed striated border. In Crocodylus niloticus the columnar cells lying near the openings of the oviducts are ciliated. The high columnar epithelium contains throughout many goblet cells and gradually becomes two-layered near the fold separating the colon and cloaca (Taguchi, 1920). The circular fold between the colon and cloaca of snakes bears columnar epithelium containing scattered goblet cells (Junqueira *et al.*, 1966).

In the terminal part of the intestine, near the cloacal

opening (the coprodaeum of Gadow, 1887) four morphologically distinct cell types are identifiable histologically. The first type is represented in all lizards, and characterized by the presence of a simple epithelium, the cells of which are of two types: cells with neutral mucosubstances as secretory product and less numerous cells with an acid mucin (Gabe, 1971b).

The second morphological type (characteristic of the Colubroidea among the ophidians), consists of a stratified surface epithelium with two cell types: columnar mucocytes with a PAS-and alcian blue-positive mucin and goblet cells with an acid mucin. Xenodon merremii belongs to the second type (Junqueria *et al.*, 1966).

The third type, found in Sphenodon punctatus, is characterized by a stratified columnar epithelium in which all the cells of the superficial layer produce a PAS-positive, acid mucin, which stains with alcian blue.

The fourth type is represented by the Typhlopidae, Leptotyphlopidae and Boidae among the lepidosaurians and differs from the third type solely in the simple form of the epithelium, the mucous cell layer lying directly on the basal membrane (Gabe, 1971b).

(b) *Connective tissue and musculature*. In all reptiles the colon differs from the smell intestine in having a clearly developed muscularis mucosae which usually consists of an inner circular and outer longitudinal layer (Gianelli and Giacomini, 1896). Only in Emys is the musculature of the mucosa reported to be so thin that the double layering is no longer clearly apparent (Machate, 1879).

The tunica propria varies in its development. In Lacerta agilis, Platy-dactylus, and Cordylus it is clearly differentiated into a central, lymphatic, reticular zone and a fibrillar zone. Lymphatic follicles show some individual variation in their occurrence (Jacobshagen, 1937).

The tunica muscularis is thick in comparison to the mucosa.

As in the small intestine, it consists of two layers, usually with the circular layer the thicker. However, the longitudinal musculature of Anolis and Trachydosaurus is thicker, and it is as thick as the circular musculature in Ophisaurus and Anguis.

The serosa is a simple squamous epithelium lining the coelom.

(c) *Histochemistry.* The striated border of the columnar marginal cell yields a positive PAS reaction indicating neutral mucopolysaccharides. In general, the supranuclear zone of the marginal cells in the colon shows similar histochemical reactions to that in the small intestine. The supranuclear region of the marginal cells shows a very strong activity of non-specific acid phosphatase and non-specific esterase. The activity is located in granules and distributed quite widely in the paranuclear and infranuclear cell regions. The tests for oxidoreductases generally show activity similar to that in the marginal cells of the small intestine. The strong activity of the glucose-6-phosphate dehydrogenase is notable.

The goblet cells give strong, uniformly positive reactions with PAS and with astra blue and alcian blue stains. Thus the secretion probably contains neutral and acid mucopolysaccharides with the fraction of mucopolysaccharide clearly being the greater. The reaction for the acid mucopolysaccharides is also stronger than that of the goblet cells of the small intestine.

The subepithelial and intermuscular connective tissue shows a strongly positive PAS reaction. Non-specific acid phosphatases and non-specific esterases here appear to the diffuse; they react in an uniformly strong positive fashion in the serosa and the circular muscles. Cellular elements of the tunica propria, which are presumably mast cells, show a positive reaction for granules with these enzymatic tests. The serosa is generally characterized by the strong activity of various oxidoreductases.

(d) *Fine structure of the epithelium.* The fine structure of the marginal cells of the colon resembles closely that of the marginal cells of the small intestine. The aggregated appearance of lipid droplets in the supranuclear region in the colon is an important difference. Tubular and vacuolar indentations in the plasmalemma are occasionally visible between the microvilli. Small, serially arranged vacuoles and canals that intersect in various planes occasionally occur in the region of the terminal network.

The goblet cells have isolated microvilli, but they occur only in the marginal zone of the cellular apex. The supranuclear region of the cell contains bell-shaped enlargements when it is full of mucous secretory granules of varying sizes. The mucous granules are surrounded by thin membranes and embedded in a grainy ground substance, which contains the Golgi apparatus, ergastoplasm and free ribosomes. Cristate mitochondria occur between these, and multivesicular bodies and lipid droplets are scattered among them. Gradual transitions exist between the marginal cells and goblet cells.

Coecum

There are no detailed studies on the microscopic anatomy of this section of the intestine, but it is generally assumed that the caecum shows approximately the same structure as the large intestine. The region of the caecal tip is supposed to contain some individual lymphatic follicles.

Urodaeum

The complexity of the cloacal region of reptiles is well known since the classical investigations of Gerhardt (1937). The important work of Gabe and Saint Girons (1965) contains comprehensive information about the anatomy, histology and histochemistry of the cloaca of 57 species of reptiles, while Whiting (1969) reports on 84 species of reptiles. A detailed study of histogenesis and histochemistry of the cloaca of Anguis fragilis is given by Raynaud *et al.*, (1963, 1968).

(a) *Histology*. The reptilian cloaca is lined with columnar

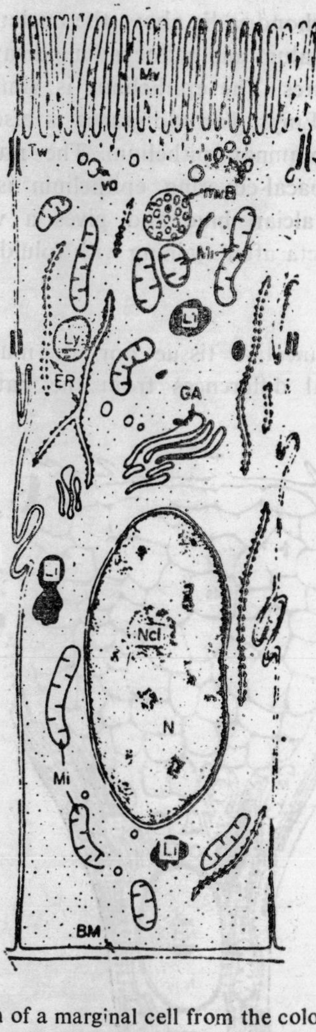

Fig. 7.4. Diagram of a marginal cell from the colon of Natrix natrix. BM, Basement membrane; De, desmosome; ER, ergastoplasm; GA, Golgi apparatus; Li, lipid; Ly, lysosomes; Mi. mitochondria; Mv, microvilli; mvB, multivesicular body; N, nucleus; Ncl, nucteolus; Tw, terminal web, Va, vacuole.

epithelium, the marginal and goblet cells of which resemble those of the colon. The narrow marginal cells have ovoid to elongate ovoid basal nuclei. A well-developed Golgi apparatus and numerous mitochondria lie above the nucleus. Preparations stained with chrome alum and gallocyanin indicate the presence of basophilic structures. The epithelium is simple in Lacerta but is compound in Emys; Testudo graeca is also supposed to have a compound columnar epithelium. The mucin formed by the cells of the cloacal-covering epithelium is PAS-positive, strongly takes up alcian blue and gives a very clear metachromasia of type beta after staining with toluidine blue (Gabe, 1971b).

Neither the connective tissue nor the musculature shows significant structural differences from the pattern seen in the

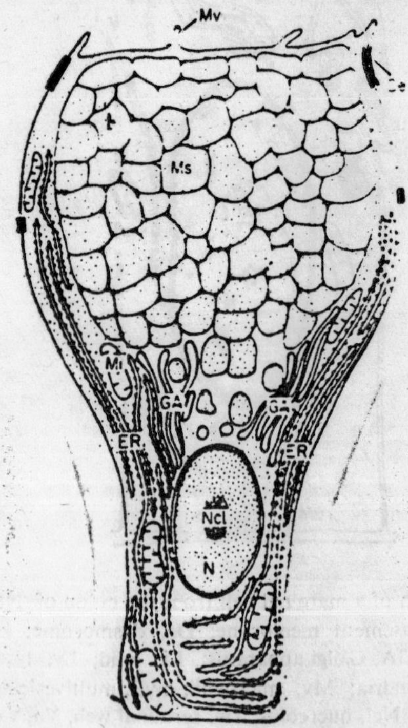

Fig. 7.5. Diagram of a goblet cell from the colon of Natrix natrix. De, Desmosome; ER ergastoplasm; GA, Golgi apparatus; Mi, mitochondria; Ms, mucous secretion; Mv, microvilli; N, nucleus; Ncl, nucleolus.

colon. The muscularis of the cloaca is thicker than that of the small intestine.

(b) *Histochemistry*. The relatively few columner epithelial cells of Lacerta and Agama have PAS-positive apical zones. Non-specific acid phosphatase and non-specific esterase show positive reactions that resemble each other in the localization and strength of reaction. The supranuclear and infranuclear regions of the cells show strongly positive reactions for isocitrate dehydrogenase, lactate dehydrogenase, and α-glycerophosphate dehydrogenase. A similar positive reaction is obtained for succinate dehydrogenase and monoamine oxidase, with that for monoamine oxidase being somewhat weaker. The activity of glucose-6-phosphate dehydrogenase is primarily concentrated in the basal region of the epithelium.

The supranuclear secretory zone in the goblet cells of the cloaca resembles that in the goblet cells of the colon, reacting strongly with alcian blue stain; the supranuclear region remains unstained with PAS, but the infranuclear region shows a positive reaction. The zone in which the secretion collects reacts negatively with all the enzymatic tests used here.

The connective tissues show a uniformly strong, PAS-positive reaction. The musculature shows weakly positive reactions for non-specific acid phosphatase and non-specific esterase (with naphthol-AS-acetate as substrate). If 5-bromoindoxylacetate is used, the test for esterase gives totally negative results. Lactate dehydrogenase, which is generally involved in anaerobic respiration, shows a strong positive reaction in the muscularis of Lacerta and Agama. The test for isocitrate dehydrogenase gives equally positive results, while those for α-glycerophosphate dehydrogenase, succinate dehydrogenase and monoamine oxidase yield only very weak reactions.

Glands of the Cloaca

A variety of glands has been reported in the cloaca of reptiles (Rathke, 1866; Gunther, 1867; Osawa, 1897; Voeltzkow, 1902;

Disselhorst, 1904; Taguchi, 1920; Schaffer, 1930; Eggeling, 1931; Woepke, 1931; Gerhardt, 1937; Volsoe. 1944; Guibe, 1948; Barone, 1954; Fox, 1956; Gabe and Saint Girons, 1964; Fox, 1965; Watkins *et al.*, 1970; Kroll and Reno, 1971). Two types of glands are found after Gabe and Saint Girons (1965):

1. Urodeal glands; only simple epithelial crypts.
2. Cloacal glands; anatomically limited monostomatic glands.

Holocrine, strongly lobular scent glands have been observed in different crocodilians (Rathke, 1866; Voeltzkow, 1902; Disselhorst, 1904; Taguchi, 1920). Simple or branched mucous glands are located in the cloaca of Crocodylus (Taguchi, 1920). The strongly lobular scent glands that form a large ovate body between the subepithelial connective tissue and the striated cloacal sphincter of crocodilians develop only with sexual maturity and are related to sexual stimulation (Disselhorst, 1904).

Simple tubular glands, independent of one another, occur in Sphenodon, Xantusiidae and Chamaeleonidae. The epithelium of the glands and the ducts is equivalent. Female lizards of numerous species have independent tubular cloacal glands. The tubes are branched and the epithelium is differentiated. The cloacal glands of males of the same species are more developed, but they are monostomatic in both genera. Differentiation is generally missing in Hoplodactylus pacificus, in Gekkonidae and also in all Typhlopidae and Amphisbaenidae (Whiting, 1967). In Typhlops and Leptotyphlops the cloacal glands are most concentrated.

In Leputyphlops dulcis, the single median cloacal gland is tubuloalveolar. Similar unpaired cloacal glands as well as paired sacs are noted in certain colubrid snakes (Kroll and Reno, 1971). The cloacal sacs ("and sacs" after Gabe and Saint Girons, 1965) of Leptotyphlops dulcis are nonglandular, posterior evaginations of the cloaca.

Carbohydrates are the most important component of glandular secretion. Mucous glands possess a large amount of acid mucopolpsaccarides. In some species acid mucopolysaccharides cannot be demonstrated and neutrol mucopolysaccharides and glycoproteins are probably present. Histological and histochemical peculiarities of the cloacal glands in female squamates are given

The paired glands found in the dorsal cloacal wall of Agama agilis sanguinolenta were investigated histochemically. Among the oxidoreductases, isocitrate dehydrogenase and lactate dehydrogenase show a uniformly positive reaction in the glandular cells, while monoamine-oxidase gives a relatively weak response. The reactions for α-glycerophosphate dehydrogenase and succinate dehydrogenase always show a topagraphically different response, part of the glands being strongly positive and another part almost entirely negative.

Non-specific acid phosphatase and non-specific esterase show corpuscularly distributed reactions of comparable intensities.

Proctodaeum

The epithelium of the generally smooth proctodaeal mucosa consists of high columnar cells, only some of which have a clear striated border while most contain mucous material. The simple, high columnar epithelium becomes double-layered, and thus compound, within the proctodaeum. Close to the cloacal aperture, the epithelium becomes compound and keratinized. Osawa (1897) reported that many reptiles show an intermediate epithelial state consisting of a compound columnar epithelium without goblet cells; it contains additional layers and the cells are flatter.

In agamids and lacertids the proctodaeal epithelium resembles closely that of the cloaca in its histochemical reactions for carbohydrates and enzymes. The marginal cells may easily be differentiated from the adjacent goblet cells by their more intense PAS-reaction. During the transition from columner to stratified squamous epithelium the histochemical appearance of the epithelium also changes. The stratified squamous epithelium is PAS-negative and does not stain or with alcian astra blue. Only the middle and superficial layers of cells contain lipids, as indicated by their staining with Sudan black B. These same cells are the only ones to show even a weakly positive reaction to non-specific acid phosphatase, while the test for non-specific esterase gives negative results in the basal layer of cells and moderately positive ones in the remainder of the epithelium.

Tests for several oxidoreductases confirm the suggestion that the stratified squamous epithelia near the cloaca may be divided into a basal zone, which generally shows little esterase activity, and a middle and superficial zone, which shows greater activity. Thus tests for monoamine oxidase, isocitrate dehydrogenase, and lactate dehydrogenase give moderate to very strong reactions in the superficial layer of cells, while the basal layers show almost no reaction.